POWERING UP CANADA

MCGILL-QUEEN'S RURAL, WILDLAND, AND RESOURCE STUDIES SERIES

Series editors: Colin A.M. Duncan, James Murton, and R.W. Sandwell

The Rural, Wildland, and Resource Studies Series includes monographs, thematically unified edited collections, and rare out-of-print classics. It is inspired by Canadian Papers in Rural History, Donald H. Akenson's influential occasional papers series, and seeks to catalyze reconsideration of communities and places lying beyond city limits, outside centres of urban political and cultural power, and located at past and present sites of resource procurement and environmental change. Scholarly and popular interest in the environment, climate change, food, and a seemingly deepening divide between city and country, is drawing non-urban places back into the mainstream. The series seeks to present the best environmentally contextualized research on topics such as agriculture, cottage living, fishing, the gathering of wild foods, mining, power generation, and rural commerce, within and beyond Canada's borders.

Powering Up Canada

A History of Power, Fuel, and Energy from 1600

Edited by

R.W. SANDWELL

McGill-Queen's University Press
Montreal & Kingston • London • Chicago

ISBN 978-0-7735-4785-8 (cloth)
ISBN 978-0-7735-4786-5 (paper)
ISBN 978-0-7735-9952-9 (ePDF)
ISBN 978-0-7735-9953-6 (ePUB)

Legal deposit second quarter 2016
Bibliothèque nationale du Québec

Printed in Canada on acid-free paper that is 100% ancient forest free `
(100% post-consumer recycled), processed chlorine free

McGill-Queen's University Press acknowledges the support of the Canada
Council for the Arts for our publishing program. We also acknowledge the
financial support of the Government of Canada through the Canada Book
Fund for our publishing activities.

Library and Archives Canada Cataloguing in Publication

Powering up Canada: a history of power, fuel, and energy from 1600/edited by
R.W. Sandwell.

(McGill-Queen's rural, wildland, and resource studies series; 6)
Includes bibliographical references and index.
Issued in print and electronic formats.
ISBN 978-0-7735-4785-8 (cloth). – ISBN 978-0-7735-4786-5 (paper). –
ISBN 978-0-7735-9952-9 (ePDF). – ISBN 978-0-7735-9953-6 (ePUB)

1. Power resources – Canada – History. I. Sandwell, R.W. (Ruth Wells),
1955–, editor II. Series: McGill-Queen's rural, wildland, and resource studies
series; 6

TJ163.25.C3P73 2016   333.790971   C2016-902003-7
                                                 C2016-902004-5

This book was typeset by Interscript in 10.5/13 Sabon.

Contents

Illustrations

TABLES

PLATES

Acknowledgments

We gratefully acknowledge the financial assistance with publishing this volume provided by the following:

The Network in Canadian History and Environment
   (Niche – http://niche-canada.org/)
The Dean of Arts and Social Sciences, Simon Fraser University
The Book and Creative Work Subvention Fund, University of Victoria
The University of British Columbia
The Department of History, University of Toronto
The Wilson Institute for Canadian History, McMaster University
Geoff Cunfer, Department of History, University of Saskatchewan, PI,
   "Sustainable Farm Systems Project," supported by the Canadian
   Social Sciences and Humanities Research Council, Partnership Grant
   895-2011-1020

The authors gratefully acknowledge as well the Social Sciences and Humanities Research Council Connection Grant that allowed us to get together in Bloomfield, Ontario, for a two-day symposium to discuss draft chapters of the book, and the project as a whole, in May 2013.

The contributors at the Bloomfield Symposium in May 2013.

POWERING UP CANADA

1

An Introduction to Canada's Energy History[1]

R.W. Sandwell

WHY A CANADIAN ENERGY HISTORY?

Human beings, like other living organisms, require energy to live; without food energy, without heat energy, and without the energy needed to fabricate such goods as clothing, shelter, and tools, people cannot survive. Over the last two hundred years, people in industrialized countries have experienced an unprecedented shift in the nature and scale of the energies they consume. In Canada, as in other industrialized countries, the shift from the organic energy regime of muscle power, wood, wind, and water power to the mineral regime of coal, oil, gas, and electricity has allowed people to do a wide range of things previously difficult or even impossible. The cheap and abundant energy created by the energy revolution has made it possible to have more heat and more light and more transportation, reducing enormously the amount of hard physical labour that so many people endured throughout their lives. New forms of energy have not only powered new machines and facilitated new industrial processes, they have generated completely new objects and activities – making instant communication possible over vast distances, for example, and creating sound and moving images for people's entertainment and edification. They have made it possible to exist for long periods of time in previously uninhabitable environments – in the air, underwater, underground, and in outer space. By creating previously unknown levels of economic growth, the new energy regime has also increased the standard of living in many nations to new heights. "Electricity will let you live like a King!" proclaimed a typical advertisement from the 1920s, and it was not far wrong: only a tiny minority of people throughout history have previously had access to the kind of wealth now provided by modern forms of energy. These influence every aspect of daily life, from the food people eat to the places they live and work, from the way they get from place to place to the way they relate to others.

But if Canadian society has followed the general pattern of industrializing economies during the great energy transition, it has demonstrated variations from the norm set by Britain, the first country to industrialize. Canadians consume more energy per capita than almost any other peoples in the world and have done so in almost every decade since European resettlement of this vast country, and before. They have, however, been later than many other industrializing countries in entirely replacing traditional or organic forms of energy with those of the new mineral energy regime. It was not until 1955 that Canada reached the 90 per cent level of modern versus traditional energy use that Britain had attained by 1845, more than a century earlier.[2] Canadians are also unusual in the variety of energy sources they have used at any given time over the last two hundred years, which might include, depending on the decade, wood (and now wood pellets), peat, coal, hydroelectricity, oil, propane, natural gas, or nuclear power. And unlike most other countries in the Western world, Canada has emerged as a large net exporter of natural gas, oil, and electricity, almost all of which goes to one country: the United States.[3] As consumers and producers of vast amounts of energy, and in a country where subsurface rights to vast reserves of oil and gas generally belong to federal or provincial governments, Canadians today are profoundly influenced by the politics and economics of energy, and they live daily with its social, cultural, and environmental implications.

If the energy revolution has both dramatically shaped and been shaped by human and environmental history, it is a revolution largely taken for granted by many of those in industrial and so-called post-industrial societies. Or this was the case until recently. The pollution from the extraction, processing, transportation, and use of the newer forms of energy has troubled human and other ecologies for decades, but in the early twenty-first century Canadians are among those worrying about increasing costs, declining reserves, and political instability of energy supplies. Even more urgently, world populations are confronting new warnings from scientists about the larger contexts of the energy revolution. Given the "great acceleration of impact on our environment" caused by fossil fuel burning, it now seems likely that we are facing "the possibility that our economy is transgressing the planetary boundaries that provide a safe operating space for humanity, threatening functioning of our ecosystems and threatening rapid climate change."[4] For the first time since the "oil shock" of the early

1970s, when oil prices tripled within six months as immediate shortages were predicted in the wake of the Arab Israeli war, Canadians are again becoming more conscious of the ways in which their lives are deeply intertwined with the energy systems that provide the food, water, light, heat, freedom from heat, shelter, medical services, communications, transportation, entertainment, waste disposal, as well as a new world of goods – indeed almost all aspects of daily life upon which people in industrialized societies depend for their living and their lives. Realizing that this high-energy mineral regime comes at a very high cost and may not be sustainable, people and some corporations and governments around the world are beginning to think differently about our energy future.

These changes are also encouraging some people to think differently about energy in the past. Generally, historians have tended to naturalize the changes long associated with the Industrial Revolution with such concepts as urbanization, development, and technological change, paying less attention to the changing energy regime that provided their foundation.[5] If, as now seems possible, the twentieth century emerges in the historical record *not* as the century in which industrial economies "took off" towards unlimited growth, but instead as the first and last Age of Abundant Energy, historians need to re-evaluate the changes associated with modern forms of energy. The early twenty-first century, therefore, provides a good vantage point from which to re-evaluate earlier assumptions about the sustainability of high-energy use and to examine particular ways that energy history has been deeply entwined with the social history of Canada.

Resisting engagement with eco-apocalyptic discourses that can accompany a burgeoning energy awareness, this book seeks to highlight, problematize, and probe very specific ways in which Canadian people and places, and kinds of people and places, produced and consumed energy in the past, and how they made the transition from relatively low to high energy consumption, and from an organic to a mineral energy regime. The book was conceived as a collective project responding to the near invisibility of most aspects of energy in the country's historiography, before and after industrialization. Chapters here explore the material conditions of the production/extraction/nurturing, harvesting/processing, transportation, consumption, and waste issues relating to fuels, powers and energies. Separate chapters on food (a case study of pemmican), animal muscle power, water power, wind power, firewood, coal, oil, gas, and nuclear power generally focus on the period 1800–1990. Appendices provide specific technical and statistical information on energy history in

Canada, beginning with one explaining the most common energy terms. Some readers might, indeed, find it useful to begin with this Primer on Terminology; words explained in that appendix are rendered in italics when they first appear in the chapters. Indeed, the book as a whole and the small symposium it came out of were conceived as a kind of historical energy primer, a place where people not well-versed in energy studies could find in one place some material details of how energy, fuel, and power work in the science, technologies, and infrastructures involved.

The book in its final form emerged as something more than we initially envisaged. Part energy primer, it is also part energy sampler – albeit a necessarily preliminary and brief one – that articulates different ways of thinking historically about myriad and profound ways that energy and the increasing scale of energy use worked out in Canada, in the broad contexts, causes, and consequences in which earlier energies were used and new ones developed. Exploring the history of energy, the authors reveal, also means exploring its intimate ties with a wide range of issues, topics, and events, including those in politics, business, economics, culture, society, environment, racialized populations, gender, various sciences (geological, biological, and mechanical), geography, and everyday life. Individual chapters in the volume, therefore, necessarily cut across many disciplinary and subdisciplinary categories. If the volume as a whole reaches across disciplinary boundaries, it also seeks to reach across the particularities of distinct energy histories, drawing them together into a broader synthetic whole. It is the authors' hope that this preliminary foray into energy history will demonstrate, therefore, why and how energy brings many aspects of Canada, past and present, into new focus. We also hope that it will provide a platform that Canadians can use to insert themselves into the growing, urgent international conversation about the larger implications of energy in human society, past, present, and future.

UNDERSTANDING ENERGY IN HISTORY

Given the enormous role that fuel, power, and energy have always played in human societies, and the huge recent changes in human energy use and their far-reaching effects, historians' silence about energy as a discrete force or phenomenon is more than a little surprising. Perhaps part of the explanation can be found in the fact that economic historians have been most interested in energy over the past half-century, and their interest in the relationship between energy and economic growth has focused on economic models and laws rather than on the kinds of human behaviours, beliefs, and ideas that have occupied social historians' interest in recent years.[6] Or perhaps, as E.A. Wrigley has argued, the energy revolution was so gradual, and so tied up with other changes associated with "modernization" that the significance of the shift to fossil fuels and electricity was missed: a case of not seeing the forest for the trees.[7] Or perhaps historians' recent interest in cultural history, micro studies, and discursive interpretations has made irrelevant or uninteresting the material connections between energy, environment, and diffuse social change. Historians of the future will doubtless have their own interpretations of why what is arguably the biggest story of the twentieth century went largely unremarked at the time.

Many historians in Canada and around the world have, of course, written about specific energy forms and technologies, particularly about their economic and political contexts, some in a comparative framework.[8] Others have written about specific effects, and stimulators, of energy use – material, cultural, and environmental[9] – and on different kinds of people. Energy studies have also emerged as important components of the sub-field of "envirotech," which fruitfully brings together environmental history with science and technology studies.[10] Many of these topics and approaches are discussed and referred to throughout this volume. In Canada, Richard Unger and John Thistle's recently published *Energy Consumption in Canada in the Nineteenth and Twentieth Centuries* provides for the first time statistical data on Canadian energy use in an international context, while Peter Sinclair's *Energy in Canada* gives readers a useful political and economic overview of energy's pivotal role in the country's history in the last hundred years or so.[11] Both have been particularly useful to our authors here.[12] Taken together, however, these histories and approaches provide us with multiple "energy silos," stopping short of providing a larger, interpretive framework for understanding the relationship between energy and society in general.[13] What lens is big enough to bring into focus the vast, entangled, and complex interactions and massive changes that are here being attributed to changing energy use? This introduction will suggest that E.A. Wrigley's concept of the solar or organic versus the mineral energy regime provides a good place to start understanding the role energy has played in mediating the relationship between the environment and human societies.

THE ORGANIC AND THE MINERAL ENERGY REGIMES

In a series of lectures published in the late 1980s and early 1990s, economic and demographic historian E.A. Wrigley argued that energy use was the key factor in the great transformation occurring over the last two centuries. Using detailed economic and demographic data, he explained that the vast and sustained changes associated with the Industrial Revolution and the rise of modern societies can be found in the monumental shift from the organic energy regime that previously characterized all human societies, to the new mineral energy regime ushered in by coal, electricity, oil, and natural gas in the nineteenth and twentieth centuries.[14] While England was experiencing dramatic changes in production techniques and economic, social, and political organization from the sixteenth century onward, changes that created unusual levels of economic growth, Wrigley argued that it was the massive nineteenth-century shift to fossil fuels and electricity that, unbeknownst to economic theorists of the time, would eventually transform almost every aspect of life and work, and in ways that would dramatically increase opportunities for economic growth for the first time in history.[15]

Until quite recently in human history, Wrigley argues, the relationship between people and the environments on which they relied for their lives and livelihoods was based almost entirely on an organic or solar energy regime. Land (and waters) met most of people's energy needs in the form of food, fodder, and fuel, and provided as well almost all the animal and vegetable raw materials for construction and industry, including "wool, flax, silk, cotton, hides, leather, hair, fur, straw and wood."[16] Economic support in organic regimes was derived directly and indirectly from the sun through the process of *photosynthesis*; people tapped into "renewable flows of solar energy, largely by growing plants that can be eaten, fed to animals, and burned."[17] It in no way disparages pre-industrial cultures to note that most people spent most of their time in wresting energy from their environment, and processing it in various ways to support life, for in the endosomatic (human-powered) world of the organic energy regime, human muscle fibres provided more than 70 per cent of mechanical energy involved in doing so.[18] People supplemented the energy provided by food, muscle-power, and biomass fuels (fires from wood and its derivative charcoal, but also peat, used to heat food, human and animal shelters, and also to work metals and glass) with the kinetic energy they wrested from wind and water (both products of complex solar hydrological systems).[19] They used this energy directly for transportation by sail and river boats, and to provide energy indirectly by turning water wheels and later turbines that more efficiently turned kinetic energy to mechanical for a variety of grinding, pressing, twisting, and milling tasks. Ian G. Simmons, and other contemporaries of Wrigley, began to speak of human epochs in terms of the increasing intensification of energy inputs required in foraging, pastoral, traditional, and then modern agricultural societies.[20] More recently historians like Manuel González de Molina describe human epochs as "metabolic regimes," characterized by the increasing appropriation of material and energy flows, and David Christian demonstrates the consistency of human energy intensification in the very long run, noting that humans, like all organisms, are "antientropy machines."[21]

The solar or organic energy regime supported life, but in ways that placed very specific constraints on human societies. Most significantly, people, like other living things, had to tap into *flows* of energy from the sun. Organic energy flows on a highly variable seasonal basis in places like northern Europe and Canada, which are far from the equator and where the solar energy (light and heat) available changes dramatically over the course of the year. Inuit and Cree hunters, fishers, and gatherers, for example, move across land and water to catch the energy from plants and animals at seasonally appropriate times. Pastoralists, including even Alberta cowboys, move so that they and the animals on whom they rely can do so, while agricultural people, including Canada's first European settlers, plant and harvest particular foods and remove others in their attempts to tap more effectively into the flows of available energy. People and their animals followed the diurnal as well as seasonal rhythm of the solar energy regime, using the relative warmth and light of daylight hours to perform most tasks, and hoping for moonlight for nighttime activities requiring sight. Energy availability was, therefore, directly linked to energy flows available in their changing and variable local environments. As many of the chapters in this collection confirm, capturing and using these flows of organic energy not only occupied most of people's time, but involved layers of complicated knowledge about their local environment, and a wide range of difficult but widely diffused skills, from hunting, gardening, and fishing to cooking, building, and toolmaking.

Relying on flows of energy, pre-industrial peoples were obliged to make hard decisions about how much land to allocate for competing energy and land-use needs throughout the year: food and fodder, fuel, and direct power. This had to be set against the amount of land needed to produce the raw materials for the major industries of the organic economy: textiles, leather, brewing, and construction. Relying on a fixed amount of land meant that "all basic human needs, for food, clothing, housing and fuel, inevitably meant mounting pressure on the same scarce resource" – land.[22] Furthermore, because in pre-industrial agricultural societies "greater productivity tends to provoke earlier marriage, reduced celibacy and lower mortality,"[23] economic growth meant increasing populations that put yet more pressure on limited energy and resources. Economic growth in one area could be gained only by a loss in another: horses could be used to plow more land to grow more food, but only if more land was used to provide fodder for the horses. J.I. Little estimates in this volume that it took about two and a half acres of "fertile corn belt land" annually to support a horse.[24] As Wrigley emphasized, "Preindustrial economies were fundamentally circumscribed in the nature and extent of growth that was possible by the limitations on the output of organic materials. Photosynthesis is an inefficient process that extracts only about one-tenth of one per cent of the energy that reaches the earth in the form of insolation."[25] It also creates the largest carbon store we know, the biosphere. But capturing energy from its flows of energy is often difficult. In pre-industrial societies, land and energy were directly linked and limited, and many have been characterized by poverty and periodic famine.[26]

The energy available for human use was limited not only by the carrying capacity of the land (the amount of energy that the land could provide for the people and animals relying on it) but by the difficulty of transporting forms of energy or power. Before fossil fuels, organic energy carriers such as trees or oxen were bulky and typically expensive to transport, while wind and water power could be employed only where, when, and if they appeared. As Christopher Jones summed up, "Because of these two factors – the limited carrying capacity of the land and the expense of transporting bulky fuel sources – Organic Economies tend to be self-sufficient in, and relatively small consumers of, energy."[27] As we will see, the fact that much of Canada possessed both vast stores of organic energy in the form of trees, and the means of transporting them in an abundance of flowing rivers, helps to explain the country's high organic energy use before and after the rise of the mineral economy. The existence of large organic stocks of energy indeed was one factor that made the "New World" so different from European, Asian, and African pre-industrial societies.[28] In general, however, pre-industrial societies were forced to tap into very limited seasonal flows of energy daily for their survival.

Fossil fuels such as coal, oil, and gas were originally the products of solar energy; only nuclear and geothermal energy have no connection to the organic energy regime. Over millions of years, however, decaying plant matter was compressed and transformed into dense, underground supplies of coal, oil, and natural gas. Identified by Wrigley as the mineral energy regime, these new (that is, newly exploited) energy carriers, including electricity, differ from those of the organic energy regime in two society-changing ways. First, fossil fuels and electricity are uniquely focused forms of energy, dense and often capable of fine control. Using them increased dramatically the scale of work – making things, moving things, rendering and fashioning things – that can be accomplished with the energy they provide. Britain was the first country to witness, if only slowly for the first few hundred–odd years, such changes associated with coal use.[29] By the early nineteenth century, coal was providing a cheap source of heat for most English homes; by the mid-nineteenth century, it was transforming the way people travelled and worked. For there is a transformative difference of scale between what a person can move in a day by using a wheelbarrow, a horse-drawn wagon, or a coal-burning steam-powered locomotive. There are vast differences of scale between preparing land for planting with a shovel, working with a horse-drawn plowing team, and driving a steam tractor with multiple plows. Railway trains travelled significantly faster and farther than stage coaches.[30] The new energy available from the mineral economy makes people capable of doing exponentially more than they could in the organic economy. As a recent energy history of Europe summed up, humans have tried to "augment their power through tools and animals for many thousands of years. The wide availability of steam engines and fossil fuels allowed them to do so, and on a much greater scale. In other words, it allowed the augmentation of capital per worker."[31]

Even more significant than the remarkable labour efficiency provided by modern energy carriers, however, is the fact that fossil fuels and electricity present as "massive *stocks* of energy available for immediate use."[32] Unlike the relatively dispersed stocks of energy in the form of forests or winds, fossil fuels are concentrated and more easily transported to human settlements. The implications of having massive dense stores of energy to draw on, though only slowly realized at first, have been many and profound. Because stocks of coal, oil, and natural gas naturally exist in large, consolidated areas, people no longer need to tap daily or seasonally into organic flows (or trickles); instead, energy carriers can be captured and physically transported to where they are needed. For the first time, it made economic sense to invest in massive infrastructure projects – first more and larger canals, then railways and later roads, drilling rigs, massive hydroelectric dams and plants, pipelines, refineries, and tankers – to extract, process, and transport more of such highly useful forms of energy over long distances. The sites of energy production, no longer constrained by the carrying capacity of the land or prohibitive transportation costs, could now be separated even farther from the sites of energy consumption.[33] The limits of the local, in other words, were effectively transcended for the first time in human history as economic growth was fuelled by reserves of energy that, for the first time, existed as commodities that could be bought and sold across vast distances.

The transition to new energy carriers – from coal to electricity, oil, and natural gas – occurred differently in different times and places, typically occurring more quickly in later-industrializing areas, which could draw on techniques, processes, and hardware developed and often produced elsewhere. But whenever they happened, they grew up as parts of new growth complexes or "development blocks" in industrial societies, each accompanied by new technologies, physical and legal infrastructures, and human practices.[34] Coal needed skilled miners for extraction, then canals and railways to transport it. Coal-fuelled mining pumps, elevators, and fans in turn increased the production and safety of mines, and, for a while, the strength of mining unions.[35] More coal meant more steel, more transportation, and more trains to carry the coal, and all reduced the price of coal, making it an increasingly popular choice of fuel. The internal combustion engine made gasoline usable as a fuel for the first time, a use that was premised on transportation by railways, then roads and pipelines, then tanker ships, and its consumption was dependent as well on gas stations and an auto industry. Manufactured gas and then electricity required networks that were in some ways even more complex and multifaceted than the railways. As Wrigley and Christopher Jones emphasize, "the zero sum game of the organic energy regime"[36] was eclipsed by an upward spiral of positive energy feedback loops, each of which made the next transformation easier, more productive, and less costly. Real per capita incomes soared with the unprecedented economic growth that the new, moveable energy stocks provided. For all these reasons, Wrigley argues that "the most significant triumph of 'capitalism' was not that celebrated in Marxist or liberal historiography, but that which saw energy stock substituted for energy flow as the foundation of modern economic activity."[37]

Wolfgang Schivelbusch summed up the revolutionary spread of modern, system-based, de-localized energy regimes in one word: distance.[38] As he emphasized, the mineral energy regime meant that for the first time people could live, mine, manufacture, and move goods and people wherever it made economic, political, or aesthetic sense to do so. And yet while fossil fuel and electricity use have erased the local into "geographies of nowhere," the extraction, processing, transportation, and consumption of the new forms of energy, and the new world of goods whose production and movements they have facilitated, really have occurred somewhere, if far from the gaze of the urban majorities.[39] The ready portability of the new super-energy sources and the resultant freedom from the local was, paradoxically, accompanied by a sharp intensification and concentration of land use, or re-localization. No longer dependent on a particular place to support a range of energy uses – food, shelter, fuel, manufacturing – industrial societies have been free to dedicate large areas of land to a particular exclusive use, such as a city, a suburb, a massive farm, a landfill site, or an open pit mine. By the last third of the nineteenth century, as the new energy regime made possible the concentration of population in cities on a scale previously unimaginable, industrial areas, and particularly those associated with fossil fuel processing, had already "earned a reputation for their distinct sights, sounds and smells: cavernous buildings, tangles of railway tracks, towering cranes, clattering machines and, above all else, dense clouds of smoke and pungent aromas of industrial waste."[40] Dedicated "zones of sacrifice" appeared, expanded, and moved out of cities, becoming designated single-use areas. These are now as characteristic of the mineral energy regime as huge conurbations and Canadian wilderness parks.[41] The mines, the oil and gas wells, the pipelines, the dammed rivers, and the hydro lines needed to support the mineral energy regime created massive wealth, leading to increases in living standards for many. But the costs and the benefits of energy-at-a-distance have been distributed spatially in distinctly unequal terms. The net global effects of climate change, while unclear as to details, are clearly transcending the limits of the local as well, reworking the politics of energy use in new ways.[42]

The mineral regime is characterized by another kind of concentration: that of power, both literal and figurative, needed to make the new industrial networks function. The amount of capital needed to build such massive, modern, and integrated infrastructures as an electrical grid, a pipeline, or a liquefied natural gas plant far exceed that required by earlier forms of workshop-based and artisanal industrial investments, such as a grist mill or even an early locomotive factory. The massive infrastructure necessarily disrupted earlier patterns of economic logic, reaching as they did far beyond simple notions of small-scale businesses or workshops, or the balance between supply and demand. Electricity, for example, with its unusual characteristics (invisibility, death-dealing electric shocks) and new powers (to move highly sophisticated machinery in minute ways, to create light without flame, and to create new substances through its role in chemical reactions), while celebrated early on for its myriad and profitable industrial applications,[43] was necessarily created well in advance of generalized popular demand: "a loose network of power companies, consulting firms, public agencies, and electrical device manufacturers had to invent the electrical power consumers that their institutions and systems seemed to require."[44] Inventing the new energy consumer, indeed, was a significant task within industrializing societies in the late nineteenth and early twentieth centuries.[45]

A new form of capitalism grew up with the massive, capital-intensive projects that defined the new energy regime.[46] Governments were necessarily involved, as the extraction and processing of new, highly profitable (mostly) underground sources of energy required much more money than individual entrepreneurs or companies normally had access to in those days. The new forms of energy also provided a source of revenue through sales (when government owned), taxes, fees, and royalties, and required new legal instruments for ownership, export, and regulation. Concentrated and easily transportable stocks transformed energy from a local concern to a national one; energy security became a key political issue for both exporters and importers of energy for the first time. Energy issues became national issues. And the new forms of capital involved in the massive infrastructure changed the nature of capitalism itself. As Schivelbusch observed, "The transformation of free competition into corporate monopoly capitalism confirmed in economic terms what electrification had anticipated technically: the end of individual enterprise and an autonomous energy supply." As he went on to conclude, "The concentration and centralization of energy in high capacity power stations corresponds to the concentration of economic power in the big banks."[47] Unlike the variable, dispersed, and often unpredictable energies that people tapped into daily and seasonally from their local environments, the new superabundant energies were distributed from a central source and distributed in a controlled, metered, and generally predictable way, through highly sophisticated networks that few understand, and no one has free access to.

In what may be the final decades of the mineral economy, rising prices, decreasing supply, unintended environmental consequences, and political pressure to reduce fossil fuel use to mitigate climate change combine to make the world's energy future uncertain. While some countries are beginning to recognize the need to drastically reduce fossil fuel consumption and provide alternative energies, no country (with the possible exception of Germany) has yet embarked on the next major energy transition, from fossil fuels to renewable energy sources.[48] The energy specialists already anticipating the transition, however, note that the shift from the mineral regime will probably differ from the shift to it, and in some important respects. Roger Fouquet is one of many energy historians to observe that in the past two hundred years, "the main economic drivers identified for energy transitions were the opportunities to produce cheaper or better energy services." If local conditions, whether environmental, geographic, cultural, or economic, did not facilitate the transition to a cheaper and better form of energy, people did not make the change.[49] Will cheap substitutes for fossil fuels be found? If not, will industrial populations willingly reduce their energy consumption and pay more for it? If the next energy transition is as transformative as the shift from organic to mineral, exciting times lie ahead.

CANADA'S ENERGY HISTORY: AN OVERVIEW

The chapters in this collection suggest that Canada's energy history has followed the broad outlines charted by Britain in a shift from a relatively low-energy organic regime to the high-energy mineral regime in the nineteenth and twentieth centuries, though as we will see, with some distinct national characteristics. Canada is far less homogeneous geologically and economically than Britain and played a vastly different and smaller role in geopolitics. But the shift from organic to mineral transformed both societies and provides, therefore, the organizational structure of this book. Chapters on food, muscle, water, wind, and wood comprise Part One, the Organic Energy Regime, and chapters on hydroelectricity, coal, petroleum, gas, and nuclear power comprise Part Two, the Mineral Regime. A short concluding chapter highlights many of the themes and issues running through this volume.

It is unlikely, however, that later editions of this volume will include a Part Three: the Post-Carbon Regime. In 2010, Canada ranked as the second-largest per capita energy consumer in the world, about 80 per cent of it from fossil fuels.[50] Canada stands out in global terms today not only for its very high per-capita energy consumption from modern energy carriers, but also because it is one of the few industrial nations firmly entrenched as a net exporter of energy. In 2006, it was exporting energy "equal to over half of domestic consumption, and the proportion was rising."[51] Annual export revenues from oil, natural gas, and hydroelectricity were $107.6 billion in 2012, comprising almost a quarter of the country's export revenues.[52] Keeping in mind that estimates (see plate 1.1) suggest that total energy consumption of Canadians for home use, industry, transportation, and internal commerce in 1800 was 44 petajoules, 279 in 1850, 631 in 1900, and increasing more than a tenfold to 7,929 in 1980, current production statistics are staggering. In 2012, Canada produced almost 17,000 petajoules from modern energy carriers: 8,021 petajoules from oil (as the world's fifth-largest oil producer[53]), 5,500 from natural gas (the world's third-largest producer[54]), and 1,416 from hydroelectricity (the world's second-largest producer[55]). Only in producing nuclear power (328 petajoules) and coal (1,521petajoules) does Canada rank below the top ten world producers. Other sources of energy, such as solid wood waste, spent pulping liquor, wood, and other fuels for electrical generation provided 527 petajoules for Canadians. The total of wind, tidal, and solar electricity in 2012 was 43 – up more than 300 per cent from 2008, but still only the tiniest fraction of 17,000 petajoules of energy provided by the mineral regime. Some of the "green" fuels, such as agricultural ethanol and methane, have received substantial support from Canadian taxpayers in agricultural subsidies, but their "greenness" has been subject to sustained criticism in the face of evidence that they actually consume more fossil energy than the green alternatives they produce.[56] Without substantial government investment in energy from solar, geothermal, small-scale hydro, and tidal power, it is likely that the "the innovation chain" – a phrase used to describe "the time from invention to uptake to dominance" of a new energy carrier – will remain stalled at the "invention" phase.[57] Attention to climate change is growing, but Canada's role as a fossil fuel–exporting nation makes serious investment in alternatives seem politically unlikely, at least in the short term. Few think that Canada's energy security is a matter of national concern.[58]

Canada's transition from the organic to the mineral regime is marked by the characteristic dramatic scaling-up of energy consumption (see plate 1.1). Per capita energy consumption doubled in Canada between the early years of the nineteenth and the early twentieth centuries; by the 1960s, it had increased three-fold from 1800, and by the end of the millennium, almost six-fold.[59] Energy consumption overall rose sharply to 2001. Even while manufacturing was increasingly moving offshore, overall industrial consumption (much in the fossil fuel sector) accounted for an increasing share of the energy used over the last forty years, rising from about 22 per cent in 1972 to almost 50 per cent in 2011. Household consumption, though still increasing in absolute and per capita terms, fell from about 27 to 13 per cent of the total.[60] Plate 1.2 outlines Canada's shift from reliance on wood, to coal, electricity (created first by coal then hydro and then nuclear energy), oil, and then natural gas. As energy historians in this volume and elsewhere remind us, at the same time that the amount of energy consumed was increasing dramatically, so was its intensity: stoves and later steam engines and lighting sources became increasingly efficient in the nineteenth century, a trend that only accelerated in the twentieth.[61]

Chapters on hydroelectricity, coal, oil, and gas explore how Canada's large reserves of high-energy stocks of energy were developed, used, and exported in the twentieth century. They transformed important aspects of the economy and society, changing it from a resource-poor, energy-importing nation to a resource-rich, energy-exporting nation, with huge boosts during both world wars. Manufacturing and transportation grew rapidly in response through a series of positive feedback loops. In the later nineteenth century, coal (about half of it imported[62]) powered trains, lowered transportation costs, and increased the number of factories and workshops. Indeed, on the basis of abundant wood and rapidly expanding coal use, "Canada enjoyed a pace of growth in economic activity from 1870 which was among if not the highest in the world."[63] Resource extraction expanded and became vastly more profitable. Mines (silver, gold, nickel, copper, uranium, molybdenum, and asbestos on the Shield, and various minerals including coal in western and eastern Canada) became safer and exponentially more productive as the result of electric lighting, ventilation, motors, furnaces, pumps, and improved transportation by rail, steamships, roads, and (after the First World War) cargo aircraft.

Chapters emphasize as well that older forms of energy within Canadian resource industries coexisted with the new. Lumber had long been a staple of the Canadian economy, and harvesting increased in the North and West with the development of steam-powered "donkeys" and later bulldozers and truck logging in the 1940s and as a growing network of railways, roads, and pipelines spanned the country. Transforming trees into paper depended on the right chemical mix, abundant forests, and a supply of cheap labour, as we will see below, much of it still performed in the forests by horses and men cutting and moving trees with the same tools and vehicles in use a hundred years before.[64] Also of key importance to the Canadian economy were highly sophisticated new mills for processing metals that needed massive amounts of electricity. For example, early in the twentieth century multinational copper and aluminum interests relocated to Canada, enticed in some cases by explicit promises of cheap and abundant hydroelectric power, and during the Second World War, industrial demand from the military boosted production and consumption dramatically.[65] Hydroelectricity had been developed early in central Canada, where the absence of coal meant importing high-quality stocks from mines in Pennsylvania. As Andrew Watson shows in this volume, the country's energy vulnerability caused increasing concern to central Canadians after a protracted miners' strike in 1897 threatened vital supplies, and in the early twentieth century the Ontario government created one of the world's earliest publicly owned electrical utilities.[66] The smaller stations of the pre–Second World War period gave way to hydroelectric mega-projects dotting the country by mid-century, a trend that was in turn facilitated by a massive surge in the use of liquid petroleum as a fuel for military-industrial activity during the war, as well as for internal transportation and home heating. By the 1940s, when the internal combustion engines in cars, trucks, buses, and bulldozers were well-established, oil and hydroelectricity had largely replaced coal, including steam-powered machines, for most applications. It would not be until the 1980s, when thermal energy from coal was used to supplement nuclear and hydroelectric power for generating electricity, that coal burning would again reach the pre-1940 levels. While southwestern Ontario had exploited some relatively small stocks of oil and natural gas from the 1850s, oil production moved to southern Alberta with the first major Turner Valley finds in 1914. As Steve Penfold describes, the huge Leduc Strike of 1947 put Canada on the map as a significant producer, and national and international pipelines soon linked Alberta to profitable markets in the United States, as well as in eastern and western Canada.[67]

The Canadian oil and gas industry was developed almost exclusively as a US subsidiary, with the United States owning, extracting, and consuming much of Canada's oil supplies as its own dwindled. Between 1928 and 1947 the famous privately owned conglomerate, the Seven Sisters, controlled global oil supplies. Between 1950 and 1970, about half of Canada's surging supplies of oil were sent to the United States, which, as the result of the Cold War, was increasingly interested in securing "domestic" (i.e., North American) supplies.[68] It was in this context that the National Energy Board was created in 1959, after a royal commission of 1958 recommended that Canada reject protectionist energy policies. In the 1960s, when foreign ownership in oil and gas operations had reached 77.3 per cent, policies were implemented to encourage exports of oil, gas, and hydroelectricity; it "suited the Canadian government's general orientation to growth based on resource exports; private enterprise produced and marketed oil and gas."[69] While electricity (hydro, coal, and nuclear) has tended, by contrast, to be publically owned across the country, provincial governments have remained in control of most energy resources, creating in Canada a patchwork of regulatory and legal structures for extracting and transporting oil, natural gas, and electricity, except for export, which falls largely under federal jurisdiction. The nuclear industry in Canada, however, has been tightly regulated at the federal level, and until recently state owned. As Laurel Sefton McDowell argues in this volume, it is distinguished from oil, gas, and other electric developments in other interesting ways as a result. Overall, oil- and gas-producing provinces have profited considerably from these key resources, which are closely guarded by governments and corporations alike.

For a brief moment between 1973 and 1984, OPEC's decision to dramatically increase oil prices in light of war in the Middle East created such an "oil shock" that the federal government considered actively monitoring and managing Canada's supplies of energy, using a series of tariffs and price controls. It even worked out a federal National Energy Policy in the attempt to guarantee Canada's access to energy supplies ("energy security"), particularly in light of Canada's very high rates of foreign ownership of oil and gas. The moment passed, and the legislation was never implemented.[70] Over the last forty years, there has been a massive expansion of oil and gas extraction in Canada, mainly from the tar sands of Alberta and off-shore drilling off Newfoundland and Nova Scotia. As Duncan and Sandwell show in this volume, natural gas production and consumption have soared along with exports. Governments have been reluctant to impose regulations, let alone limits on production, even in the wake of growing concern about environmental risks of extracting, transporting, and burning fossil fuels. Exploration in the North continues, and British Columbia is considering removing its moratorium on oil and gas exploration off the West Coast. Canada is already heading away from a path of cheap and available energy, to costly and less available energy, but this change may be mitigated for a few more decades by a strange brew of factors: increased global supplies, novel efficiencies in energy use, more efficient methods of extracting oil and gas from difficult sources, and a political will heavily invested (mired?) in what has been the highly profitable status quo.[71]

By 1980 Canada was displaying all the characteristics of "first world" mineral economies: dependence on centralized sources of energy from the mineral economy, complex and highly specialized networks of energy distribution, and standards of living much higher than at any other time in history.[72] The new reliance on easily transportable huge stocks rather than flows of energy tended to separate sites of production and consumption, often by vast distances. Canada was sorted into single-purpose areas: cities, parks, suburbs, highways and other transportation corridors, some industrial areas, particularly pulp and paper harvesting and production, mining, flooded lands for hydroelectric development and cleared corridors for transmission lines, zones for gas and oil extraction, refining, and waste. Also apparent was the unequal division of monetary benefits and costs so characteristic of the new energy regime.[73] By the late nineteenth century, proximity to the new industries of the mineral regime, particularly those related to manufactured gas and oil refining, had already become new markers of poverty and ill-health within Canadian cities and towns.[74] Across the country, rural areas, particularly provincial norths, were becoming identified as the resource- and/or energy-abundant places designated for massive resource development projects. The resulting pollution and ecological degradation often guaranteed that many areas remained uninhabitable, even after the resource industry had moved on to more profitable ventures, ensuring that the mineral economy would leave a long environmental and social legacy.[75] The fact that most residents of Canada's North are Indigenous peoples with treaty rights adds another dimension to the inequalities imposed by the new regime. Many rural people continue to pay a relatively high price, in their economies, environments, and health, and receive relatively few benefits of lucrative oil, gas, uranium, and hydroelectric sales. Extreme regional variations continue to define Canada, east and west, north and south, urban and rural.

Chapters in this collection point to the highly uneven process of the change, or rather, the series of changes that marked the transition that Canadians made to near-total reliance on fossil fuels and hydroelectricity by the end of the twentieth century. Electricity, for example, was available in homes in Victoria and Toronto from the 1880s, while those living in rural areas across the country typically had to wait another seventy years for electric light and power. Coal was used for a variety of industrial purposes from the 1840s, but it was only in areas close to railways or canals that it displaced cheap and abundant stores of firewood for home heating and not until the early decades of the twentieth century. Many Canadians made the switch from firewood to natural gas or electricity, without ever using coal or oil, and it is only in southern Ontario and New Brunswick that nuclear power generates significant quantities of electricity. Some people took up and then abandoned new energy carriers, particularly in times of economic hardship, or when frustrated by the difficulties and unexpected expenses involved in making a change.[76] While uneven uptake marked the energy transition from organic to mineral in most countries, Canada remained unusual in the variety of energy choices that remained available to many over the last two hundred years.[77] Certainly the coincidence of the country's vast size, its geological history, and the juxtaposition of this lightly populated country next to a large and powerful industrial neighbour to the south does much to explain these distinctive patterns both in energy use and in the development of new energy carriers for export.

This volume highlights two other energy characteristics of Canada that are distinctive and related: high per capita consumption and relatively late adoption of modern fuel carriers, for Canadians were very high consumers of energy per capita long before the mineral regime turned Canada into a net energy exporter. Evidence suggests that they consumed more than two and a half times the energy (mostly from fuel-wood) of their counterparts in coal-burning England in the 1820s, a leading place – if such a term can be used – that they lost by the late nineteenth century and regained in world terms by the 1970s.[78] Large consumers of energy, diverse in their use of fuels, Canadians also differed overall from their industrializing counterparts in other countries by being much later in making the transition from the organic to the mineral regime. As plate 1.2 illustrates, well into the twentieth century Canadians were still relying heavily on flows and stocks (in the form of trees) of energy from the organic economy. By contrast, England and Wales were meeting 90 per cent of their energy needs with fossil fuels (coal) by 1845, the United States by 1915; it was only in 1955 that Canada reached the 90 per cent mark, and it was only in 1980 that Canada joined other industrialized countries in obtaining 98 per cent of its energy from modern carriers. In this volume MacFadyen recalculates the amounts required in Canadian homes and finds that estimates of wood consumption have been short by at least 50 per cent in the nineteenth century and 120 per cent in the early twentieth century. Wood consumption doubled in Canada between 1871 and 1911. Throughout the nineteenth century, most energy used in Canada was coming from firewood; as late as the beginning of the Second World War, more energy still came from burning wood than crude oil.[79] Chapters in this volume find Canadians using organic energy in many ways seldom recorded or quantified in other studies or other countries. Similarly underlining Canada's distinctiveness is the fact that an early massive resort to hydroelectricity in many parts of the country was followed by the relative eclipse of hydro power at the national level as petroleum products began their ascent.

Both of these trends – high energy use and late adoption – have been explained by Canada's geography and geology, which provided both the demand for and abundant supplies of energy, organic and otherwise.[80]

Canadians have a high demand for energy: more energy, quite simply, in the form of food, heat, and light is needed to support life in this vast, northern, dark, and cold country. Unger and Thistle use these factors to explain their estimates that Canadians consumed about six times as much energy as Europeans around 1800.[81] Abundant stocks of organic energy in the form of trees, food, and rushing water provided abundant energy long before coal, oil, natural gas, and electricity were used. Later, with an economy rooted in extracting and processing these and other natural resources abundantly available in this country, particularly in mining and pulp and paper, Canadian industries consumed huge amounts of energy, amounts that increased exponentially as fossil fuel–powered machines replaced human and animal labour.[82]

Another factor accounting for Canadians' high per capita energy use has been distance. Within the organic energy regime, Europeans first settled on large (by their standards) pieces of land, often far from each other, creating, as Cole Harris famously described, an archipelago of settlements rather than integrated settlement across the country. As chapters in this volume document, moving people, food and fuel energy, and goods required large amounts of energy. Transportation absorbed increasing amounts of energy as the new mineral regime made it possible for the first time to transport fuels and other goods, particularly iron ore, lumber, grain, and manufactured goods across the vast distances. Transportation became cheaper almost everywhere. And it now made sense to transport people, along with the energy supplies they needed, to work in remote extractive industries across the country, if only seasonally, and for a few months at a time. As we have seen, freed from the imperative of sustainability by huge stores of cheap energy, people were able to live and work just about anywhere, expanding economic growth dramatically. In Canada, most of the land area is now dedicated to resource extraction, machines use most of the energy, and people are stored in cities, where they can purchase an increasing array of goods transported from around the world.[83] The implications of the transition to the mineral economy have, therefore, been profound for the history of Canada.[84]

Canada's geography, geology, settlement patterns, and resource-based, energy-intensive industrial development go a long way to explaining the country's high per capita energy consumption rates, but their role in slowing the transition to modern energy carriers is a little more problematic.[85] Given this sparsely populated country's well-known difficulties with labour shortages, as well as its early and vigorous exploitation of some modern energy carriers in certain parts of the country (particularly coal, steam, and hydroelectricity in Ontario and Quebec), and its quick adoption of mechanization within the organic regime (new horse-powered machinery was widely adopted in the nineteenth century[86]), it is more than a little surprising that Canada was so much later than Britain and the United States in using "modern" proportions of mineral energy. Canada's relatively late adoption of tractors, cars, and trucks, and indeed electric stoves, home appliances, and coal or oil furnaces is notable. Notwithstanding high energy use in a few sectors, in an important sense, as Joy Parr has observed, "the industrial revolution, though presaged in the National Policy days of the late nineteenth century, really came to Canada after World War II,"[87] a point well-supported by Unger and Thistle's energy data. Abundant supplies of organic energy carriers are not sufficient to explain what Unger and Thistle call Canada's "lag."

Chapters in this collection suggest that the relationship among apparently competing energy carriers might have played a role in Canada's late transition. As the mineral energy regime increased the amount and the pace of energy used in manufacturing, agriculture, and transportation, one initial consequence was to increase, not decrease, amounts of energy from Canada's rich organic economy. More trade, more equipment, and better transportation meant more manufacturing workshops, from barrel-makers to brewers, and more work in Canada's vitally important resource extraction industries, including mining, fishing, and logging. All relied heavily on horses, oxen, human labour, and firewood until the end of the nineteenth century and beyond. Even the electrical dams of the early twentieth century relied on draught animals to excavate sites and to carry in the first electrical machinery, wires, and cables. The number of people moving about increased dramatically with growing use of steam-powered trains, but the risk of fire associated with steam trains limited their use in cities for many decades in the late nineteenth and early twentieth centuries. The number of urban horse-powered vehicles, therefore, increased dramatically. Horse-drawn (i.e., horse-powered) machinery such as seed drills and gang plows transformed Canadian agriculture in the nineteenth century. As Little, as well as Dean and Wilson show in this volume, notwithstanding the rapid development of cars, trucks, and tractors in some areas, the number of horses kept rising in Canada until the 1920s, and Canada's largest railway-building era occurred over four decades after the driving of the "last spike."[88] In 1921, only 6 per cent of farms had tractors, a figure that rose to only 21 per cent in 1941. As Clayton and van Huizen demonstrate, the mineral economy first boosted and did not initially reduce water-power use in some areas of the country, and Sager emphasizes the well-known positive role that competition from steam power had on the growth of sailing ship trade in the late nineteenth century.[89] The chapters in this collection document the long persistence of the organic energy regime in Canada and invite more research into the complex interrelationships among energy carriers to explain rates and kinds of energy transitions.

The country's distinctive political economy before the Second World War may provide another important context for understanding Canada's energy history, for not only did Canada possess abundant supplies of organic energy, but a large proportion of the population – the rural – also had relatively free access to that energy. While historians have tended to emphasize the country's urbanization and industrialization from the late nineteenth century, Canada remained a rural country in some important respects as well until the 1940s, including in employment and population distribution. The number of farms in Canada peaked in 1941, and it was not until 1951 that the number of men working in manufacturing exceeded those in agriculture.[90] Notwithstanding the fact that in 1921, a majority (51 per cent) of the population was designated "urban" for the first time, it was not until 1941 that most people were living in towns and cities with a population over 1,000, and it was not until 1961 that a bare majority were living in urban areas with more than 5,000 people. The rural population grew continuously in the century after 1871, more than doubling from 3 million to about 7.5 million in 1971. It fell for the first time ever in 1976, and has been stable ever since.[91] Canada was a rural place much longer than other industrializing countries.

Late nineteenth- and early twentieth-century Canadian rural households, furthermore, differed from those in industrializing European countries in some important respects. While rural economies varied considerably across the country and over the 1850–1950 period, they shared some common and distinct features. Rural Canadians were not all farmers, or at least not all the time. An abundance of land, and a generous system of land granting (itself premised on a basic principle of the organic economy: a country needed to produce its own food locally) meant that many Canadians had much greater access to the energies available on their own and neighbouring lands "for free" than did European populations. They needed it; marginal lands and a short growing season often limited the production and sale of agricultural commodities. Even as Canada was becoming a major international exporter of food energy by the early twentieth century (wheat), the vast distances implied by large land holdings and low population density in this vast country meant that transportation to and from markets, or indeed the railhead, was often difficult and expensive before year-round roads became common in the 1940s. With flows of inexpensive food and wood fuel readily available through the organic economy, many rural Canadians were used to supplementing their energy needs with a wide range of self-provisioning activities, including hunting, gathering, fishing, gardening, and, perhaps most significant from the point of view of energy consumption, cutting wood for heating homes, barns, and food.[92]

This is not to suggest that twentieth-century Canadians were backward peasants excluded from the mineral economy and capitalism's growing international reach. Most post-contact Canadians were never completely self-sufficient in commodities, relying on a variety of purchased and traded goods characteristic of the organic regime – from leather harnesses to imported silks, coffee, and rum, to commercially produced iron cookware and machinery – transported through wind and water power of the organic regime until the 1860s.[93] As early as the 1840s, products and processes emerging from the mineral economy in Britain, including gas lighting, steam engines, and steel, were already being imported into colonial Canada and beginning to change the ways people worked, travelled and lived, particularly in the towns and cities. Many rural Canadians sold their agricultural produce and purchased a range of goods within the expanding markets of the globalizing mineral economy and its transportation networks. And they also sold their labour and that of their draught animals, usually in a seasonal, part-time, and intermittent way, to rural industries – particularly the global resource extraction industries of forestry and mining growing up so rapidly in rural Canada in the early years of the twentieth century.

Off-farm wages, indeed, added a third pillar of support to rural households, providing a welcome supplement to self-provisioning activities and commodity sales, and allowing many families to remain on their own lands, with their own access to abundant energy and other resources, well into the twentieth century. In this hybrid energy regime, organic energy continued to play a vital role in agriculture and the rural extractive industries alike throughout the first half of the century. "The great bulk of power used in most production processes was derived from organic sources. Human or animal muscle power ... was the prime mover for most agricultural, industrial and transport operations."[94] Rural dwellers, Native and non-Native alike, often working with their draught animals, comprised an important reserve army of labour, working in bush camps across the country, sustaining the nascent industries of the mineral economy in the years before modern infrastructures provided the better working conditions, more stable employment opportunities, state-supported employment insurance, and higher wages that eventually tempted rural dwellers off the land and provided a more stable investment for industry.[95]

Until the Second World War, rural and many urban people were still deeply embedded in organic energy regimes. Patterns of work continued to be determined by seasonal variations within local environments, not the regular and standardized patterns of indoor modern industrial processes.[96] Every year before 1940, millions of farmers, loggers, road crews, and construction workers across the country laboured on a seasonal basis, during daylight hours only, using their labour and that of their animals to move trees, or gravel, or grain.[97] Women continued to tap into the organic energy regime by growing and raising food and other materials for consumption and sale, well into the mid-twentieth century. One man summed up its importance in his description of growing up in rural Canada in 1930s and 1940s (in what Neil Sutherland has called "the age before television"): "Arduous, endless and generally mindless work remained a central element" of childhood; rural people worked within "two endless wheels of labour ... one within the other: a daily round of chores spinning inside the greater circle of seasons' tasks."[98] The organic regime was still very much part of Canadian life well into the twentieth century.

Access to abundant energy from the organic regime limited the rural appeal of fossil fuels and hydroelectricity; even when these were available, they were typically much more expensive, and the service was much poorer than in cities. Many of the daily practices of rural Canadians, furthermore, simply did not fit well with the demands of the mineral regime, which relied on dense population nodes for the economies of scale to make them cheap and convenient. Farmers were reluctant to pay year-round for a relatively expensive electrical service that would be used only a few days a year to the full capacity they were paying for daily. Women were reluctant to pay up front for the expensive wiring required for an electric stove, for the unfamiliar stove "hardware," and on an ongoing basis for a supply of energy that had previously been free or relatively inexpensive (in cash, if not in labour) for most rural families. And this for an energy source that provided only a single function; unlike the wood stove that cooked, and heated water, the kitchen, and often the entire house, that dried wet and snowy clothes, and also provided the heat energy used for ironing, the electric stove was effective in serving really only one energy function: heating food.[99]

Loggers and farmers alike also found it difficult to pace their work to the exigencies of the steam engine, which required hours to "get up steam" and could be operated only by experienced engineers, which caused fires, and whose cumbersome transportation to fields and forests often created much more trouble than readily available horses and oxen. It was not until the internal combustion engine, with its flexibility in mobility and start-up time, that the mineral economy first met the needs of rural people.[100] Later still, small electric power tools further transformed their possibilities. The poor fit between the rhythms of the organic regime and the rigid demands of the mineral, and not simply the "supply-side" problems of getting electrical and other networks to low-density areas certainly helps to explain why wood, as Joshua MacFadyen details in his chapter in this volume, still provided the dominant fuel for almost half of all Canadian homes in 1941, and over 80 per cent of rural homes.[101] The late adoption of the mineral energy regime speaks to this distinctive political economy of the country until the Second World War, and some profound changes to it after.

These and other issues raised in the following chapters warrant much more exploration in Canada's energy history. This volume will also facilitate a better accounting of the Canadian path to massive reliance on modern energy carriers. And it is our hope that it will stimulate much more research and discussion about the role of energy in Canada's past, present, and future.

NOTES

1 My thanks to Eric Sager, Steve Penfold, Sean Kheraj, Josh MacFadyen, and the anonymous reviewers of McGill-Queen's University Press for their extremely helpful and insightful comments on earlier versions of this chapter.

2 Unger and Thistle note that it was not until the 1970s that Canada's dependence on traditional forms of energy fell to the negligible levels reached by England in the early decades of the nineteenth century. Unger and Thistle, *Energy Consumption in Canada in the 19th and 20th Centuries*, 77, 53.

3 In 2001, 62 per cent of annual oil production and 55 per cent of annual gas production was exported, primarily to the United States. Canada, Statistics Canada, *Human Activity and the Environment: Annual Statistics 2004*, 2.

4 Kander, Malanima and Warde, *Power to the People: Energy in Europe over the Last Five Centuries*, 2.

5 The purpose of E.A. Wrigley's *Continuity, Chance and Change: The Character of the Industrial Revolution in England* is to urge a re-evaluation of the Industrial Revolution from the perspective of energy use. See as well

Wrigley, "Reflections on the History of Energy Supply, Living Standards and Economic Growth," esp. 19–21, and Wrigley, *Energy in the English Industrial Revolution.*

6 Kander, Malanima, and Warde, *Power to the People,* take this argument a little further, saying that most economic historians, however, have no interest in energy history because they believe that (1) energy occupies too small a proportion of the income of modern countries to be significant; (2) economic growth is entirely explained by knowledge, technical progress, and capital; (3) and (most remarkably) that economic growth was created by increases in economic efficiency that will continue indefinitely, with the result that "the economy can dematerialize, or get rid of its dependence on energy or other material resources almost entirely" (10). As the authors note, "We show that this is a false belief" (7–10).

7 Wrigley, *Continuity, Chance and Change,* 5–6.

8 A number of books draw on international comparisons to explore energy transitions, but while their work can be extremely important to historians, they often assume a straightforward economic relationship between energy use and historical change. See, for example, Darmstadter, Dunkerley, and Alterman, *How Industrial Societies Use Energy;* Hausman, Hertner, and Wilkins, *Global Electrification: Multinational Enterprise and International Finance in the History of Light and Power, 1878–2007;* Schurr et al., *Energy in the American Economy, 1850–1975;* and Fouquet, *Heat, Power and Light: Revolutions in Energy Services.* Thomas Hughes provided an early comparative social history of electricity in Berlin, Chicago, and London in *Networks of Power: Electrification in Western Society, 1880–1930.* Most recently, Kander, Malanima, and Warde, *Power to the People,* explores the history of energy in western Europe over the last five centuries, providing new comparable datasets and new explanations of the relationship between energy and society. Most countries have a national or provincial/regional history of one or more of the new energy carriers, and its influence. In Canada, the work of the following authors has been particularly important in economic and political analyses of various forms of energy: Nelles, *The Politics of Development: Forests, Mines and Hydro-electric Power in Ontario, 1849–1941;* Nelles and Armstrong, *Monopoly's Moment: The Organization and Regulation of Canadian Utilities, 1830–1930;* Lower, *Settlement and the Forest Frontier in Eastern Canada* and Harold A. Innis, *Settlement and the Mining Frontier;* Drummond, *Progress without Planning: The Economic History of Ontario from Confederation to the Second World War;* Mochuruk, *Formidable Heritage: Manitoba's North and the Cost of Development, 1870 to 1930;* Piper, *The Industrial Transformation of*

Subarctic Canada; Wynn, *Canada and Arctic North America: An Environmental History*. The "oil shock" of the 1970s prompted some historical consideration of Canada's relationship to oil and natural gas, as well as reflections on its future. See, for example, McDougall, *Fuels and the National Policy*; Laxer, *Canada's Energy Crisis*; and Laxer, *Oil and Gas*.

9 Internationally, the cultural study of energy history has perhaps been a little more developed, as in the works of Bruno Latour, David Harvey, and Henri Lefebvre. See also Hecht, *The Radiance of France: Nuclear Power and National Identity after World War II*; Ross, *Fast Car, Clean Bodies: Decolonization and the Reordering of French Culture*. In Canada, histories exploring the impact of technology on women's work, lives, and status comprise an important contribution. For a recent summary of the work of such historians as Joy Parr, Bettina Bradbury, Dianne Dodd, Emanuela Cardi, and Dorotea Gucciardo, about gender and energy, see Sandwell, "Pedagogies of the Unimpressed: Re-Educating Ontario Women for the Mineral Economy, 1900–1940."

10 See, for example, Russell et al., "The Nature of Power: Synthesizing the History of Technology and Environmental History"; Macfarlane, *Negotiating a River: Canada, the United States and the Creation of the St Lawrence Seaway*. There are two recent general collections on the envirotech theme: Reuss and Cutcliffe, eds., *The Illusory Boundary: Environment and Technology in History*; Jørgensen, Jørgensen, and Pritchard, eds., *New Natures: Joining Environmental History with Science and Technology Studies*. Other environmental historians have been conducting important research on the material impact of changing energy carriers on the environment. Many of these are discussed in the relevant chapters here, but see, for example, Manore, *Cross-currents: Hydroelectricity and the Engineering of Northern Ontario*; Loo, "People in the Way: Modernity, Environment and Society on the Arrow Lakes"; Parr, *Sensing Changes: Technologies, Environments, and the Everyday, 1953–2003*; Armstrong and Nelles, *Wilderness and Waterpower*.

11 Sinclair, *Energy in Canada*.

12 Unger and Thistle, *Energy Consumption in Canada*; Sinclair, *Energy in Canada*. F.R. Steward also compiled a statistical account of Canada's energy consumption between 1867 and 1970, which provides estimates similar, though not identical, to Unger and Thistle's. Steward, "Energy Consumption in Canada since Confederation." Also useful is the chapter "Changing Energy Regimes" in MacDowell, *An Environmental History of Canada*.

13 Important exceptions that have provided invaluable models, approaches, and resources for the authors of this volume are Crosby, *Children of the Sun: A*

History of Humanity's Unappeasable Appetite for Energy; Smil, *Energy in Nature and Society: General Energetics of Complex Systems*; and Smil, *Energy at the Crossroads: Global Perspectives and Uncertainties*. Williams's *Energy and the Making of Modern California*, with its emphasis on the social and environmental history of energy use in a state with about the same population as Canada, also provided an important model for this book.

14 Wrigley, *Continuity, Chance and Change*; Wrigley, "Reflections on the History of Energy Supply."

15 Wrigley argues that England had characteristics in the early modern period that created an "advanced organic economy" – including appropriate legal frameworks, protection of property rights, secure enforceable contracts, removal of constraints on use of capital and freedom of labour, ensurance that governments refrained from arbitrary taxation, encouragement for freedom of trade, promotion of specialization of labour, mechanization, and restriction of births through late marriage – so that "societies could liberate powers of production long frustrated and suppressed by the ineptness of feudal or mercantile states." These measures could advance economic growth within the organic economy, but could not escape its fundamental limitations – the finite nature of land. Wrigley, *Continuity, Chance and Change*, 4.

16 Ibid., 18.

17 Jones, *Routes of Power: Energy and Modern America*, 3.

18 The expression "endosomatic" is McNeill's, described in *Something New under the Sun: An Environmental History of the Twentieth Century World*, 11. See Colpitts, this volume, chapter 2. Wynn, *Canada and Arctic North America*, 113.

19 There is a debate over the classification of peat as a biomass or fossil fuel. The United Nations Framework Convention on Climate Change calls it a fossil fuel, but the Intergovernmental Panel on Climate Change calls it a "slow-renewable" biomass fuel. Apparently it regenerates at only 1 mm per year, and only in 30–40 per cent of peatlands.

20 Simmons, *Environmental History: A Concise Introduction*, 37.

21 De Molina and Toledo, *The Social Metabolism: A Socio-Ecological Theory of Historical Change*, 155, 267–75; Christian, *Maps of Time: An Introduction to Big History*, 80. Thanks to Joshua MacFadyen for drawing these to my attention.

22 Wrigley, *Continuity, Chance and Change*, 5.

23 Ibid., 30.

24 Little, this volume, chapter 3.

25 Wrigley, "Reflections on the History of Energy Supply," 8.

26 Sahlins has famously argued in "The Original Affluent Society" in *Stone Age Economics* that this is a condition peculiar to agricultural, and not hunting and gathering societies, where, he argues people concern themselves not with capital accumulation, but only with subsistence and their families. As a result, they negotiate their relationship with the environment in ways that allow them an excellent standard of living with a minimum of work (energy).

27 Jones, "A Landscape of Energy Abundance: Anthracite Coal Canals and the Roots of American Fossil Fuel Dependence, 1820–1860," 453.

28 Why Canadian Indigenous peoples did not exploit their forests in the ways that Europeans did is a field of energy history ripe for historical examination. As Weaver argues, the movement of millions of people into "new Europes" had the effect of dramatically increasing the world food (a.k.a. energy) supply, particularly of wheat, thereby raising the world's population dramatically. Weaver, *The Great Land Rush and the Making of the Modern World, 1650–1900*.

29 For a fascinating overview of the slow and irregular pace of change in Britain's transition to the mineral economy, see Fouquet, "The Slow Search for Solutions: Lessons from Historical Energy Transitions by Sector and Service."

30 Paraphrased from Wrigley, *Continuity, Chance and Change*, 27.

31 Kander, *Power to the People*, 158.

32 Wrigley, *Continuity, Chance and Change*, 27. As we will see in Josh MacFadyen's chapter, Canada long had relatively small-scale infrastructure for harvesting and transporting wood, but once riparian supplies had dwindled, the cost of transporting wood over land rose substantially.

33 This paragraph is paraphrased from Jones, "A Landscape of Energy Abundance" 453; and Jones, *Routes of Power*, 18.

34 Energy transitions have occupied the attention of numerous economic historians, including Schurr et al., *Energy in the American Economy*; Schurr, *Electricity in the American Economy: Agent of Technological Progress*; and Kander, Malanima, and Warde, *Power to the People*. The term *development blocks* is from the last. For a thoughtful historical overview and analysis of energy transitions generally, and to see a handy table depicting this erratic process in Britain across various centuries and energy carriers, see Fouquet, "The Slow Search for Solutions," 6591.

35 Mitchell, *Carbon Democracy: Political Power in the Age of Oil*, provides a sweeping overview of the ways in which fossil fuels have transformed global politics and social relations.

36 Jones, *Routes of Power*, 18. Jones avoids the technological determinism trap and shows how these "social and technological worlds are

co-produced" (8). Other historians have called this "technological momentum." See especially Hughes, "Technological Momentum." See also
Christian, *Maps of Time,* 252–3.

37 Wrigley, "Reflections on the History of Energy Supply," 20.

38 Schivelbusch, *Disenchanted Night: The Industrialization of Light in the
Nineteenth Century,* 44.

39 Kunstler, *The Geography of Nowhere: The Rise and Decline of America's
Man-Made Landscape.* On the inequality of the distribution of costs and
rewards of the mineral energy regime, see Jones, *Routes of Power,* esp.
10–14. For a reflection on the spatial dimensions of energy use in Canada,
see Sandwell, "Mapping Fuel Use in Canada: Exploring the Social History
of Canadians' Great Fuel Transformation."

40 Hurley, "Creating Ecological Wastelands: Oil Pollution in New York City,
1870–1900"; see as well Tomory, "The Environmental History of the Early
British Gas Industry, 1812–1830"; Tarr, "Toxic Legacy: The Environmental
Impact of the Gas Industry in the United States."

41 The phrase *zones of sacrifice* is from Kuletz, *The Tainted Desert: Environ-
mental and Social Ruin in the American West.* Parr, "'Lostscapes': Found
Sources in Search of a Fitting Representation"; Parr, *Sensing Changes;*
Lecain, *Mass Destruction: The Men and Giant Mines That Wired America
and Scarred the Planet;* Keeling and Sandlos, "Claiming the New North:
Mining and Colonialism at the Pine Point Mine, Northwest Territories,
Canada."

42 Jones, *Routes of Power,* 18–19.

43 Schurr, *Electricity in the American Economy,* is among those arguing
the pivotal role of electricity in transforming the modern economy.

44 Frost, review of *L'Élecricité et ses consommateurs. Actes du Quatrième
colloque de l'Association pour l'histoire de l'électricité en France.*

45 The role of "cultural factors" in British energy transitions is a matter of
some debate, one taken up enthusiastically by Gooday, *Domesticating
Electricity: Technology, Uncertainty and Gender, 1880–1914;* and
Clendinning, *Demons of Domesticity: Women and the English Gas
Industry, 1889–1939;* as well as by many of the authors in this collection.
For a thoughtful cultural history of electricity in Canada, see Bruce
Stadfeld, "Electric Space: Social and Natural Transformations in British
Columbia's Hydro-electricity Industry to World War II."

46 The implications of these concentrations of power are at the heart of
Kander, Malanima, and Warde's *Power to the People.* For an excellent
overview and analysis in the Canadian context, see Nelles, *The Politics
of Development,* and of the global context, Mitchell, *Carbon Democracy.*

47 "It is well known that the electrical industry was a significant factor in bringing about these changes." Schivelbusch, *Disenchanted Night*, 74. As Thrift observed, electricity played a "critical role in the change to a new form of more integrated capitalism. Society was tied together and made subservient to the new networks of power." Thrift, "Inhuman Geographies: Landscapes of Speed, Light and Power," 206.

48 See Fouquet, *Slow Transitions*, for an overview of some of the approaches and research now directed at the problems of the next energy transition. 6586–8.

49 Fouquet, *Slow Transitions*, 6591.

50 See, for example, EnergyRealities.com, "Per Capita Energy Consumption." In 2010, 80 per cent of Canadian energy consumed was from coal, oil, and gas, and 18 per cent from electricity. In 2007, 27 per cent of that electrical energy was being produced by coal. See Unger and Thistle, appendix; and Statistics Canada, *Electric Power Generation, Transmission and Distribution*.

51 Unger and Thistle, *Energy Consumption in Canada*, 87.

52 Canada, National Energy Board, *Energy Briefing Note: Canadian Energy Overview 2012*. Unless otherwise noted, all 2012 figures presented in this paragraph are from this source. https://www.neb-one.gc.ca/nrg/ntgrtd/mrkt/vrvw/2012/2012cndnnrgvrvw-eng.pdf.

53 Basov, "Top 10 Oil and Gas Producing Countries in 2012."

54 McFarlin, "Top 10 Natural Gas Producers by Country." Russia and the United States ranked first and second.

55 Epstein, "Top Hydroelectric Producing Nations."

56 There is a substantial literature on this topic. See, for example, Pimentel and Patzek, "Biofuel, Ethanol Production Using Corn, Switchgrass, and Wood: Biodiesel Production Using Soybean and Sunflower."

57 Fouquet, "Slow Transitions," 6592; Sager, this volume, chapter 6.

58 Notwithstanding Sinclair's impassioned argument in its favour, and mounting public pressure on the federal government to implement a national energy strategy. Sinclair, *Energy in Canada*.

59 Unger and Thistle, *Energy Consumption in Canada*, 103.

60 See appendix 4a and 4b; Canada, Statistics Canada, *Energy Use Data Handbook*, 1990–2008, table 1, 5–6; Darmstadter, Dunkerley, and Alterman, *How Industrial Societies Use Energy*, table 3.2, 26.

61 See Kander, Malanima, and Warde, *Power to the People*, 9–13; and Darmstadter, Dunkerley, and Alterman, *How Industrial Societies Use Energy*, esp. 9–10 for a discussion of the broader implications of this trend.

62 See Watson, this volume, particularly figure 8.2.

63 Under and Thistle, *Energy Consumption in Canada*, 83.

64 On the persistence of energies from the organic regime in Canada's logging industry well into the twentieth century, see Radforth, *Bushworkers and Bosses: Logging in Northern Ontario.*

65 See Evenden, "Aluminum, Commodity Chains, and the Environmental History of the Second World War"; Evenden, *Allied Power: Mobilizing Hydroelectricity during Canada's Second World War*; Evenden and Peyton, this volume, chapter 9; Lecain, *Mass Destruction.*

66 Fleming, *Power at Cost: Ontario Hydro and Rural Electrification, 1911–58*; Nelles, *The Politics of Development*; Armstrong and Nelles, *Monopoly's Moment*; Watson, chapter 8, and Evenden and Peyton, chapter 9, this volume.

67 Recent studies urge the importance of manufactured gas as not only a significant energy carrier for lighting in nineteenth-century urban Canada, but a major urban polluter, as the gas chapter summarizes. See particularly Tarr, "Toxic Legacy"; and Tomory, "The Environmental History of the Early British Gas Industry."

68 Sinclair, *Energy in Canada*, 25.

69 Laxer, *Oil and Gas*, 7–8; Sinclair, *Energy in Canada*, 25.

70 Laxer, *Canada's Energy Crisis.*

71 Sinclair's *Energy in Canada* is an argument against this probable course of events, as is Klein, *This Changes Everything: Capitalism vs the Climate.*

72 For a detailed discussion of the unequal impact of oil production in different countries, see Mitchell, *Carbon Democracy.*

73 Jones, *Routes of Power.*

74 See, for example, Cruikshank and Bouchier, "Blighted Areas and Obnoxious Industries: Constructing Environmental Inequality on an Industrial Waterfront, Hamilton, Ontario, 1890–1960"; Evenden and Peyton, chapter 9, Duncan, appendix 1, and Penfold, chapter 10, this volume.

75 See, for example, Keeling, "'Born in an Atomic Test Tube: Landscapes of Cyclonic Development at Uranium City, Saskatchewan"; Keeling and Sandlos, "Claiming the New North"; Sandlos and Keeling, "Zombie Mines and the (Over)burden of History"; Duncan and Sandwell, chapter 11, Penfold, chapter 10, and Sefton McDowell, chapter 12, this volume.

76 See Sandwell, "Pedagogies of the Unimpressed," for an overview of some of the problems experienced by Canadians making the shift to the new energy regime.

77 In 1821, Canadians obtained three times as much energy from wood as did people in England and Wales from coal. Unger and Thistle, *Energy Consumption in Canada*, 39, 81.

78 Ibid., 94–6; Darmstadter, Dunkerley, and Alterman, *How Industrial Societies Use Energy*, 186.

79 Unger and Thistle, *Energy Consumption in Canada*, 84, 77–8, 105.

80 This seems to be the explanation favoured by Unger and Thistle, ibid., 105–7.

81 Unger and Thistle, *Energy Consumption in Canada*, 53. They also note that Indigenous people in the sub-Arctic "enjoyed higher levels of food consumption than their European counterparts, as much as 40 per cent greater than English workers of mid-century" (24–5).

82 In 2008, Canadian mining alone used 827 petajoules of energy, manufacturing 641, pulp and paper 612, and petroleum refining 337, with these top four uses accounting for 75 per cent of all industrial energy, and 28 per cent of all energy used in Canada. Natural Resources Canada, *Energy Use Data Handbook: 1990–2008*, table 1, 5–6.

83 Jones, *Routes of Power*, 233–4, 116–20, 176–7.

84 Sandwell, "Notes towards a History of Rural Canada, 1870–1940." The other chapters in that volume are, in one way and another, essays that in effect explore the impact of the mineral energy regime on rural Canada.

85 As Unger and Thistle note, "The question of why British North America lagged behind is one which bothered politicians at the time and historians ever since." *Energy Consumption in Canada*, 105.

86 See J.I. Little, chapter 3, and Dean and Wilson, chapter 4, this volume, on increases in horse power throughout the nineteenth century.

87 Parr, "Modern Kitchen, Good Home, Strong Nation," 666.

88 See Wilson and Dean, esp. note 4.

89 As Fouquet has argued, energy transitions are complex and varied, but two clearly observed variations are the "sailboat effect," whereby competition with a new energy carrier improves an older form (not only in the case of sailboats, but also gas lighting with the invention of the Welsbasch Mantle), vs the stage coach effect, where an older carrier dramatically increases its prices in order to cash in as much as possible before obsolescence. Fouquet, "Slow Transitions," 6592.

90 Demonstrating a pattern very different from England during the nineteenth and twentieth centuries, where only a tiny proportion of people ever owned land, and most rural dwellers had moved to the city by 1850, the number of farm households in Canada kept increasing until 1941. The number of acres being farmed increased steadily between 1871 and 1971. Sandwell, *Canada's Rural Majority, 1870–1940*; Statistics Canada, Historical Statistics of Canada, series A67–69, *Population, Rural and Urban, Census Dates 1871 to 1976*, http://www.statcan.gc.ca/access_acces/archive.action?l=eng&loc=A67_69-eng.csv; M12–22, *Farm Holdings, Census Data, Canada and by Province, 1871 to 1971*, http://www.statcan.gc.ca/access_acces/archive.action?l=

eng&loc=M12_22-eng.csv; M23–33, *Area of Land in Farm Holdings, Census Data, Canada and by Province, 1871 to 1971*, http://www.statcan. gc.ca/access_acces/archive.action?l=eng&loc=M23_33-eng.csv; M34–44, *Area of Improved Land in Farm Holdings, Census Data, Canada and by Province, 1871 to 1971*, http://www.statcan.gc.ca/access_acces/archive. action?l=eng&loc=M34_44-eng.csv. In 1931, more Canadians were still employed in agriculture – almost one out of every three men – than in any other single occupation. According to the 1931 census, agriculture employed 1.1 million of the 3.2 million "occupied" men, and when combined with other rural occupations of logging and fishing, employed twice as many men as manufacturing across the country. *Seventh Census of Canada, 1931*, tables 20 and 27, 7:30 and 7:38. In 1941, there were more than twice as many men (213 per cent) working in in the rural resource occupations of agriculture, logging, fishing, and trapping as there were in manufacturing. *Eighth Census of Canada, 1941*, table 2, 7:7.

91 Sandwell, "Notes towards a History of Rural Canada."

92 Ibid.; Sandwell, *Canada's Rural Majority, 1870–1940*; Sandwell, chapter 13, this volume; Sandwell, "Rural Households, Subsistence and Environment on the Canadian Shield 1901–1941."

93 Craig, *Backwoods Consumers and Homespun Capitalists: The Rise of Market Culture in Eastern Canada*; McCalla, *Consumers in the Bush: Studies of Consumer Buying in Rural Upper Canada, 1808–1861*.

94 Wrigley, *Continuity, Chance and Change*, 27.

95 On the role of First Nations in the formal waged economy, see Lutz, *Makuk: New History of Aboriginal-White Relations*.

96 For a thorough discussion of the changing role of seasonal labour in industrializing cities, see Baskerville and Sager, *Unwilling Idlers: The Urban Unemployed and Their Families in Late Victorian Canada*, chap. 4.

97 Sandwell, *Canada's Rural Majority*, chap. 2, for a detailed description of the work of Canada's large part-time resource-working labour force.

98 Sutherland, *Growing Up: Childhood in English Canada from the Great War to the Age of Television*, 142.

99 Sandwell, "Pedagogies of the Unimpressed"; Sandwell, "Mapping Fuel Use in Canada."

100 Meyer, "The Farm Debut of the Gasoline Engine." For the gendered effects of tractors, see Jellison, *Entitled to Power: Farm Women and Technology, 1913–63*, 184–5.

101 "Table 18: Occupied Dwellings with Specified Conveniences, 1941," *Census of Canada, 1941*, vol. 9, *Housing*.

Plate 1.1   Energy consumption in Canada by energy carrier, 1800–2010 (petajoules)

	1801	1811	1821	1831	1841	1851	1861	1871	1881	1891	1901	1911	1921	1931	1941	1951	1961	1971	1981	1991	2001	2010
Electricity											1	4	17	59	123	214	399	765	1 246	1 706	1 809	1 791
Natural gas												13	16	29	48	92	439	1 260	1 723	2 577	3 718	3 448
Crude oil									2	4	4	13	52	172	325	745	1 637	2 641	3 604	3 018	3 694	3 481
Coal	1	1	2	3	6	10	18	33	60	151	260	648	834	665	1 012	1 178	577	766	1 039	1 336	1 721	1 005
Wind and water						1	1	1	2	1	1	0	0	0	0	0	0	0	0	0	0	0
Firewood	41	56	67	117	153	268	331	365	390	363	303	301	254	292	246	291	85	29	43	60	17	17
Food for people	1	2	2	4	5	11	15	17	20	22	25	33	40	45	55	63	82	104	117	136	172	171
Working animals	3	4	5	7	13	27	37	37	48	61	70	99	132	119	106	48	19	13	14	16	17	17

Source: Unger and Thistle, *Energy Consumption in Canada in the Nineteenth and Twentieth Centuries*, appendix 1.

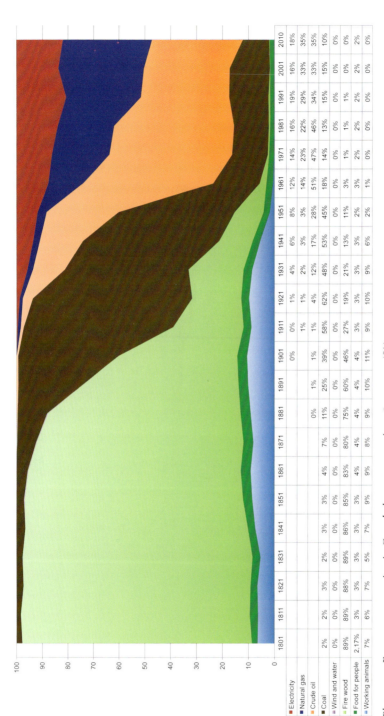

Plate 1.2 Energy consumption in Canada by energy carrier, 1800–2010 (%)

Source: Unger and Thistle, *Energy Consumption in Canada in the Nineteenth and Twentieth Centuries*, appendix 1.

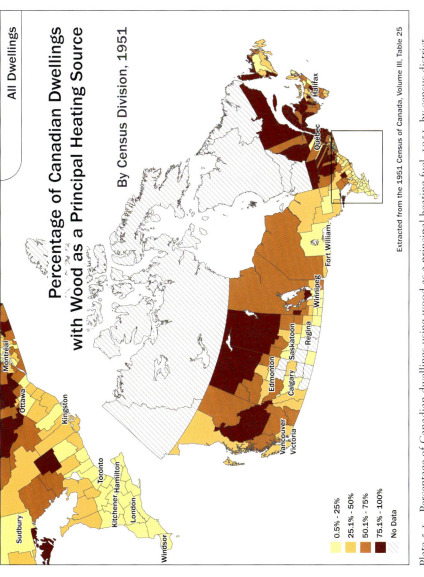

All Dwellings

Percentage of Canadian Dwellings
with Wood as a Principal Heating Source

By Census Division, 1951

0.5% - 25%
25.1% - 50%
50.1% - 75%
75.1% - 100%
No Data

Extracted from the 1951 Census of Canada, Volume III, Table 25

Plate 5.1   Percentage of Canadian dwellings using wood as a principal heating fuel, 1951, by census district

Source: Data from 1951 Census of Canada, volume 3, *Housing and Families*, table 25. Map by Ruth Sandwell, with thanks to Marcel Fortin and his HGIS team.

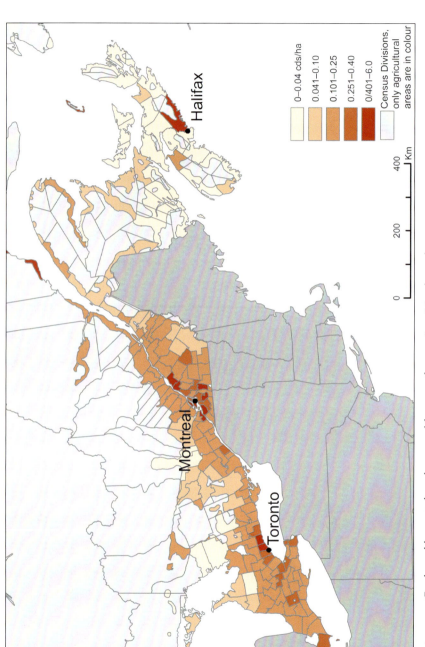

	0–0.04 cds/ha
	0.041–0.10
	0.101–0.25
	0.251–0.40
	0/401–6.0
	Census Divisions, only agricultural areas are in colour

Halifax

Montreal

Toronto

0   200   400
Km

Plate 5.2a   Cordwood harvested per hectare of forest on farms, 1871. Map by author.

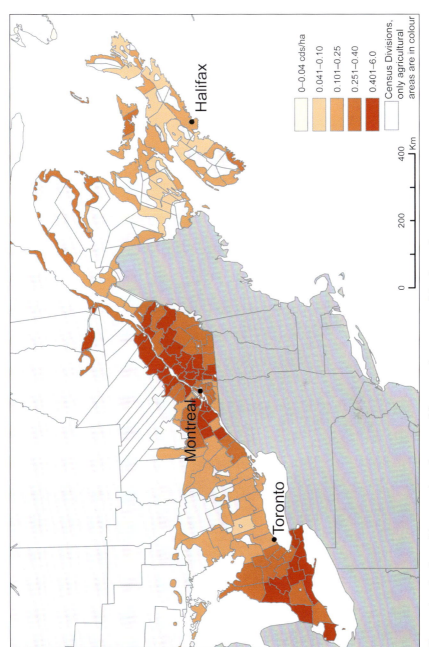

Plate 5.2b   Cordwood harvested per hectare of forest on farms, 1931. Map by author.

Legend:
- 0–0.04 cds/ha
- 0.041–0.10
- 0.101–0.25
- 0.251–0.40
- 0.401–6.0
- Census Divisions, only agricultural areas are in colour

Halifax

Montreal

Toronto

0   200   400

Km

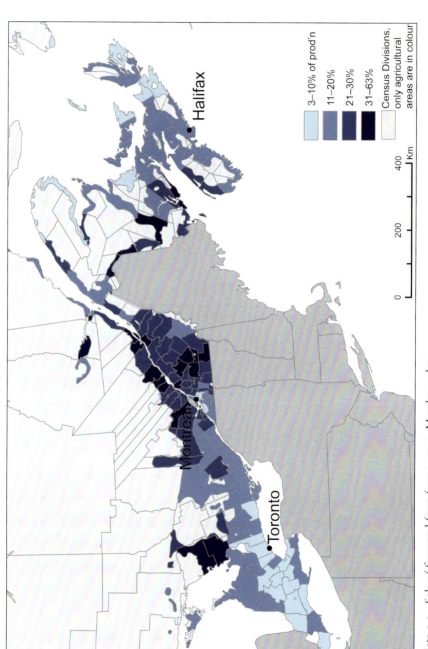

Halifax

Montreal

Toronto

3–10% of prod'n

11–20%

21–30%

31–63%

Census Divisions,
only agricultural
areas are in colour

0   200   400
Km

Plate 5.3   Sale of firewood from farms, 1931. Map by author.

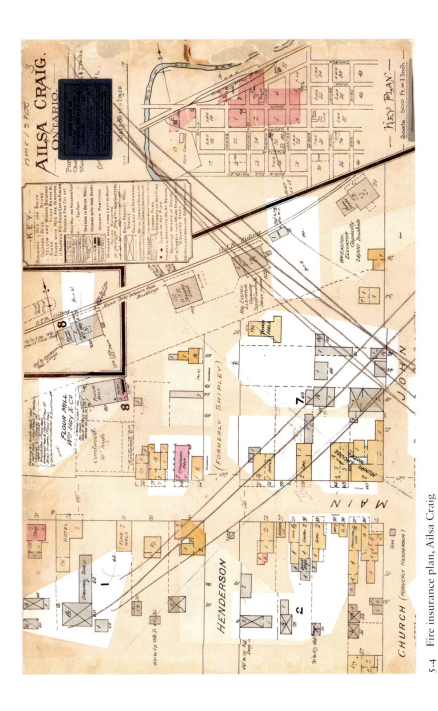

5.4   Fire insurance plan, Ailsa Craig

The William Hay & Company flour mill in Ailsa Craig, Middlesex County, contained three woodpiles, and the largest held over 400 cords, possibly for multiple businesses.

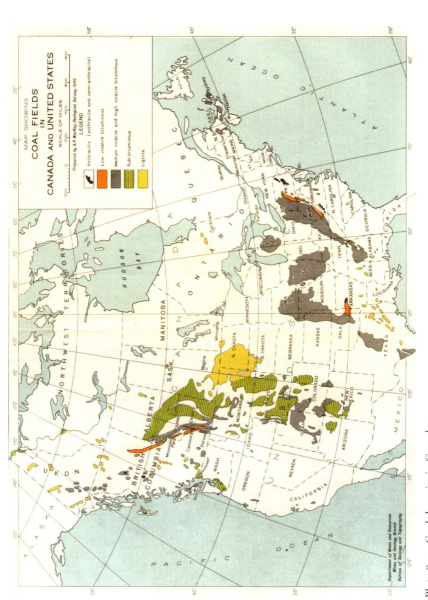

Plate 8.1 Coal deposits in Canada

Source: Canada, Royal Commission on Coal, *Report of the Canada Royal Commission on Coal, 1946.*

PART ONE

The Organic Regime

2

Food Energy and the Expansion of the Canadian Fur Trade

George Colpitts

As a recent global history of eating and feasting has pointed out, "Historians are newcomers to the study of food."[1] Although historians have engaged in questions of diet and nutrition, recently exploring the centrality of food in its social and cultural contexts in "edible histories" of Canada, they have generally overlooked the particular ways that carbohydrates, fats, and proteins actually combusted at a molecular level to drive colonialism, form linkages between rural landscapes and aboriginal territories, or fuel a brawny environmental overreach in Canada's staple resource economies.[2] But however invisible to historians, food has been the foundation of much of the energy and power in Canada. In pre-industrial settings, food energy animated muscle power. Much of the work in back-country farming, logging, fishing, and proto-industry was powered in the hoppy conviviality of countryside taverns, the fatty tourtières of Lower Canada, or even the seasonal clockwork of binge religious feasting. Bursts of caloric energy, in periods of better food production or purchasing power, arguably provided catalysts for historical economic and social change. Food is an energy fact of life, and the history of energy and power in Canada requires its own food energy tally.

The *calorie* unit, invented in the nineteenth century to measure the potential energy of food,[3] seems especially relevant in Canada, given the preponderance and grinding nature of the endosomatic (or human-powered) energy regime dominating much of its earlier past.[4] This chapter will argue that in the case of that quintessentially Canadian endeavour, the fur trade, food energy was probably *the* driver of newcomer and First Nations' relations, unsustainable fur production, and, ultimately, rapid commercial expansion. Food energy in the fur trade changed at critical moments, offering higher orders of potential labour productivity. Indeed, if Harold Innis saw transportation innovations and the opportunities of Canadian Shield river systems allowing for the trade's geographic expansion – a mainstay of his classic "staples" thesis – it was surely food energy that provided a motive force for the national story arc he saw arising from it.[5] This chapter suggests that the Canadian fur trade, centred upon a caloric-poor subarctic prime furbearer habitat, expanded at a key moment when a new food energy regime, based on bison fat, allowed St Lawrence valley and Hudson Bay fur traders to manifestly extend their extractive activities across far larger spaces, indeed across the continent.

THE OLD FOOD REGIME OF THE CANADIAN FUR TRADE

That energy shift took place deep in the interior of North America. In the swaying tall grass prairie of present-day southern Manitoba, in 1793, fur trader Duncan M'Gillivray took note of no fewer than twenty-one trading posts busily operating in the Qu'Appelle, Assiniboine, and Red River valleys alone.[6] Their major trading item was not fur, but food. Companies around him in the North Saskatchewan basin, too, were shifting fur trader diets into the high caloric energy offered in bison flesh and fat. So frantic was food purchasing that these prairie outposts could seem more like abattoirs than commercial way stations. Rafters hung with hundreds, even thousands, of drying tongues. Cauldrons of flammable, occasionally exploding tallow fats littered courtyards. Blood-soaked timbers lined meat sheds. As M'Gillivray observed, the fur trade's very success now hinged on the uninterrupted supply of plains foods, and within them, critical bioenergy for work. If the bison hunt was stopped, it "would prevent progress & stop the Trade," he said. "It is wisely ordered by Providence that this should be so."[7]

Just prior to M'Gillivray's making these observations, the fur trade had reached its absolute geographic limits. Certainly by the 1770s, company men dependent on corn carbohydrates to muscle their canoe voyaging were quite simply running out of fuel.[8] Travelling by canoe from the St Lawrence, fur traders, and their Hudson's Bay Company (HBC) rivals from James and Hudson Bay, began to starve in the difficult transits across the Rainy Lake–Lake Winnipeg corridor or the uplands of the Canadian Shield. Given that most of the fur trade sprawled across the subarctic, an ecological biome with one of the lowest carrying capacities in North America, the fur trade from the start depended on food carried by brigades, not traded from dispersed and small populations of boreal hunting and gathering people.[9] In the next decades, Europeans who would have otherwise starved, harnessed a revolutionary new bioenergy source in bison fat. Offered in the form of pemmican, this food source proved far more effective and converted to unheard of quantities of energy to be released in the heavy labour driving fur trade brigades and York boat crews.

The reason why food energy was so critical to the fur trade was the comparatively lower returns on labour made in canoes and other water craft used to move commodities and traders. Canoe design borrowed from Algonquian allies in New France was admirably suited to subarctic travel. However, it lent itself to none of the technological innovations of the modern age, except that the French, for commercial purposes, were able to enlarge Algonquian canoes and further organize labour in brigades to race time coming and going from their ever-larger trading hinterlands. These limits were not uncommon in pre-industrial settings where relatively small inputs of wind and water power meant that human muscle fibre accounted for up to 70 per cent of all the mechanical energy transforming the environment – perhaps 80 per cent if plant and animal energy inputs are also considered.[10]

But unlike farm economies freed up by breakthroughs in labour-saving devices (the cradle scythe, the fanning mill, or the first seed drills drawn by horse, to name a few),[11] the expanding fur trade reached the limits of its potential work efficiency relatively early. Carolyn Podruchny has described the almost inconceivably difficult human effort to drive the fur trade, one giving rise to a distinctive and "manly" voyageur culture.[12] As Claiborne Skinner simply observed, men employed driving canoes "did physically brutal work in an absurdly long work day."[13] Certainly, the manpower driving the fur trade staggered the imagination of contemporaries. A late eighteenth-century traveller viewing the fur trade described it frankly: "No men in the world are more severely worked than are these Canadian voyageurs. I have known them to work in a canoe twenty hours out of twenty-four, and go on at that rate during a fortnight or three weeks without a day of rest or any diminution of labour." He added, significantly, that "they lose much flesh in the performance of such journies [sic], though the quantity of food they consume is incredible."[14]

Two environmental realities conspired against the fur trade's expansion. The Canadian Shield, although appearing flat, proved a long-distance canoeist's nightmare. It was shaped like a great Precambrian granite saucer, with its centre in the middle of Hudson Bay. Shield rivers flushed seasonally down the "rock" or "upland" roil areas along its rim, and then slowed into flattened lowland stretches. Montreal companies tended to stay high "along the rim" of the shield and use the concentric circles of river routes that followed on or around it, to only occasionally cut across water basin portage points. The Hudson's Bay Company, unfortunately located at the bottom of the shield, had to literally climb rivers. Its traders ascended the Hays, Nelson, Albany, and other Shield rivers on foot, dragging watercraft behind them. Most brigades timed their journeys across the Shield to periods when river draw-down occurred in the later summer and avoided them completely in the spring flush. Wind-battered sections of lakes, baldly exposed big-current rivers, and reed-tangled marshes were skirted as much as possible. The other problem was biotic: the Shield sprawls across the large transcontinental sweep of the subarctic biome, where horticulture was largely non-existent, big game easily hunted out, and aboriginal populations sparse. For that reason, fur companies constantly reconnoitred new routes, bettered seasonal ascents, and quickened their return voyages in order to economize very finite food supplies as best they could.[15]

Even finessing and enlarging Algonquian canoe technology to the needs of commerce in the seventeenth century did not give impressive returns on scale. The Montreal cargo canoes emerging in the 1680s were intentionally supersized, but still ungainly and limited to deep water routes. These *canots du maître* plied the St Lawrence–Great Lakes corridor with crews numbering as few as six but often as many as twelve, and averaging eight to ten; these craft could carry up to 8,000 pounds. Larger canoes by the mid-eighteenth century allowed for the 50-pound *pièce* to be packed in 100-pound packs, sixty of which could be hauled in these larger canoes, along with 1,000 pounds of provisions.[16] When companies spread farther inland on the Shield or into the smaller, more difficult waters of the Hudson Bay watersheds, the "north west canoe" was adopted: smaller, carrying about two tons of cargo, perhaps twenty-five feet in length, and manned by an average of five or six men. Even paddling in unison, and well-disciplined by emergent voyageur traditions – syncopating their paddling to the songs they sang; regular rests about on the hour to have smoke breaks; and ritualized rewards of brandy after extraordinarily difficult portages and dangerous rapids – the canoe frontier reached the outer limits of its productivity and geographic extent by the mid-eighteenth century.[17]

One insurmountable problem of the canoe frontier was the stout and relatively strong narrow-bladed paddle almost universally used by voyageurs. Snout-nosed and blunt to withstand shocks on rock, the voyageur paddle did not displace a great deal of water, and paddlers had to use a short stroke of forty counts a minute to gain any speed.[18] Given the staggering dead weight of a Montreal canoe in water – well over four tons – and the fact that these craft did not glide well, voyageurs paddled almost constantly to keep a canoe in motion. Meanwhile, the relatively short season in northern Canadian latitudes and need to cover sometimes thousands of miles before ice-up forced voyageurs to work almost ceaselessly, twelve- to fourteen-hour days typically. Men, then, would launch canoes often by 4 a.m. and put up them up at 9 or 10 p.m., or with the sun's final setting. In lake country, the working shifts lengthened considerably because ideal voyaging occurred at night, when the wind fell and trips became safer on open water. In such circumstances, either to make up for lost time or to speed up the journey, voyageurs paddled throughout the night, and twenty-four-hour voyaging was not uncommon, especially through the dangerous *traverses* of the Great Lakes.[19]

French fur traders learned from First Nations to smoke tobacco as an appetite suppressant,[20] but their work ultimately produced significant caloric deficits. The common estimate is that voyageurs burnt between 4,000 to 6,000 calories a day.[21] Unlike the images from 1920s silent movies showing strapping and muscular French-Canadian voyageurs, these chronically underfed and overworked labourers were likely seriously underweight and debilitated by personal injuries, bone deformities, and stressed constitutions.[22] A better image of voyageurs is hinted in modern-day transatlantic rowing competitions. Despite consuming up to 8,000 calories a day (65 per cent carbohydrate, 10 per cent protein, and 25 per cent fat), paddlers still suffer daily calorie deficits ranging from 1,200 to 1500 kcal; in one case, these super-athletes on average shed between twenty and twenty-five pounds in fifty-eight days of intensive rowing.[23]

The voyageur's diet in the St Lawrence–Great Lakes corridor was limited in a number of other ways. Since the Shield areas did not provide much food en route, their brigades took on agricultural supplies back in New France or at points supplied by the corn-growing centres around Detroit and the southeastern marsh-tracts of the Great Lakes. The "custom of the voyageur," a phrase likely borrowed from unwritten customary law traditions,[24] was pretty meagre, however: it consisted of leached corn (Indian corn first dried out and boiled before eating to allow its sugars to be better digested), and/or dried peas, and a portion of fat in the form of lard (the Montreal men going to Grand Portage, hence, were called the "lard-" or "pork-"eaters for that reason).[25] Corn supplies from the St Lawrence corridor, however, quickly ran out en route. Supplies later derived from Detroit were prohibitively expensive, as were tallow pork fats available to voyageur crews from that location.[26] The cargo of a three-man canoe in 1694 heading for Illinois country was likely typical, its voyageurs consuming more carbohydrates than fats and very little protein. Their diet was made of a biscuit-to-grease ratio of 6:1. For their entire voyage, a man had forty pounds of biscuit and seven pounds of grease. He also had three gallons of dried peas.[27] From a caloric point of view, these outfits were woefully under-stocked. The diets of men going on to Detroit in the same period were much the same: one hundred pounds of biscuit and twenty-five pounds of lard per man.[28] How this might have been improved, even if there were not such limits on cargo space, is not certain. Skinner's estimate of the *livre* cost of even a late seventeenth-century three-man canoe suggests that food already constituted the greatest expenditure for the outfit: some 55 per cent of total expenses, in fact.[29]

Given the poor carrying capacity of the subarctic, it became a gruesome truism of the trade that the farther the destination, the greater the difficulty in feeding the men going there. Le Sieur de La Véréndrye, who perhaps moved the fur trade the farthest in the early eighteenth century, left Montreal in 1731 to reach Kaminisquia, the farthest point on the other side of Lake Superior after 79 days, to cover about 1,100 miles.[30] Later, outbound voyages, first to Rainy Lake, then Lake of the Woods, and finally, Lake Winnipeg, took 104 days, and were slowed considerably by the deteriorating conditions late in the season. In such extremities of the French fur trading empire, traders searched for Ojibwa and Cree to trade meats, fats, and wild rice. The only and very uncertain supply to be added beyond Lake Superior and the Grand Portage, however, were high-starch wild rice stocks harvested by Ojibwa near Lake of the Woods. In the case of Alexander Henry the Elder, a relatively short voyage for twelve of his men between Michilimackinac and the end point of Lake Superior in the mid-1770s, of about a month, required fifty bushels of corn. The ration of just over four bushels per man would have given about 230 pounds of corn for each. A generous estimate would give each voyageur, then, about 7 pounds of corn a day, or about 3,680 calories, likely well below their own energy needs.[31]

The human body consumed food energy in a variety of ways in the course of a voyageur's travels. Transforming energy into work required the release and combustion of adenosine triphosphate (ATP), which is broken down into chemical energy for muscle contraction. In the fur trade, combinations of muscular contraction dominated work: short-term anaerobic systems provided energy for short bursts of effort for sudden, short-duration exertions – to respond, for instance, to the sudden, temperamental changes in a river's currents and velocity, to wild rapids that imperilled life and limb, and, of course, to the portage. With a 100-pound *pièce* suspended on one's back by tumpline, men would portage in adrenaline-hopped half-mile runs, doubled over. This ATP energy, however, came with a cost. Anaerobic metabolism left a residue of lactate acid that built up and quickly debilitated an individual doing such maximal exercise.[32] James Sutherland clearly reached this maximal muscle exertion when he travelled with Ojibwa over a fifty-six-day period across Shield country. On one day paddling and carrying a canoe over a nineteen-hour stretch, he confessed in his journal that night to being "very unhappy" and "not able to lift my arms."[33]

Greater food energy was more typically released from aerobic metabolism where ATP was created through the burning of glucose. Oxidative systems, sustaining the greatest part of the voyageur's workday of monotonous but tiring paddling, importantly, took energy from fats and carbohydrates to create comparatively vast amounts of ATP. Most paddling, at forty strokes a minute, would have constituted medium-exertion exercise in which this form of muscle energy was released.

Whatever way they metabolized energy, voyageurs rapidly consumed whatever food they could carry inland. They came up against the absolute caloric limits of the canoe frontier as early as the 1770s.[34] After the conquest of New France, British capital prompted the movement of many more Europeans inland, 1,500 to 2,000 miles from Montreal. Starvation awaited many traders upon entering the Lake Winnipeg region and the distant Athabasca drainage. In 1775, Peter Pond and Alexander Henry "barely reached their wintering place" across Lake Winnipeg; their 100 men in twenty canoes came very close to starving to death en route, and they lived miserably on fish during the winter, their corn and wild rice long before exhausted. Another Montreal trader, Thomas Frobisher, led crews that almost starved crossing Lake Winnipeg, and also ended up subsisting only on fish in their push to distant Île-à-la-Crosse.[35]

In the 1770s, fur trade journals more commonly reported cases of exhaustion, hunger, starvation, and even cannibalism in the farther extremes of the Canadian Shield.[36] Even if they arrived at their wintering camps, traders faced worst conditions during the cold season when food supply virtually ended altogether. Benjamin Frobisher in 1775 admitted to his HBC rivals the "great Destress he ware in for Provisions, which ware realy shocking" in what became known as the English River district, just north and east of Lake Winnipeg. One or two of Frobisher's men starved to death during the winter, one had apparently resorted to cannibalism, and Frobisher himself had survived only by eating the post's moose skins and many of the furs traded from Indians – "and even a few garden seeds" imported from Canada.[37] By the time traders in a last gasp reached Lake Winnipeg, swatting mosquitoes and cursing blackflies, food figured as their greatest concern. Europeans learned in a hurry how to fish the vast but capricious waters of the lake, but winter hunting in forests nearby was always difficult, or game populations easily exhausted.[38]

NEW FOOD, NEW ENERGY

It was fortunate that, immediately after their arrival in Athabasca territory, Montreal traders made a critical discovery in native food production and transferred it to their business operations. In the 1770s they made a critical leap of harnessing pemmican energy – a combination of dried, pounded meat, joined with melted fat in almost equal ratios, sewn in bags. In so doing, they decisively shifted diets from their carbohydrate base on one side of the Grand Portage, on the east side of Lake Superior, to one of protein and fat on the other.[39]

Already by the 1780s, local Chipewyan, Cree, and Beaver were providing supplies of bison, elk, and deer during the winter to these newcomers in the North.[40] Europeans systematically converged Dene fats and dried meat of wood bison (*Bison bison athabascae*), converted to pemmican, to the portal of the La Loche carrying place for incoming crews. They, in effect, joined northern winter pemmican to the needs of crews going on to the nascent post system on the Athabasca, Slave, and Mackenzie Rivers. At Fort Chipewyan, Alexander Mackenzie coordinated the first relay system between camps of Beaver and Cree by 1788, when he sent "provisions for the canoes in their voyage out in the Spring."[41] A few years later, a single canoe plying the Peace River would collect about two tons of dried meat for that purpose.[42]

The true takeoff occurred as fur companies began developing a longer-distance food system with which they could feed the fur trade in the barren grounds on pemmican produced far to the south in the parkland and the plains. There plains bison herds (*Bison bison bison*) roamed in seemingly limitless numbers. In 1775, the Montreal trader Alexander Henry the Elder noted the quantities of food already being traded by the plains Assiniboine to Nor'westers for the new purpose[43] (see map 2.1). Indeed, by the mid-1780s Montreal companies were deriving so much pemmican

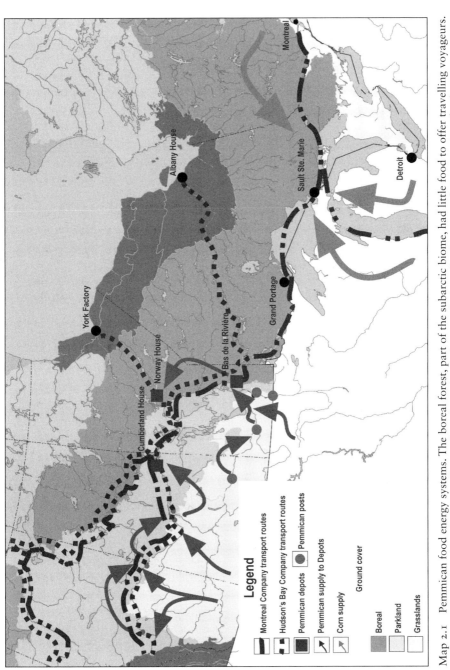

Map 2.1 Pemmican food energy systems. The boreal forest, part of the subarctic biome, had little food to offer travelling voyageurs. The grasslands and parklands areas, offering plains bison herds, became the breadbasket region for a pemmican food energy system to support the Canadian fur trade. Map by author.

that they could trade it cheaply to boreal forest people who, in turn, used it to visit farther areas of the Canadian Shield to trade. John Sutherland, travelling through north central Ontario in an Ojibwa brigade in 1785, noted that the twelve Ojibwa he accompanied, led by Newitchcanisium, were using buffalo pemmican to make marathon-like passages by canoe over the Shield from their visits to HBC posts. Instead of stopping and cooking to eat, they had a little fat "which they mix with Ruahaggan [a form of buffalo pemmican] and eat while they paddle."[44] They canoed through the daylight hours without pause, their fuel source "Buffalo Rucheggan mixed with fat." Sutherland expressed in wonder, "What may seem surprising they traded most of it from the [Montreal] pedlers who has more of it than they can use, as the fire [or plains] country through which they pass is so plentiful of that kind of provision."[45]

Sutherland was witnessing a significant shift in food energy driving the fur trade both on the plains and now the boreal forest. The numbers speak for themselves: a single pound of pemmican might have contained as much as 3,200–3,500 calories. What became the standard bag of the stuff, weighing 90 pounds, likely had the caloric value of between 288,000 and 315,000 kilocalories. That almost doubled the food energy value of dried corn or wild rice. According to Alexander Mackenzie, who was a pemmican-eater, the "custom of the voyageur" offered "wholesome, palatable food, and easy of digestion … sufficient for a man's subsistence during twenty-four hours," but he pointed out that "it is not sufficiently heartening to sustain the strength necessary for a state of active labour."[46] At most, the *custom* offered 510 calories from the lard, and the 1.43 pounds weight in corn another 2,676, for a total of 3,186. But that was still far below the needs of a man burning 4,000 to 6,000 calories a day. If a Montreal canoe usually carrying 1,000 pounds of provisions was carrying corn, it might have availed about 300 man-days of energy, in an eight-man canoe, perhaps thirty-eight days of actual journeying. If they were fortunate, like Alexander Henry the Younger was, to take on 100 bags of wild rice in Lake of the Woods (at about 1,619 calories per pound), the same canoe could carry energy for a further thirty-three days. By contrast, a Montreal canoe hypothetically carrying 1,000 pounds of of pemmican would have food energy of about 533 man-days, or almost sixty-seven days of journeying, almost double that of corn and rice (see figure 2.1)

Comparing carbohydrate-rich diets on the eastern side of the divide to the protein- and fat-heavy diets on the other side, however, is fraught with problems from the perspective of physiology. Simply having more

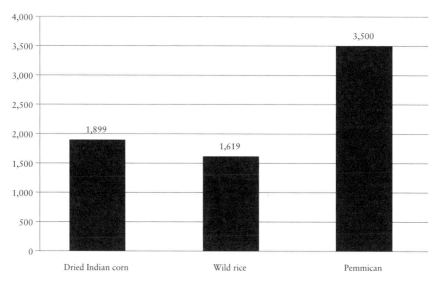

Figure 2.1   Caloric energy of dried Indian corn, wild rice, and pemmican (cals/lb)

Source: Author, based on "USDA National Nutrient Database for Standard Reference"; and Wentworth, "Dried Meat: Early Man's Travel Ration," 8.

calories of energy available in their food did not necessarily mean that it could be used by muscle tissue to do work. Fats and proteins contain greater number of calories; however, body metabolism derives that energy less efficiently than it can from carbohydrates. Aerobic exercise probably better converts carbohydrates into ATP than it does fat. The fur trade, creating such extremes in diet, for that reason, continues to raise numerous questions about its ultimate sustainability in the present day.[47] The switch from high carbohydrates into an almost exclusive diet of fat and protein, whatever its long-term effects on voyageur bodies, availed significant gains for commercial expansion all the same.[48] Exercise physiology studies, indeed, have recently suggested that carbohydrates provide "most of the energy for short-term maximal endurance exercise, whereas lipid [fat] makes substantial contributions to the energy requirements of more prolonged exercise," essentially making these fats most appropriate for the extraordinary long days and almost unlimited outlays of exercise undertaken by voyageurs. Modern-day elite athletes, in fact, after adapting themselves to a high-fat diet for a number of days, are able to undertake ultra-endurance events with the higher energy now available to them without apparently deleterious effects to the body.[49]

More than simply providing an alternative food energy source, then, fats in pemmican likely constituted a revolutionary energy shift to prompt commercial expansion. They undoubtedly increased the very pace of voyaging, allowing paddlers to significantly lengthen their work in time and intensity, especially now into the farthest reaches of the subarctic. They most certainly bolstered the constitutions and body masses of voyaging crews.[50]

The switch to a higher-energy food source in the fur trade was apparent by the 1790s. By 1792, Montreal companies built their first depots at the head of the Winnipeg River, christened Le Sieur's, after its builder, but more commonly known as Bas de la Rivière Winipic. With Cumberland House, it became the NWC's key supply base to what was developing as the jumping-off to a vast commercial empire. Eventually the HBC, too, established a pemmican depot at Cumberland House as well as Norway House to service its own incoming transport crews. In the case of the Montreal companies, initially, a modest 100 to 250 bags of pemmican annually were delivered from the Assiniboine and upper Red River posts to Bas de la Rivière, but that quickly ballooned to thousands of bags by the early nineteenth century when a "bandwagon" effect – so common in the story of other energy sources in this volume – brought companies into pemmican production and competition with each other to procure this energy source.[51] By 1809, Alexander Henry the Younger was mass producing pemmican and shuttling it back from Fort Vermillion, near present-day Edmonton. He used 66 bags of pemmican to move a single brigade of eleven canoes (each paddled by five men and one woman) all the way to Bas de la Rivière – about a single bag allotted to each person for the entire trip. Another 40, going the other way, could get the entire contingent to the salty airs of the Pacific on the Columbia.[52]

The wealth of food energy now available was made manifest at Le Sieur's post by the turn of the century. Hardly a starvation zone, Lake Winnipeg now constituted a meeting point for Montreal crews who, after regrouping, competed with each other in great races across the lake to reach the Athabasca district. The first crew arriving won. Sometimes up to a hundred canoes took part. The forty-stroke paddle that set pace in corn country was here quickened considerably. As historian Marjorie Wilkins Campbell described it, voyageurs eating pemmican now increased their speed from forty to sixty, then to sixty-five strokes per minute, with guides in the front of each canoe who "hacked off hunks of pemmican for the straining voyageurs who chewed as they paddled till they verged on exhaustion."[53] The record at such speed was forty hours of continuous paddling. Officials finally stopped that contest, fearing for the men's lives.[54]

NEW ENERGY REGIMES: NEW HUMAN
AND ENVIRONMENTAL DYNAMICS

Pemmican presented a solution to a fundamental logistical problem of the British fur trade. Ironically, it also raised a host of new problems. As rivals competed to gather pemmican, price escalated. In general, fur traders paid First Nations hunters progressively more for bison products to make more pemmican by the 1790s. In places like the Red River valley, where competition was intense, pemmican was rated at a shilling a pound by the trader Alexander Henry the Younger. Their rising spending power prompted some aboriginal groups, such as the Assiniboine, Cree, and Ojibwa, to specialize or at least hunt more for the pemmican market forming around them.[55] The HBC's own accounts from Cumberland House show that pemmican, fat, and dried meat warehoused at that post was rated by 1810 at four pence per pound. That climbed to five and six pence by 1815 and 1816, and then reached a critical cost by 1821–22 of eight-and-a-half and nine pence per pound.[56] The upwards spiral on prices for pemmican led eventually to the great "pemmican wars" of the region by 1814 when the HBC and NWC attempted to corner the supply on regional pemmican so necessary for their operations. There is no coincidence, too, that the most violent, competitive period of the northwest fur trade, approximately from 1780 to 1821, coincided with the release of comparatively vast amounts of this food energy. Pemmican freed cargo space and allowed for greater energetic output from crews. They could now deliver more goods to Native people, more otherwise heavy and bulky alcohol supplies, more guns, and, because they were competing with each other, use more violence, over-trapping and over-hunting as a strategy among nations inland.[57]

The Nor'Westers, in particular, gained their reputation as brazen bullies in the 1780s and 1790s. Their food supplies could allow them to impose upon, physically threaten, and intimidate Cree, Chipewyan, and other Athapaskan people. Their behaviour more than suggests that fur trade reciprocity, long respected at posts vulnerable to starvation and always needing provisioning from large numbers of hunters, had been decisively upended in the new energy regime. Pemmican, used in transport, or stockpiled for long-term storage at a post, gave posts far greater independence from Native people and their support inland.

Furthermore, it was after and not before pemmican began to be used that fur trade companies, enjoying greater food security, could embark on environmental scorched-earth policies against their rivals. Given the low carrying capacity of much of the subarctic, better-fed traders confidently hired First Nations hunters to trap out and overhunt areas to create game and fur "deserts" to discourage competitors from moving into their hinterlands. By the late eighteenth century, large game deserts emerged in Ojibwa territories in northwestern Ontario and present-day Manitoba. By the early 1800s, Montreal companies, and then HBC companies, hired Iroquoian hunters from the Montreal area to create such deserts in Peace River country, the east slopes of the Rockies, and the Athabasca district.[58]

Directly affecting Dene and Algonquian seasonal rounds, scorched earth policies all the same manifested the power gained for Europeans in pemmican energy. That power expanded as companies improved larger and more efficient food systems that delivered pemmican from points of supply on the plains, to district depots, and finally by circuitous routes to where it was needed. What might be called "peak production," however, came after 1821, when the HBC gained monopoly in British territories. At that point, the company fine-tuned a massive food system to support its brigades and post system, and had supply available to support British naval exploration in the Arctic. Food allowed the company to span the continent. Most key in that respect, monopoly allowed the company to suppress prices on its pemmican, dried meat, and fat purchases, and set quotas from "districts" it set across buffalo territories. Indeed, when the HBC was created as a monopoly in 1821, in large measure to reduce provisions costs in the fur trade, it was able to drive prices to four pence per pound by 1825 and by 1833 to two pence a pound.[59] It kept prices on the product low throughout the bison era by using a district ordering system. Since it could buy pemmican products from plains First Nations and Metis groups from any of its three main districts, it could play aboriginal people against each other to cheapen prices.

But typical of moments of cheap energy supply in other settings, the company did what might be expected: it increased, not decreased, consumption. If in 1821, it typically procured 800 to 900 bags of pemmican for its transportation system, it progressively increased purchases in the 1840s to 1,200, and by 1869 to 2,500. Extraordinary increases in consumption were seen in the now more efficient York boat system: at Norway House, pemmican stocked for crews grew from 7,180 pounds in 1822, to 8,840 pounds in 1827, to almost 36,720 pounds in 1835. In years when pemmican had been so cheaply rated, in 1832 and 1833 Norway House crews received 47,520 and 62,509 pounds respectively.[60] With such an abundant supply of cheap energy, the HBC mobilized a fantastic commercial expansion northward and westward to the Pacific. Well-fed brigades could now trade extensively in the Athabasca and Cordilleran stretches of British Columbia. They could also wage war against American rivals north of the Missouri, and especially in the Columbia.[61]

Cheapened food energy had vast environmental consequences. Not only did it prompt a constantly expanding hunt of the herds themselves by First Nations who were paid low prices for their product. It also gave the company's traders the purchasing power to amass pemmican beyond their own immediate needs and the district quotas they had to fill. They commonly exceeded quotas to amass "discretionary" stock to be kept on hand in parkland and boreal areas to support First Nations in increasingly exhausted territories to continue to trap and hunt out their areas. Already by 1800 the HBC was stockpiling so many discretionary funds in pemmican that it could support Wood Cree hunters in the environs of Cumberland House. These otherwise generalist hunters and gatherers, in already over-hunted territories, were now trapping intensively for the market, rather than their subsistence, supported by pemmican kept on hand for them.[62] By the 1860s, so large were HBC discretionary funds that posts like Fort Ellice in the western edge of present-day Manitoba could allocate some 3,000 to 4,000 pounds of pemmican to the Egg Lake and Riding Mountain areas for a single winter to feed destitute Ojibwa on near-exhausted territories. There they continued to trap and hunt on behalf of the company[63]

Perhaps more grimly, however, excess food energy ended up unravelling, not enhancing, human relations in the fur trade. Food fights became common between traders who had better access to a larger supply. In 1800, Ferdinand Wentzel, a Nor'Wester employed in the Lake Athabasca area, contended with the XY Company and HBC. He used his own large pemmican surpluses to deliberately starve out his rivals. After hunting out the environs of the area and stopping First Nations from trading with the XY men, Wentzel waited until they were reduced to eating leather used for window covering, and one nearly starved to death, before purchasing up their stock of goods at a fantastic profit, in exchange for a few pounds of pemmican, enough to supply the traders out of the country.[64]

There is no better example of the new human and environmental relations in the energy regime than in 1816, when Montreal traders in the Fort Vermillion area deliberately starved to death sixteen of their HBC rivals. That year, the inexperienced and poorly supplied John Clarke was sent to move the HBC into the Athabasca district. Without enough provisions, he decided to relocate his fifty men, in eight canoes, up the Peace

River to present-day northwestern Alberta. His Montreal Nor'Wester rival, William MacKintosh, sent his own men ahead of the contingent to fire guns along the river to chase away game and pay off Chipewyans who might otherwise feed the HBC contingent. When Clarke's men arrived in late fall and the main body took up at a place called Loon River "where the ice stopped them," they were already out of food. Over the course of the winter, the group desperately broke up in search of support. Given the scantiest aid from their rivals, one of Clarke's men died of starvation before the men regrouped at the trader's main post. Three more men died before Clarke finally sold his entire trading supply to the Nor'Westers for a mere 700 pounds of dried meat and pemmican, enough, it was hoped, for an ill-fated evacuation to Fort Chipewyan. Tramping to that centre, all but three of the twelve perished.[65] Ferdinand Wentzel, hearing of the incident, reported with satisfaction, "No less than 15 men, 1 clerk with a woman and child died of starvation going up Peace River." The HBC's Colin Robertson, looking back on "the starving of Mr Clarke's Men in Athapascow," cited it as "the most deliberate and wanton acts of cruelty towards the Company's servants" ever.[66]

CONCLUSION

The high calories offered in pemmican, delivered in ever-larger food systems, significantly shifted the endosomatic energy regime of the fur trade by the late eighteenth century. Having reached the end of their food supply by the 1770s, fur trading companies from Montreal or Hudson Bay discovered the vast energy sources to be tapped from seemingly super-abundant bison herds on the plains. Metabolically, however, the caloric wealth of pemmican differed from that offered in the corn-based "custom of the voyageur." Representing a far more compact and higher order of energy, this food fuel allowed for massive and rapid expansion of British commercial frontiers, especially after 1821, when the monopolized Hudson's Bay Company fine-tuned its food system and used this fuel to entrench itself, and the empire, in the farthest reaches of the Arctic and Northwestern Pacific watersheds. But the shift into a new energy source presented as many new problems as it did solutions to the logistic nightmares of the fur trade. Pemmican energy shifted more power into European hands and allowed for not only a more intense pace of trade, but increased competition and violence. More importantly, Europeans could use greater food security to change their relationships with First Nations and the environment itself. The expanding market for cheapened pemmican supplies prompted greater hunting "overreach" on the plains.

It is clear that the pemmican trade fatefully tipped the balance against the bison herds north of the Missouri River.[67] The supressed price on this cheap energy source also meant that surplus stocks could be kept at posts in parkland and boreal areas to subsidize more intensive trapping and hunting and, eventually, completely exhaust large regions where over-hunting and trapping had already occurred. By the early 1870s, when the bison herds were quickly disappearing from the plains, the HBC perceived the shift into a settlement and missionary frontier. It turned to steamship technology in a bid to rid itself of the dependency on bison fats for its transport system. The first steamship in the west, the *Northcote*, began plying the Saskatchewan in 1874, just as a pemmican energy crisis fully enveloped the company's entire transport and post system.[68] The pemmican era ending decisively in 1879, when the last bison herds were seen in Canadian territory, constituted a remarkable early chapter in the larger environmental history in Canada, and certainly the history of energy and power in its past.

NOTES

1 Kirkby and Luckins, *Dining on Turtles: Food Feasts and Drinking in History*, 3–4.

2 See Iacovetta, Korinek, and Epp, *Edible Histories, Cultural Politics: Towards a Canadian Food History*. Over the past forty years or so a number of historians have explored the history of food. See, for example, Otter, "The British Nutrition Transition and Its histories," 818; Grigg, "The Nutritional Transition in Western Europe"; also Collins, "Dietary Change and Cereal Consumption in Britain in the Nineteenth Century"; Muldrew, *Food, Energy and the Creation of Industriousness: Work and Material Culture in Agrarian England, 1550–1780*, 260–3, 319; Vernon, *Hunger: A Modern History*, 18–21, 42–4; Burnett, *Plenty and Want: A Social History of Diet in England from 1815 to the Present Day*, 7–13; Kiple, *A Movable Feast: Ten Millennia of Food Globalization*, 214–25. On developing food systems and colonialism, Belich, *Replenishing the Earth: The Settler Revolution and the Rise of the Anglo-World, 1783–1939*; and Perren, *Taste, Trade and Technology: The Development of the International Meat Industry since 1840*.

3 Vernon, *Hunger*, 84–6; Drouard, "Reforming Diet at the End of the Nineteenth Century in Europe," 216.

4 The expression is McNeill's, described in *Something New under the Sun: An Environmental History of the Twentieth Century World*, 11.

5 Innis, *The Fur Trade in Canada*, 383–402.

6 Morton, *The Journal of Duncan M'Gillivray of the North West Company at Fort George on the Saskatchewan, 1794–5*, 58; Robert Goodwin to Charles Isham, 6 February 1795, B.22/a/1, Brandon House Journal, Hudson's Bay Company Archives (hereafter HBCA).

7 Morton, *Journal of Duncan M'Gillivray*, xlviii.

8 Yudkin, "Some Basic Principles of Nutrition."

9 Krech, "On the Aboriginal Population of the Kutchin"; Vale, "The Pre-European Landscape in the United States," 11.

10 Wynn, *Canada and Arctic North America: An Environmental History*, 113.

11 Clarke, *The Ordinary People of Essex: Environment, Culture, and Economy on the Frontier of Upper Canada*, 219–20; Bleakney, *Sods, Soil, and Spades: The Acadians at Grand Pré and Their Dykeland Legacy.*

12 Podruchny, *Making the Voyageur World: Travelers and Traders in the North American Fur Trade*, 86–134.

13 Skinner, "The Sinews of Empire: The Voyageurs and the Carrying Trade of the *Pays d'en Haut*, 1681–1754," 232; and Skinner, *Regional Perspectives on Early America: Upper Country: French Enterprise in the Colonial Great Lakes.*

14 Podruchny, *Making the Voyageur World*, 88.

15 Morse, *Fur Trade Canoe Routes of Canada, Then and Now*, 27–32.

16 Skinner, "Sinews of Empire," 208.

17 Morse, *Fur Trade Canoe Routes of Canada*, 7–9; Jennings, *The Canoe: A Living Tradition*, 20–3, 40–5.

18 Skinner, "Sinews of Empire," 230.

19 Podruchny, *Making the Voyageur World*, 114–18.

20 Skinner, "Sinews of Empire," 224–7.

21 Morse rates the calories as 4,000 to 5,000, without attribution (*Fur Trade Canoe Routes of Canada*, 24). Skinner counts 4,000 to 6,000 calories, without attribution ("Sinews of Empire," 59).

22 Lai and Lovell, "Skeletal Markers of Occupational Stress in the Fur Trade: A Case Study from a Hudson's Bay Company Fur Trade Post;" Lovell and Dublenko, "Further Aspects of Fur Trade Life Depicted in the Skeleton."

23 Clark et al., "Food for Trans-Atlantic Rowers: A Menu Planning Model and Case Study," 229.

24 Osborn, *A Concise Law Dictionary for Students and Practitioners*, 66.

25 Nute, *Voyageur.*

26 Henry, *Travels and Adventures: In Canada and the Indian Territories*, 56.

27 Skinner, "Sinews of Empire," 61; the pound-weight of corn is derived from US Department of Agriculture data provided in Patzek, "Ethanol from Corn: Clean Renewable Fuel for the Future, or Drain on Our Resources and Pockets," 3.

28 Skinner, "Sinews of Empire," 109.

29 See table, ibid., 61.

30 Ibid., 219.

31 Henry, *Travels and Adventures* 184.

32 Goktepe, "Energy Systems in Sport," 26. On these energy systems, see Hargreaves and Spriet, *Exercise Metabolism*; and Brown, Miller, and Eason, *Exercise Physiology: Basis of Human Movement in Health and Disease.*

33 16 June 1785, James Sutherland's journal entry, B.78/a/15, Gloucester House Journal, HBCA.

34 Innis well anticipated this crisis, in his description of the strategies of Peter Pond and his partners sharing food to get him into the Athabasca district. Innis, *The Fur Trade in Canada,* 93–111.

35 Sloan, "The Native Response to the Extension of the European Traders into the Athabasca and Mackenzie Basin, 1770–1814."

36 Colpitts, *Game in the Garden: A Human History of Wildlife in Western Canada,* 19.

37 Tyrell, *Journals of Samuel Hearne and Philip Turnor between the Years 1774 and 1792,* 190.

38 The poor winter conditions are best illustrated in B.254/a/1, Blood River Journal 1794–95, HBCA.

39 Innis, *Peter Pond: Fur Trade and Adventurer,* 83–4; Mackenzie, *Voyages from Montreal,* 152.

40 See my references to fur traders using the term *custom of the country,* in my *Game in the Garden,* 36–7.

41 McKenzie, "Réminiscences," 24.

42 Sloan, "Native Response to European Traders," 285n10.

43 Henry the Elder, *Travels and Adventures in Canada and the Indian Territories between the Years 1760 and 1776,* 275–6.

44 12–17 June 1785, Gloucester House Journal, B.78/a/15, HBCA.

45 25 July 1785, Gloucester House Journal, B.78/a/15, HBCA.

46 Mackenzie quoted in Innis, *Fur Trade in Canada,* 224.

47 McArdle, Katch, and Katch, *Essentials of Exercise Physiology,* 1:94.

48 Noakes, "Physiological Models to Understand Exercise Fatigue and the Adaptations That Predict or Enhance Athletic Performance," 134–5.

49 See Hawley and Hopkins, "Aerobic Glycolytic and Aerobic Lipolytic Power Systems"; Lambert et al., "Enhanced Endurance in Trained Cyclists during Moderate Intensity Exercise Following 2 Weeks Adaptations to a High Fat Diet"; and discussion in Noakes, "Physiological Models."

50 Hamilton, *Incidents and Events in the Life of Gurdon Saltonstall Hubbard 1802–1886,* 20–1.

51 Morton, *A History of the Canadian West,* 429

52 *The Journal of Alexander Henry the Younger,* 2:395.

53 Campbell, *The North West Company,* 44.

54 Ibid.

55 On their specialization, see Ray, *Indians in the Fur Trade: Their Role as Trappers, Hunters, and Middlemen in the Lands Southwest of Hudson Bay, 1660–1870,* 131–3; as well as my argument that these groups selected product from their traditional pounding activities, in Colpitts, "Environment, Pemmican, and Trade in a Northern Great Plains Bioregion."

56 See a full price series in my "Provisioning the HBC: Market Economies in the British Buffalo Commons in the Early Nineteenth Century," 189–92.

57 Ens, "Fatal Quarrels and Fur Trade Rivalries: A Year of Living Dangerously on the North Saskatchewan, 1806–07."

58 Ray, "Periodic Shortages, Native Welfare, and the Hudson's Bay Company, 1670–1930"; Colpitts, "Moose-Nose and Buffalo Hump: The Amerindian-European Food Exchange in the British North American Fur Trade to 1840"; Kirkby and Luckins, *Dining on Turtles,* 64–81; Davies, *Peter Skene Ogden's Snake Country Journal, 1826–27;* and Hammond, "Marketing Wildlife: The Hudson's Bay Company and the Pacific Northwest, 1821–1849," 206–7.

59 Colpitts, "Provisioning the HBC," 192.

60 Ibid., 197.

61 Ibid., 179–203.

62 26 August 1808, *The Journal of Alexander Henry the Younger,* 2:345.

63 Cited in Peers, *The Ojibwa of Western Canada, 1780–1870,* 194.

64 Colpitts, "Moose-Nose and Buffalo Hump," 64–7.

65 See Goldring, "MacKintosh, William."

66 Vernon, *Hunger: A Modern History,* 19–20.

67 Ray, "The Northern Great Plains: Pantry of the Northwestern Fur Trade, 1774–1885"; and Dobak, "Killing the Canadian Buffalo, 1821–1881."

68 On steamship technology, see Innis, *Fur Trade in Canada,* 343.

3

Ox and Horse Power in Rural Canada[1]

J.I. Little

The Industrial Revolution was characterized, to use E.A. Wrigley's terms, by the shift from an organic economy based on energy captured from water, wind, and plants to a mineral economy based on fossil fuels, notably coal.[2] For agriculture, however, the major sources of energy long continued to be people, animals, wind, and water. The mechanization of farming proceeded more slowly than in other industries, P.K. O'Brien notes, because "the spatial dimension in farm work implies that machines must be mobile and travel to crops." Furthermore, the requirement for specialized machines to prepare the soil, sow crops, then cultivate and harvest them, means that those machines remain underutilized for much of the year.[3] As a result, the agricultural revolution – which was marked by a significant growth in output per acre between 1600 and 1800 – was based not on fossil fuels but on improved techniques such as the increased use of manure as fertilizer, which was in turn made possible by the trend towards pastoral farming.[4] Wrigley also stresses the growing role of draft horses, noting that "one horse provided as much energy as five men."[5] He neglects to mention oxen, however, or that horses continued to be essential sources of power for the production and marketing of food crops until well into the twentieth century. A crucial element of Wrigley's organic economy therefore persisted much longer than he suggests.

Indeed, recent studies argue that horses played an important role in the history of energy and technology in the industrial era, and that as "living machines" they challenge the nature-technology divide.[6] An obvious example of this point is that the eighteenth-century designer of the steam engine, James Watt, experimented with draft horses in order to devise a measure of performance that he termed "horsepower." Furthermore, while there is a tendency to see machines as ever improving while nature is static, farm horses were changed over time in response to human needs and desires. The same was not true of oxen, which are simply castrated bulls of any breed, but, prior to the introduction of agricultural machinery in the latter half of the nineteenth century, they provided much of the draft power required on Canadian farms, and they long continued to be used in the woods. The main role of horses was originally to provide transportation beyond the farm, thereby improving access to markets as well as strengthening rural community bonds by facilitating social interaction.

HORSES VS OXEN: THE PRE-INDUSTRIAL ERA

Despite being introduced to Port Royal near the mouth of the Annapolis River as early as 1609, horses did not gain a permanent foothold in Acadia until the 1680s, nor is there evidence of oxen used as draft animals,[7] suggesting that the agricultural land reclaimed from the sea was farmed intensively by hand labour. The more important role played by horses in the early St Lawrence valley settlements reflected the fact that boats could not be used for communication and transportation when the river was frozen during the winter. Eighty-two horses arrived in Canada from France between 1665 and 1671, when Intendant Talon declared that there was already a large enough number to supply colts to all who needed them. Although these state-owned horses were not meant for the use of peasants, the *censitaires* (peasants who had been granted land by seigneurs) owned half the ninety-six head that were reported in the colony in 1681.[8]

The number of horses in Canada increased rapidly and steadily thereafter, reaching 400 in 1692 and 1,872 in 1706, when there was an average of nearly 1 horse per family.[9] This ratio represents a sharp contrast to that of the English colonies, where horses were rare at that time.[10] Horses were not evenly distributed in the St Lawrence valley , however, for there were more in the towns than in the countryside, and ownership was sufficiently concentrated to spur the governor of the Montreal district in 1709 to forbid anyone from possessing more than two mature horses and a foal. Dependence on horses was said to be causing cattle to be undernourished, and to be softening the militia, but the chief concern of the authorities may well have been the threat to social class barriers that peasant ownership of horses represented.[11]

Because most heavy farm work was done during the summer by oxen, four of which were generally needed to pull one of the cumbersome wheeled wooden plows then in use,[12] horses were used primarily during

Figure 3.1   Many habitants took their produce directly to urban markets long after the railway era had begun.

the winter when hard-packed snow on the frozen St Lawrence facilitated movement over relatively long distances. It was during this agriculturally dead season that the habitants used horses to haul their grain to the seigneurial mill and their produce to market, and that they also had ample time to socialize as well as to scandalize their social superiors by racing their sleighs on the way home from Sunday Mass.[13] Efforts to limit the numbers of horses were obviously rather futile, for they increased to 5,603 in 1721, when the average was 1.4 for each family and the cost of an ordinary farm horse was less than that of an ox (see table 3.1).[14]

The horse/household ratio had not changed significantly by the time of the Conquest, but the British authorities nevertheless shared the concerns of their French predecessors, for General Murray proposed a tax on horses rather than dry goods to support his administration, arguing that it would "restrain a piece of luxury the people of this Country are too apt to run into."[15] This it failed to do, however, for the number of horses enumerated in 1784 increased to 1.6 for each of the 18,924 households (see table 3.1).

Table 3.1   Horses per household in Canada (French Regime)
and Lower Canada

1721	1765	1784	1831	1844
1.4	1.3	1.6	1.4	1.3
			(2.3/farmer)	

Sources: *Journals of the Legislative Assembly of Lower Canada*, vol. 41
(1831–32), appendix oo; *Journals of the Legislative Assembly of the
Province of Canada*, vol. 5 (1846), appendix D; *Canada Census Report*,
1871, vol. 4.

There would be no census enumeration for Lower Canada between
1784 and 1827, but analysis of post-mortem inventories has revealed
that the average number of horses per family in the Quebec district
increased from 1.1 in 1792–96 to 1.5 in 1807–12, and in the more agriculturally advanced Montreal district from 2.0 to 2.3 during the same
period.[16] The increase in the number of horses was clearly connected to
the use of horse-drawn implements, for the number of harrows recorded
for the Montreal district increased from 1.1 per inventory in 1807–12 to
1.7 in 1820–5.[17] The more efficient iron or steel swing plow known as the
charrue anglaise (or, in English Canada, the Scotch plow) had also begun
to make its appearance.[18] As a result, the Montreal Agricultural Society
reported in 1821 that local habitants were using horses instead of oxen
for plowing. The lighter plows were practicable, however, only once the
soil had already been broken up and cultivated. They were therefore
often used as a second plow for specific tasks such as turning potato
fields, or by small farmers who did not own a team of oxen needed to pull
the heavier wooden "Canadian" plow.[19] Horses were still relatively small
in the early nineteenth century and oxen's more docile nature, lower
centre of gravity, and superior traction made them more suitable for
heavy-duty tasks.[20] Paradoxical as it may seem, then, the ownership of
oxen rather than horses is associated with agricultural progress in early
nineteenth-century Lower Canada.[21]

Oxen and locally made wheeled plows with iron tips were still in use
in the northern Saguenay region as late as the 1920s,[22] but horses were
not only much more efficient than oxen, producing 50 per cent more
foot-pounds per second in energy because of their greater speed, they
could also work one or two hours longer each day, thereby facilitating
the increase of production for the market.[23] As British travel writer
James F.W. Johnston observed while in New Brunswick in 1851, oxen
made more economic sense in England, where labourers were relatively
plentiful, than in North America, where wages were high "and quick work

Figure 3.2   When Quebec farmers turned to tractors in the 1940s, some clung to oxen, believing they were easier to manage than horses. Oxen required an extra person to guide them, however, as reins were not used.

is therefore desirable."[24] Even in Lower Canada, where the increasingly overcrowded seigneuries were pushing more and more habitants into rural wage labour, oxen gradually began to give way to horses. In 1784 there had been 1.4 oxen for every horse in Lower Canada, and they still outnumbered horses slightly in 1827, but by 1852 there were 1.6 horses for every bull, ox, and steer.[25] Given that oxen were rarely worked alone,[26] no more than half the 95,813 farmers would have had such a team, which suggests that many now relied exclusively on their horses (the number of horses per farm was recorded as 1.9) for plowing and other heavy work. It was not always an either/or situation, however, for Irish-born Lower Canadian agricultural expert William Evans observed in 1835 that many French Canadians plowed with one or two horses in front of their oxen in order to speed them up.[27] Whether or not this was effective, it did eliminate the necessity of having one person drive the team while the other held the plow.[28]

Evans was a supporter of the small French-Canadian horse, as opposed to the heavier British breeds. In his words, not only did the larger horses cost more to purchase, they consumed more feed and compacted the heavier soils when used for plowing. Furthermore, Canadians marketed much of their produce during the cold winter months when a slow English draft horse would expose them to frostbite.[29] To increase speed, the habitants commonly used two horses – hitching one behind the other rather

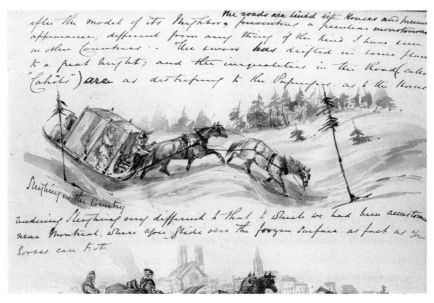

Figure 3.3   A sketch by a British military officer.

than side by side because of the narrowness of the winter roads[30] – but this was counteracted somewhat by the fact that the low runners of their sleighs (known as *carrioles*) caused the snow on the roads to be packed into large bumps. Evans argued that no improvement was "more necessary" than modification of the habitant sleigh, for good winter roads would enable two horses to haul as much produce to market as four could under the current conditions, "and in half the time."[31]

British travel writers echoed Evans's praises for the qualities of the French-Canadian horse while criticizing the care it received from its habitant owners. Thus, George Heriot wrote in 1805, "The horses are of the Norman breed, and are rather small, but stout, hardy, fleet, and well calculated for draft. Notwithstanding the little care that is bestowed on them, and the ill treatment which many of them experience, they in general possess their strength to a great age." In 1820 Edward Allen Talbot wrote that the habitants' horses would "endure all kinds of hardship, and live on any food."[32] In fact, horses can consume grasses of lower quality than any other ruminant, and they require a lower volume of food than any other large animal.[33] To survive in the wild, however, they must graze with little interruption. Even their ability (unlike cattle) to paw through snow to the dead grass underneath would obviously have been of little use during the mid-winter months of deep snow accumulation in eastern Canada. Horses were therefore fed hay and grain. Timothy hay was favoured because it is easy to grow, generally free of mould and dust, resistant to spoilage when stored, and quite rich in carbohydrates and fats. It also provides the required roughage, but the digestible protein and minerals that horses depend upon must come from grain.[34] Oats has long been the grain of choice for horses because it not only grows in most kinds of soil and matures quickly, but it is high in fibre and protein and the hull gives sufficient volume to prevent gorging.[35] This was a particular concern for horses because their relatively small stomachs and intestines are prone to twisting and blocking, namely colic.[36] As an agricultural improver, however, Evans complained in 1835 that far less oats was raised in the province than was required to feed its horses.[37]

Evans also claimed that the Canadian breed was deteriorating as a result of the habitants' failure to castrate horses that wandered at large because of the poor fences, a complaint that was echoed in the report drafted by the special committee on Lower Canadian agriculture in 1850 (the Taché Report), though this practice had not prevented the development of the hardy stock in the first place. But an even greater concern, in Evans's opinion, was the fact that Americans were buying the best Canadian horses "while we purchase and breed from their horses, which, I maintain, are every way inferior to the Canadian horses for agricultural purposes ... A tall, slender horse, well fed and groomed, and splendidly harnessed, may be very showy, and answer well for pleasure about town, but will not be the most suitable or profitable for a farmer."[38] Evans was far from alone in this opinion, for it had been voiced by agricultural societies in the province as early as the 1820s.[39] As for the export of horses, it had become so prevalent in the 1850s, and especially during the Civil War, that the French-Canadian breed – having contributed to the development of the Morgan, the Standardbred, and the Saddlebred in the United States – was threatened with extinction.[40]

In Upper Canada, early nineteenth-century assessment records indicate that there was approximately one mature horse for every seven people, a ratio similar to that in Lower Canada.[41] Not surprisingly, Robert Gourlay's Upper Canadian survey of 1822 had found that most of the colony's horses were of American origin, though some were interbred with French-Canadian and British stock.[42] A couple of years later, Edward Talbot observed that the horses of Upper Canada were more refined in appearance than the French Canadians', but less resistant to fatigue, privation, and disease.[43] By 1845 the heavy-draft Conestogas brought by the "Pennsylvania Dutch" who settled north of York and in Waterloo County were approaching extinction, having been interbred with lighter and faster British horses. Cross-bred Clydesdales and Shires

Figure 3.4   The Red River Cart, developed by the Métis and used in the fur trade throughout the west, could float across streams and carry up to 450 kg. Note the makeshift canopy for shade.

were becoming increasingly popular for the heavier work, while British light-draft horses such as the Hackney were imported to take advantage of the growing road network.[44]

Finally, on the far western frontier, First Nations war parties had introduced horses from farther south as early as the 1730s. The Selkirk settlers owned twenty-one of these Native horses in 1816, but they were all taken in a single raid. According to historian Grant MacEwan, such raids were so rampant that the man in charge of the Hayfield Experimental Farm at Red River was reduced in 1818 to hitching three horses, two cows, one bull, and a bison heifer to make up two plow teams![45] Although the first stallion to be shipped from Britain was not a draft horse but a Norfolk Trotter that arrived in 1831, MacEwan claims that its offspring crossed with "Indian" mares "were conspicuous as the best buffalo runners, the best racing horses, the best farm horses, and those commanding the premiums when sales were made."[46] Most prairie settlers nevertheless continued to rely on oxen for their farm work prior to the mechanization era (see table 3.2).

CHUNK VS HEAVY DRAFT: THE MECHANIZATION ERA

Steam power not only destroyed the stage and teaming business in many places, it also replaced horses that had been used to pull canal barges and to turn the wheels that propelled ferryboats, but the advent of the

Table 3.2 Number of oxen per farm (occupier)

	1871	1881	1891
Prince Edward Island	–	0	0
Nova Scotia	0.7	0.6	0.4
New Brunswick	0.4	0.2	0.2
Quebec	0.4	0.4	0.3
Ontario	0.3	0.1	0
Manitoba	–	1.4	0.9
Northwest Territories	–	3.3	0.8
British Columbia	–	0.8	0.4
Canada	0.4	0.3	0.2

Sources: *Canada Census Reports*, 1871, 1881, 1891.

threshing machine in the mechanization era may have actually increased the use of horse-powered treadmills and sweeps for a time.[47] The Canadian *Census Reports* include no data on these traditional power sources, but treadmills continued to be used on Quebec farms until the First World War.[48]

Furthermore, thousands of horses were required to move earth for the construction of railway beds, haul timbers for the building of railway bridges, and deliver ties, rails, and spikes for the laying of track.[49] Still more significantly, railways increased the demand for horses by stimulating the growth of the urban market for agricultural produce. Horses were now needed to haul goods to railway stations, leading historian Ann Greene to argue, "Directly and indirectly, the industrialization of transportation in nineteenth-century America centered on horses."[50] Furthermore, Greene notes, the farm mechanization spurred by the growing agricultural market was based on horse power because "most new implements required the fast-paced horse to provide enough speed to drive the ground wheels that operated the internal mechanism that propelled the machine."[51]

Although the American agricultural revolution is said to have taken place between 1830 and 1860, it would not begin to have a significant impact on Canada until after mid-century.[52] Horse-powered threshing machines had made small inroads in Lower Canada by 1844, but the Taché Report of 1850 failed to include farm mechanization as one of its recommendations to resolve the agricultural crisis caused by the failure of the wheat economy, fearing that it would lower the demand for farm labour and accelerate the exodus to the United States.[53] Perhaps not surprisingly,

Figure 3.5   A four-team sweep powering a threshing machine in the early days of
Alberta settlement.

then, the average farmer in Lower Canada had invested only $76.40 in
farm implements in 1861, as compared with $112.90 for his Upper
Canadian counterpart. The shortage of manpower caused by the French-
Canadian exodus would stimulate the purchase of farm machinery, as
would the growing American demand for agricultural produce during the
Civil War.[54] Such purchases were, however, confined largely to areas capable of meeting the export demand.

The 1871 Canada *Census Report*, which is the only one in the nineteenth century to include farm machinery, reveals that in Quebec horse-
drawn mowers, reapers, and rakes were still to be found mostly in the
area south of Montreal and the primarily English-speaking section of the
Eastern Townships.[55] Although hand raking was the first labour-intensive
process to be replaced by a machine (mowers were less practical on small
uneven fields),[56] horse rakes were almost as scarce in Quebec as a whole
in 1871 (11.4 farmers per unit) as in New Brunswick and Nova Scotia
(11.8 farmers per unit in each province), and much scarcer than in
Ontario (3.2 farmers per unit). There were more plows, harrows, and
cultivators on the average Quebec farm (1.8) than in its New Brunswick
or Nova Scotia counterpart (1.2 and 0.8, respectively), and even very
slightly more than in Ontario (1.7), but horse-drawn reapers and mowers
had yet to make significant inroads east of Ontario, where there was one
for every five farms (see Table 3.3).[57] Not surprisingly, then, there was an

Figure 3.6   The "widespread" machine had to be pulled quickly to keep the track on the wagon bed and the teeth at the rear rotating fast enough to shoot manure in a wide arc.

average increase of nearly seven acres in crop for Ontario between 1861 and 1871, as compared with less than four acres in Quebec (see table 3.4). The growth of the dairy industry is said to have accelerated mechanization in Quebec during the 1880s, but it was in the wheat-growing regions that new farm machinery would be most called for.[58]

The slower pace of mechanization in Quebec and the Maritimes was reflected in the slower increase in the number of farm horses when compared to Ontario. There was an average of not quite two horses per farm in both Lower and Upper Canada in 1852, but Upper Canada reported 2.9 per farm in 1861 when Lower Canada reported only 2.3, and that margin would persist for many years, despite the historic French-Canadian penchant for horses[59] (see table 3.5). New Brunswick and Nova Scotia reported only 1.5 and 0.9 horses per farm, respectively, in 1871, but not every farmer owned a horse, even in Upper Canada, for Douglas McCalla states that in 1861 none were reported by 32 per cent of its small farms (32–69 acres), 28 per cent of its medium-sized farms (70–169 acres), or even 21 per cent of its large farms (170 acres and more).[60] As discussed below, some farmers still relied on oxen at the dawn of agricultural mechanization in the 1860s; others, however, obviously owned substantially more horses than the Upper Canadian average of three. A likely explanation is that many were being bred to supply the growing urban demand in

Table 3.3 Farm equipment per farm, 1871

	Plows, harrows, cultivators	Horse rakes	Reapers, mowers	Threshing mills	Light carriages	Vehicles for transport
Nova Scotia	0.8	0.1	0	0	0.9	1.6
New Brunswick	1.2	0.1	0	0	0.9	2.0
Quebec	1.8	0.1	0	0.1	2.0	3.4
Ontario	1.7	0.6	0.2	0.1	1.2	1.7
Canada	1.6	0.2	0.1	0.1	1.4	2.3

Source: *Canada Census Report*, 1871.

Table 3.4 Acres in crop per farm

	1851	1861	1871	1881	1911	1961
Newfoundland	–	–	–	–	–	3.7
Prince Edward Island	–	–	–	34.3	33.7	42.0
Nova Scotia	–	–	17.1	16.9	13.4	17.0
New Brunswick	–	–	24.9	23.1	25.6	27.4
Quebec	21.6	27.8	31.5	30.8	34.3	42.6
Ontario	22.9	31.1	38.0	40.4	42.7	58.1
Manitoba	–	–	–	25.4	113.2	165.1
Saskatchewan	–	–	–	20.9[a]	94.8	245.0
Alberta	–	–	–	–	54.9	199.3
British Columbia	–	–	–	30.5	13.0	32.4
Canada	–	–	32.1	32.6	49.3	119.3

[a] Includes Alberta
Source: *Province of Canada Census Reports*, 1851, 1861; *Canada Census Reports*, 1871, 1881, 1911, 1961.

the new railway age, as well as to take advantage of the burgeoning Civil War market. In fact, horse exports were valued at $935,000 in 1863 and $1,519,000 in 1865–66, dropping sharply to $293,000 in 1868–69, but then rising again to $650,000 in 1870–71.[61] Apparently because it was believed that horses were stronger-boned when fed on oats as opposed to corn, those from Ontario continued to be more sought after on the American market than those from the Midwestern states.[62]

Despite the increased external demand as well as mechanization of the 1860s, however, the average number of horses on Ontario as well as Quebec farms was slightly lower in 1871 (2.8 and 2.1, respectively) than in 1861, and oxen had not "virtually disappeared" from Ontario farms, McCalla's claim to the contrary.[63] There were still three oxen for every ten

Table 3.5   Average number of horses per farm[a]

	1851	1861	1871	1881	1891	1901	1911	1921	1931	1941	1951	1961
Newfoundland											0.8	0.7
Prince Edward Island		1.7	2.0	2.3	2.5	2.4	2.5	2.3	2.3	2.3	2.1	1.1
Nova Scotia		0.9	1.1	1.0	1.0	1.1	1.1	1.1	1.1	1.1	1.1	0.7
New Brunswick		1.5	1.4	1.4	1.5	1.6	1.7	1.7	1.7	1.4	1.2	0.8
Quebec	1.9	2.3	2.1	2.0	2.0	2.1	2.3	2.4	2.2	2.2	1.7	1.0
Ontario	2.0	2.9	2.8	2.9	2.7	3.2	3.6	3.4	3.0	3.0	1.7	1.0
Manitoba				1.8	3.8	5.0	6.1	6.7	6.0	5.2	2.5	1.2
Saskatchewan				10.7[b]	6.6[b]	7.6	5.3	9.1	7.3	5.8	2.7	1.2
Alberta							6.6	9.8	7.5	6.5	3.1	1.5
British Columbia				9.5	4.3	5.5	3.1	2.8	2.2	2.4	1.4	1.2
Canada			1.7	2.3	2.4	2.9	3.6	4.1	4.3	3.8	1.7	1.1

Note: 1861 data for Prince Edward Island, New Brunswick, and Nova Scotia are in *Canada Census Report*, 1871, and 1871 data for Prince Edward Island are in the same report.
[a] Includes rural properties of 10 acres and under
[b] Includes Alberta
Source: *Province of Canada Census Reports*, 1851, 1861; *Canada Census Reports*, 1871–1961.

farms in that province, and four for every ten in Quebec. In fact, an average of one in five Canadian farms still reported owning a pair of oxen as late as 1891, when they remained particularly popular in the Prairies, where there was just under a pair for every two farms (see table 3.2). Perhaps many Prairie farmers could still not afford as many horses as they needed, but the main reason for the persistence of oxen elsewhere was their usefulness in logging. Thus, Greene notes that oxen "retained their niche as draft animals in logging regions such as Maine and northern New York" long after they were excluded from the 1890 census of the United States.[64] Similarly, large teams of oxen were still being used to skid the giant logs of British Columbia's coastal forests after the turn of the twentieth century, but horses appear to have outnumbered oxen, even in the woods, for equine speed was an advantage for hauling supplies to the distant logging camps and pulling double sleighs loaded with logs to river banks or frozen lakes. This helps to explain the larger production of oats, grown mainly to feed horses, in logging regions.[65] Not surprisingly, then, the average number of horses per British Columbia farm (4.3) was much greater than the average number of oxen (0.4) in 1891 (see tables 3.2 and 3.5).

But horses were still used primarily for agricultural purposes, and the number on Canadian farms had increased by 28 per cent during the

Figure 3.7   Note the size of the logs these oxen have pulled over a skid road to the water.

1880s, a decade when the first practical hay mowers were developed, and the reaper, which cut only grain, had been transformed into the reaper-binder, which bound and tied the grain with twine.[66] According to D.A. Lawr, "Before the end of the century all the horse-drawn machinery available for the next half century had been invented – manure spreaders, potato diggers, root-crop planters, cultivators, scufflers, tedders, hay-loaders, disc harrows."[67] Lawr notes that Ontario farmers eagerly purchased these machines because agricultural produce prices were declining and the price of labour was increasing as young men moved to the Prairies as well as to the cities.

More than 50,000 Ontario-made binders were sold in 1883 alone, and competition kept prices from rising.[68] As a result, according to Lawr, farm mechanization in the horse-power age "represented only a small percentage of farm capitalization, never reaching over six per cent of total farm value until well into the twentieth century."[69] In fact, the 1901 census reveals that the farm implements and machinery on the average

Figure 3.8 Invented in the late 1800s, hay loaders needed one person to toss hay to the front of the wagon, one to tamp it down, and one (often a young boy) to drive the horses.

Canadian farm were still worth less than the horses ($201.40 versus $217.15, respectively) (see table 3.6). The greater energy that horses had to expend pulling new farm machinery eventually had a major impact on horses' diet: oats production per horse jumped dramatically from 56.1 bushels in 1891 to 96.0 bushels in 1901.[70]

Furthermore, increased mechanization depended on growing numbers of horses. Those numbers may have gone up only slightly, from 1,470,572 in 1891 to 1,577,493 in 1901, but they did outpace the increase in farms, with the result that the average number of horses per farm in Canada grew from 2.4 to 2.7. In Quebec, where mechanization was proceeding relatively slowly, the average farmer continued to own 2.1 horses, while in Ontario – where half the country's horses were to be found – the average was 3.2. Once again, the Maritime ratio of 1.4 was the lowest in the country, while in the grain-producing prairies it was a remarkably high 6.1. Manitoba – where Brandon was known as the horse capital of Canada[71] – may have had 5,088 fewer farmers than New Brunswick, but it had 102,078 more horses! Farther west, where W.M. Elofson somewhat inexplicably claims that on most Alberta ranches/farms the numbers of horses remained comparatively small,[72] the region still known as the Northwest

Table 3.6   Value of farm implements/machinery and value of horses
per farm, 1901 ($)

	Implements/machinery	Horses
Prince Edward Island	187.58	152.27
Nova Scotia	54.36	65.29
New Brunswick	97.46	114.74
Quebec	179.54	160.45
Ontario	235.15	245.07
Manitoba	323.81	485.10
Northwest Territories	260.48	476.10
British Columbia	177.79	310.81
Canada	201.40	217.15

Source: *Canada Census Report*, 1901.

Territories in 1901 had even more horses than Manitoba (176,462 vs 163,867), despite having only two-thirds the number of occupiers.

Not all these horses were for draft purposes, however, for they included Thoroughbreds and Irish Hunters raised for sale to the British military during the Boer War. David Breen claims that the breeding and sale of horses "contributed substantially" to the income of many Alberta ranchers between 1900 and 1914, the return per horse being two to five times higher than that per head of cattle.[73] The First World War clearly increased the demand, though mostly for transportation purposes rather than the cavalry, for thousands of horses were required to maintain a steady flow of supplies from the railheads into the battle zones. There were approximately 24,000 horses and mules serving in the Canadian Expeditionary Force in November 1918, and many had died of disease and injury during the previous four years.[74] The cattle ranches, with their extensive grazing areas, also depended on considerable numbers of cowboys and their small tough horses of complex lineage for their spring roundups.[75] According to the historian of British Columbia's Douglas Lake ranch, "Each cowboy rode a personal string of six or seven ranch horses – five fairly gentle and one or two that were being broken."[76]

The Douglas Lake Cattle Company also raised pure-bred stock for export, and the First Nations of the province's interior were clearly involved in the horse market, as well, for 7,757 head were reported for the few hundred members of the Kamloops-Okanagan Indian Agency in 1909.[77] But the large ratio of 5.5 horses per British Columbia farm in

Table 3.7   Value of average horse ($)

	1901	1911	1921	1931	1941
Prince Edward Island	63.68	118.02	127.59	107.84	98.94
Nova Scotia	61.66	115.78	129.88	92.52	101.76
New Brunswick	69.79	123.64	145.79	105.54	111.40
Quebec	75.35	131.10	144.63	104.06	111.65
Ontario	76.17	139.79	125.08	87.52	85.81
Manitoba	96.20	168.31	134.54	61.59	54.27
Saskatchewan	76.45	174.91	126.63	53.89	49.72
Alberta	49.74	138.62	94.37	46.02	46.82
British Columbia	56.12	136.44	99.54	55.22	58.43
Canada	76.94	146.95	121.54	65.86	66.17

Sources: *Canada Census Reports*, 1901–41.

1901 must also have been due partly to the fact that West Coast logging took place mostly during the summer when farm horses were not available for lease. Prices had begun to plummet, however, for the coastal forest industry was turning to steam donkeys and railways to replace the large teams of horses and oxen used to haul the giant logs of the Pacific rainforest. Thus, one sawmill manager wrote in 1896 that the advantage of steam donkeys was that their fuel was "on the ground" and their operating expenses ceased with the end of the logging season.[78] As historian Richard Rajala points out, turning to these machines also reduced the dependency of the forest companies on skilled "bull punchers" and horsemen who had developed a special rapport with their animals.[79]

At $56.12, the average value for a horse in British Columbia in 1901 was even lower than in the Maritimes, where it was $65.27. Horses still represented a significant investment in much of the country, however, for in Canada as a whole they were valued on average at $76.94 (see table 3.7), as compared with $50.32 for the average head of cattle. Quebec's horses were close to the national average ($75.35), as were those of Ontario ($76.17), and in the Prairies it was $78.69.

The number of horses in Canada grew faster between 1901 and 1911 than it had during the previous decade, and it continued to outpace the increase in the number of farmers, so that the national average reached 3.6 (see table 3.5). Horse-drawn machinery was evidently not becoming more efficient (nor, as discussed below, were horses), for the amount of land in crop per horse remained in the thirteen- to fourteen-acre range

Table 3.8   Acres in crop per farm horse

	1851	1861	1871	1881	1911
Prince Edward Island	–	–	–	14.9	13.5
Nova Scotia	–	–	15.9	16.5	11.7
New Brunswick	–	–	17.4	16.0	15.0
Quebec	11.2	11.8	14.7	15.1	14.7
Ontario	11.3	10.9	13.4	14.2	11.9
Manitoba	–	–	–	13.8	18.4
Saskatchewan	–	–	–	2.0a	18.0
Alberta	–	–	–	–	8.3
British Columbia	–	–	–	3.2	4.2
Canada	–	–	13.4	14.3	13.6

a Includes Alberta

Sources: *Province of Canada Census Reports*, 1851, 1861; *Canada Census Reports*, 1871, 1881, 1911.

between 1871 and 1912 (see table 3.8); consequently, more horses were needed to increase production. Not surprisingly, then, the growth in the number of horses was outpaced by the growth in their value, which nearly doubled for the average horse (from $76.94 to $146.95) between 1901 and 1911. This increase reflected a national levelling out, in a sense, for it was most marked in the provinces where values had been lowest a decade earlier. The fact that the average price more than doubled in British Columbia to $136.44, despite mechanization in the forest industry, can probably be attributed to the sharp decline in the number of horses from 5.5 to 3.1 per farm (see tables 3.5 and 3.7).

One might have also suspected that the increased value nationally reflected an expansion of registered stock, yet the Canada *Census Reports* indicate that purebred horses were still only 1.3 per cent of those in the country as a whole in 1921.[80] It does not follow, however, that the stock was failing to improve, for farmers were encouraged to breed their mares to registered stallions owned by local agricultural societies. As Greene notes of the United States, "Pedigreed stock were regarded as a reservoir of quality traits on which farmers could draw to breed improved horses."[81] Thus, in the interest of improving farm efficiency, the Quebec Department of Agriculture began importing Belgian stallions for sale to agricultural societies in 1910.[82] In the Prairies, the first major step towards introducing purebred stallions was taken by Alberta's George Lane, who imported seventy-two Percheron mares and three stallions from France in 1909,

Figure 3.9   Four-horse teams pulling binders stretch nearly as far as the eye can see.

returning the following year to purchase twenty-five more stallions and fifty mares, said to be "the largest and best importation of purebred horses ever made to Canada."[83] Ten years later, in 1919, the University of Saskatchewan began to sponsor one of the most ambitious programs of Clydesdale breeding outside Scotland.[84] Not to be outdone, the Alberta government set out the same year to purchase two of the best draft horse stallions available, one Clydesdale and one Percheron.[85]

Reflecting the Canadian state's interest in promoting purebred stock, the 1921 census identifies the breeds for both that year and 1911. Even though British draft breeds had fallen out of favour in the United States because their feathered fetlocks made them prone to leg infections during winter when slush or mud froze to their lower legs,[86] the high-stepping Clydesdales remained much the most popular breed in Canada during the early twentieth century. In descending order, the main breeds in 1921 were Clydesdale (30,735), Percheron (8711), Belgian (2083), Standard Bred (1480), French Canadian (1055), and Shire (741).[87] Most of these breeds had increased significantly since 1911, with the exception of the Standard Bred, which – along with three other carriage breeds: Thoroughbreds, Hackneys, and Coaches – had declined sharply in number, obviously reflecting the rise of the automobile. Having to compete with the automobile as well as the

Figure 3.10 Imported from Scotland, Clydesdales were the most popular draft breed in Canada. Those shown here are smaller than they typically are today.

trend towards large draft horses, the all-purpose French Canadian dropped by 332, or 24 per cent, as it again moved dangerously close to extinction. Finally, mules never challenged the primacy of horses in Canada, because they were slower and said to be stubborn, but the fact that they were also hardier and more docile than horses[88] made them popular in mountainous British Columbia during the gold rush era when they were used as pack animals on the Cariboo Trail. As late as 1921 there were 158,742 of these animals recorded in the West Coast province, the average one being valued at $149.60, or slightly more than the average farm horse.[89]

As already suggested, the relatively small number of horses that were purebred as of 1921 did not mean that they were failing to change. Most horses had weighed between 800 and 1,200 pounds in 1800, but, according to Margaret Derry, the general-purpose horse known as the "chunk" began to become larger in the 1840s as a result of breeding with newly imported Clydesdales and, later, with Shires as well as Percherons.[90] Greene goes so far as to claim that in the United States "the average size of draft horses, especially in the cities, increased by possibly as much as 75 percent in the second half of the century,"[91] but Derry points out that even in the late nineteenth century North American farmers still favoured horses that were neither the specialized light nor the heavy draft type, nor associated with a specific breed. Despite the invention of heavier farm implements, the average weight of the general-purpose horse in Canada was still said to be 1,200 pounds in 1919.[92]

Figure 3.11   Known as Cataline, Jean-Jacques Caux ran his mule-packing business from 1858 until 1912. Pack trains consisted of about sixty animals, each carrying 250–300 lbs.

The Quebec agricultural expert Édouard Barnard had observed in 1895 that the vogue for the large 1,600-to-2,000-pound version of the Percheron had passed because it was too heavy for the soft country roads. The same was the case for the Shire and the Clydesdale, though there was still a market for them in the larger cities.[93] He might also have mentioned that the lighter horses were less costly to feed.[94] Even in the West, according to Derry, lighter horses were preferred for farm work once the heavy Prairie soils had been broken. Thus, a dealer appearing before a select committee of the House of Commons in 1924 claimed that western farmers preferred horses in the 1,200 to 1,400 pound range, and that his market for those weighing 1,600 pounds or more was only in the cities.[95] With electric tramways, bicycles, and horseless carriages undermining the market for lighter horses, however, farmers were encouraged to focus on the heavy drafters to meet the urban and industrial demand, and they finally became more inclined to use them on their land as well.[96]

In fact, at least one agricultural expert, E.S. Archibald of the Canadian Department of Agriculture, claimed that heavier horses were more efficient. Encouraging farmers to view their horses essentially as machines,[97] a pamphlet published by Archibald during the First World War opened with the statement that "increased production of crops on every Canadian farm is not only an economic necessity but also the patriotic duty of every Canadian farmer." Archibald's main aim was clearly to convince more farmers to purchase the "larger and better farm implements" that he felt could be operated more efficiently by heavier horses, thereby ensuring that food production for the war effort did not decline at a time when farm labour was in short supply.[98]

HORSES VS TRACTORS: THE INTERNAL COMBUSTION ERA

Archibald's obsession with the efficiency of draft horses was not new, for Greene notes that by the middle of the nineteenth century developments in biology, physics, and chemistry had blurred the boundaries between the mechanical and the organic, especially in discussions about the nature of energy. Thus, *A Textbook of the Physics of Agriculture,* published by Franklin H. King of the University of Wisconsin in 1899, presented formulas for calculating draft using (in Greene's words) "variables of horse size, weight, and build, the number of horses being used, the desired speed and distance, road grade, road surface, wheel size, width, and type, harnessing, load weight, load distribution, vehicle rigidity (vertical draft), vehicle friction, and length of time in use."[99] To take one example, the draft required to pull a wagon over a gravel road was reported to be 75 to 140 pounds per ton, but over a macadam road only 55 to 67 pounds per ton.[100]

Horses are capable of sudden bursts of great strength, as demonstrated at the state fair pulling contests that were popular in the 1970s, when double-hitch pulls of over 3,000 pounds for the standard distance of 27.5 feet (representing rolling loads of more than 100 tons) were not uncommon. These contests demonstrated that horses can exert overloads of well over 1,000 per cent for short periods.[101] But it is generally assumed that the draft or pull of a horse doing steady and continuous work for a ten-hour day should not be more than one-eighth to one-tenth its weight, namely an average pull of only 150 pounds for a 1,500-pound horse.[102] Stated another way, a team of 1,500-pound horses with equipment in good repair can be expected in a ten-hour day to either plow one and a half to two acres, harrow or plant eight to ten acres, mow seven acres, rake fourteen acres, or haul one to one and a half tons on a wagon for twenty to twenty-five miles.[103] Equine efficiency has been measured at 15 to 20 per cent, which is comparable to that of the internal combustion engine, but the fact remains that machines do not become tired, even if they eventually wear out. Horses finally lost out simply because they were less powerful,[104] but that process did take longer than many had anticipated. As McShane and Tarr point out, the efficiency of horses was not simply a matter of strength, but also of the nature of the fuel consumed, as well as initial cost and maintenance in comparison to other animals, steam engines, and, later, electric motors.[105]

The number of horses in Canada continued to grow between 1911 and 1921, though at a slower rate than in the previous decade. Because the number of farmers dropped, however, there was an increase in ratio from 3.6 to 4.1 horses per farmer (see table 3.5). Not surprisingly, given the appearance of motor vehicles on the market, the value of the average horse had declined somewhat from $146.95 in 1911 to $121.54 in 1921(see table 3.7). The exception was the Maritimes, where Prince Edward Island had banned use of automobiles in 1908 on the grounds that they caused horses to bolt, thereby endangering the lives of those driving the local roads.[106]

Gasoline tractors might have been expected to gain a quicker foothold than automobiles in rural communities, but they remained too unwieldy, undependable, and expensive to be a practical option until the First World War drove up the price of grain while making farm labour more expensive. Manufacturers took advantage of this situation to produce small tractors in the ten-to-twelve horsepower range in the hope of supplanting the horse on the half-section farm then typical of prairie agriculture. The Fordson was supposed to be capable of plowing an average of eight acres in a ten-hour day (or four times the amount of the average pair of horses, as noted above), and the Canadian government arranged for their purchase and resale to farmers at the wholesale price of $795. Sales dropped dramatically after the war, however, in part because of the postwar recession and in part because these tractors were too light and unreliable to replace the horse for farm work.[107] According to a survey made by an official with the Alberta Department of Agriculture in 1921, 80 per cent of tractor owners regretted their investment. Not only were the maintenance and repair of these machines beyond the ability of the ordinary farmer, but manufacturers had an inadequate parts supply system. Finally, horses could get on the land earlier than tractors for spring seeding, and most horse-drawn equipment could not be adapted to withstand the force of acceleration or the speeds of tractors.[108]

Figure 3.12 This harness was one of the more expensive models, and emphasis was placed on the high quality needed for western agriculture.

A booklet promoting horse breeding in 1920 claimed, "The tractor and the truck have been so perfected as to take the place of the horse in a considerable number of industries, but we now know as we did not a few years ago, that no tractor can take the place of the horse on the farm of the average size, and smaller."[109] To take one example, the fact that horses had knowledge of routine and could respond to command meant that a farmer could perform such tasks as picking stones from a field without having to get repeatedly on and off a tractor that was consuming fuel by idling.[110] Until the mid-1920s, tractors were used mostly as portable power sources for operating threshing machines, but manufacturers then began to produce more powerful, reliable, and fuel-efficient models that could pull plows with three or four blades. Most important of all was the addition of the power take-off that extended the drive shaft of the tractor beyond the rear axle to operate moving parts of the implement being towed. Given that operating with tractors was less expensive than continuing to use horses on the Prairies by the late 1920s, and that the improvement in grain yields and prices made new equipment more affordable, it is not surprising that there was a rebound in tractor sales and that the number of horses on Prairie farms began to decline.[111] In fact, the value of the average horse in that region had dropped dramatically to $52 by 1931 (see table 3.7).

Meanwhile, there were still only 800 tractors in Ontario in 1918, and still too few three years later to make a significant impact on the number of farm horses, which had continued to increase, as we have seen. Quebec farmers were slower still to embrace tractors, for there were only 7.0 units per 1,000 Quebec farms in 1921, as compared to 36.2 in Ontario, and 66.7 in the rest of Canada.[112] Clearly, then, the great majority of farmers were still reliant on horses alone in 1921.[113] And horses still represented a substantial investment to the farmer, for they were 42 per cent of the number of cattle in the country and valued considerably higher as a whole ($438,555,611 vs $343,146,320), representing 6.3 per cent of average farm value.[114] The amount of land required for their feed was also substantial. The Horse Association of America estimated that each farm horse consumed the production of two and a half acres of "fertile corn belt land" annually.[115] In Canada, 28.6 per cent of the cultivated acreage was devoted to oats in 1921 (admittedly, not all was consumed by horses), as compared with 17.7 percent for hay, and 40.8 percent for wheat.

Given that investment in horses as opposed to cattle would be less and less profitable as urban centres turned increasingly to electric streetcars and automobiles, those who had an interest in raising horses had good reason to worry about their displacement by tractors for working the land and by automobiles for travelling to town. Thus, the *Farmer's Advocate* complained in 1919, "Thousands and thousands of dollars are being expended in perfecting the tractor and making it suitable for farm work. The horse is not given a chance to demonstrate its usefulness, and is gradually being relegated to the background."[116] Quebec agronomes were still predicting as late as 1941, however, that mechanization of traction on the farm would never be more than a supplement to the use of horses.[117]

That view may have been understandable from the perspective of Quebec, where a large number of farms had steep slopes that made the operation of tractors dangerous, and where there were still only 3.8 tractors per 100 farms in 1941, but there were 19.9 per 100 farms in Ontario, and 21.8 in Canada as a whole.[118] Even though the strong western trend toward tractors had been delayed somewhat by the drought and economic depression, 38 per cent of prairie farmers owned tractors in 1941.[119] An American expert declared three years later that a farm should have more than seventy-five cultivated acres before replacing horses with a tractor,[120] acreages that had yet to be attained outside the Prairies in 1961 (see table 3.4), but engineering improvements such as pneumatic rubber tires and three-point hydraulic hitches were dramatically reducing the cost of tractor horsepower.[121] Despite restrictions in production during the Second World War, rising grain prices and increased labour costs led half the farmers of Manitoba, Saskatchewan, and Alberta to acquire a tractor by 1946. Stimulated by the implementation of government-supplied insurance against falling grain prices, tractor sales began to surge after the war. The number in Canada more than doubled between 1941 and 1951, from 159,753 to 399,686, which was enough to supply approximately two-thirds of the farms in the country.[122]

Horses did have an advantage over tractors in that they could be used for travelling to town, but farmers were also now purchasing cars and trucks. By 1951, over half of Canada's farmers reported owning automobiles, and the number of trucks on farms had more than doubled from ten years earlier. Hoping to produce a strain of general purpose horses that could serve as chore teams on farms that were otherwise totally mechanized, the University of Saskatchewan kept a French-Canadian stallion between 1942 and 1949,[123] but the timing was not propitious. Even cowboys now used pickup trucks pulling horse trailers to reach the cattle ranges more quickly, with the result that there was a dramatic drop in the cowboy remuda.[124]

Many farmers formed strong bonds with their horses, naming them and considering them to be almost members of the family,[125] but this was presumably less the case on the Prairies where large teams were used. In any case, according to MacEwan, farmers in that region had begun to lose patience "with the idle animals eating valuable grass and showing no return." As a result, the Western Horse Marketing Co-operative was organized in 1944 to take advantage of the market for canned meat in postwar Europe. The average number of horses on Prairie farms dropped by more than 50 per cent between 1941 and 1951 (see table 3.5), and the co-operative could report as early as 1952 that, having removed a quarter of a million head, its purpose had been served and its two meat-processing plants were for sale.[126] By this time, according to Ankli, Helsberg, and Thompson, "almost all serious commercial farms were operating with tractors ... The only Prairie farmers to cling to horses were small operators on the northern fringes of the park belt who still were engaged in almost subsistence farming."[127] Unfortunately for them, the fact that horse-drawn machinery came from the same corporate source as the tractor ensured that the long-range goal was to eliminate such machines rather than to improve them by adding such features as rubber tires.[128]

Meanwhile, the high-lead system of moving logs from stump to logging road had been introduced to the West Coast, with its large trees and high-density stands,[129] and gasoline-powered crawler tractors spread rapidly in the eastern forests in the 1920s, followed in the 1930s by trucks that hauled logs on iced roads. Pulpwood skidding nevertheless continued in eastern Canada to depend almost entirely on horses, many of which were rented for the winter from local farmers.[130] Eventually, though, the incentive to mechanize further became irresistible as the supply of seasonal workers declined with rural depopulation and the availability of unemployment insurance. While horses worked only 50 per cent of the time when skidding or hauling, and then only in daylight hours and seasonally, machines could be operated year-round and non-stop, if necessary. The invention of the chainsaw also meant that horses and teamsters could no longer keep up with cutters. Wheeled skidders were introduced in the 1950s, and there were over 3,500 in the woods of eastern Canada by 1965 when the Kimberly-Clark Company reported that it had doubled its man-day productivity by replacing horses with these machines.[131] When a Spruce Falls, Ontario, logging company eliminated its last horse in 1966, a supervisor commented, "We hate to see the horses go. But they are outdated. They represent a productivity factor that is just too low for us."[132]

Figure 3.13   Some horses were still used in the forest industry even after trucks took their place for hauling logs.

The mechanization of skidding had been slowed somewhat in eastern Canada by the ready availability of farm horses during the winter, but their numbers declined by more than half nationally between 1941 and 1951, from 2,788,795 to 1,306,639. And the decline would again be precipitous the following decade – down to 512,021 in 1961, by which time horses were finally outnumbered by tractors on Canadian farms.[133] Over half these machines were in the prairies, which reported an average of 1.4 for every farm, and where acres in crop had increased dramatically since 1911.[134] The increases were less notable elsewhere in Canada (see table 3.4), and the number of farms in the country as a whole in 1961 (480,903) had dropped even faster during the preceding decade than the number of horses, with the result that there was still an average of 1.1 horses per farm (see table 3.5). Many were now being raised for urban pleasure riders or racing purposes, or even kept for the collection of urine from pregnant mares to use for human hormone replacement therapy,[135] but ownership of a tractor did not necessarily preclude the use of horses.[136] In fact, the cheaper feed that resulted from the greater efficiency of tractor farming reduced the major expense of the animal mode of power.[137] While

Table 3.9   Horses on farms reporting horses

	1921	1941	1961
Newfoundland	–	–	1.2
Prince Edward Island	2.8	2.8	1.6
Nova Scotia	2.0	1.9	1.5
New Brunswick	2.3	2.1	1.6
Quebec	3.1	2.6	1.7
Ontario	4.2	3.7	2.6
Manitoba	7.9	6.3	2.5
Saskatchewan	10.1	7.4	2.7
Alberta	11.3	8.0	3.8
British Columbia	4.7	4.3	5.3
Canada	5.9	4.9	2.6

Sources: *Canada Census Reports*, 1921, 1941, 1961.

the number of horses on farms in Canada that reported horses dropped precipitously from 4.9 to 2.9 between 1941 and 1951, there was still an average of 2.6 horses on each of those farms in 1961 (see table 3.9).[138]

CONCLUSION

As early as the French Regime in Canada, horses were what Greene refers to as "prime movers," hauling produce to market as well as playing a central role in the development of sociability in a settlement zone where peasants lived on individual lots rather than in European-style villages. With the improvement of farm implements such as the plow in the early nineteenth century, horses began to take the place of the slower-moving oxen as draft animals, increasing in size with the expansion of mechanization after mid-century. They also increased in number more quickly than the human population, for the ratio of 5.3 people for every horse in the British North American colonies in 1851 had dropped to 2.8 people per horse in Canada by 1911. In Greene's words, "Horse-drawn agricultural machines cemented the relationship between the individual farmer and the economy of industrial capitalism.[139] With the rise of the internal combustion engine in the early twentieth century, horses themselves were increasingly portrayed as machines by agricultural experts. Even though British and French stallions were imported to increase the size of Canadian draft horses, however, they ultimately could not compete with the tractor as the market dictated larger and larger scales of production.

That said, draft horses have continued to be kept and bred by prosperous agribusinessmen for show purposes, and by small-scale farmers who believe they are fundamental to their economic independence. In addition, as rural sociologist Douglas Harper notes, working with horses requires a kind of intimate knowledge that is a product of experience and sensitivity, and that adds to the farmer's identity as well as the pleasure of his work.[140] The use of horses does not lend itself to large acreages or single-crop agriculture, but they are well-suited to organic production. Given that horses require pasture and hay land, one generally finds other types of livestock as well as crop rotation on such farms. Furthermore, American farmer and writer Maurice Telleen notes, a horse is "a source that reproduces itself, with good care is self-repairing, consumes home-grown fuel, and contributes to the fertility of the soil."[141] In fact, the horse's inefficiency as a food-processing machine – with only four-fifths of what it consumes being digested – means that it takes an average of only three weeks for it to produce its own weight (or a total of nine to fifteen tons per year) in high-nitrogen manure.[142] In that light, one might question the sustainability of an economic system that was once ecologically self-contained but whose off-farm energy input now exceeds its energy output by a factor of three to one, not to mention the compacting and rutting of the soil caused by the increasingly large tractors and other farm machines.[143]

Horses may never replace tractors on the farm, but MacEwan noted back in 1991, "There are progressive Canadian farmers who are determined that they can perform daily barnyard chores more economically with two heavy horses than with a tractor. The rather common sight of a $30,000 tractor doing a two-horse job makes little sense."[144] In addition, Telleen – writing in 1977 – noted a growing interest in logging with horses, "not on the basis of revoking the twentieth century or moving mechanized equipment out, but rather in a supplementary or specialized role." Not only had the investment required for large-scale logging become astronomical, but horses could be used to thin and manage a farm woodlot without scarring the landscape, skinning the bark on standing trees, or destroying the young growth.[145] Despite the long-term trend towards large-scale production, then, the sense of satisfaction many gain from working with horses, the desire for healthy food, the antipathy to clear-cutting, and the growing crisis caused by the burning of fossil fuels may ensure that small-scale farming and logging will survive into the indefinite future, thereby preserving a role for the farm horse long after the onset of its decline as a vital engine of the economy.

NOTES

1 My thanks to Ruth Sandwell, John Thompson, and Josh MacFadyen for their helpful comments.

2 In an organic economy "the productivity of the land ultimately determined all else since the material artefacts useful to man were made from animal or vegetable raw materials." Wrigley, *Poverty, Progress, and Population*, 4, 46–52, 74–6, 220–2, 225–6. Swedish research shows that energy provided by moving wind and water during the nineteenth century was rather insignificant when compared to firewood consumption or muscle energy. Ibid., 30n41.

3 O'Brien, "Agriculture and the Industrial Revolution," 171.

4 Wrigley, *Continuity, Chance and Change: The Character of the Industrial Revolution in England*, 5–6, 13–15, 35–40, 42. O'Brien notes, as well, that new crops such as clover, lucerne, and turnips were inserted into traditional rotations, reducing the amount of land left fallow, restoring soil fertility by acting as green manure, and improving the capacity of the land to carry animals. O'Brien, "Agriculture," 171.

5 Wrigley, *Continuity*, 42–3. Wrigley also states, "A man can produce only about a tenth as many foot-pounds of effort in an hour as a horse," and notes that increased agricultural productivity "yielded the fodder needed by hundreds of thousands of horses employed in mines, industrial plants and transport." *Continuity*, 36, 39, 124.

6 McShane and Tarr, *Horse in the City*, x, 2–7; Greene, *Horses at Work: Harnessing Power in Industrial America*, 7, 15.

7 Bernier, *Le cheval canadien*, 25; Griffiths, *From Migrant to Acadian: A North American Border People, 1604–1756*, 132.

8 Bernier, *Le cheval canadien*, 26–30, 38–41; Séguin, *La civilisation traditionelle de l'"habitant" aux 17ᵉ et 18ᵉ Siècles*, 531–2.

9 This calculation is based on the assumption that the 2,665 married and widowed women represented the number of families in the colony. There were more married and widowed women than married and widowed men in the colony. The census data are from Canada, *Census Reports, 1870–71*, vol. 4.

10 McShane and Tarr, *Horse in the City*, 8.

11 Dechêne, *Habitants et marchands de Montréal au XVIIᵉ Siècle*, 319; Dechêne, *Le peuple, l'état et la guerre au Canada sous le Régime français*, 214–16; Bernier, *Le cheval canadien*, 42, 60–1.

12 See Séguin, *La civilisation traditionelle*, 545–9.

13 Dechêne, *Habitants et marchands*, 319–20; Jones, "The Old French-Canadian Horse: Its History in Canada and the United States," 133–4.

European visitors were surprised to find, however, that some horses were used for plowing in Canada. Bernier, *Le cheval canadien*, 42–5.

14 Townspeople apparently still owned more horses than did farmers, on average, for evidence from post-mortem inventories suggests that no more than 10 per cent of habitants had more than two horses during this period. Dechêne, *Habitants and Merchants in Seventeenth-Century Montreal*, 190, 215. The horse/family ratio for 1721 is based on the assumption that each married woman represented one family. Canada, *Census Reports, 1870–1*, vol. 4. Séguin reports lower numbers of horses for these years, even though he cites the same census reports as I do. Séguin, *La civilisation traditionelle*, 540.

15 Shortt and Doughty, *Documents Relating to the Constitutional History of Canada, 1759–1791*, 50; Dechêne, *Le peuple*, 352–3.

16 Paquet and Wallot, "Structures sociales et niveaux de richessse dans les campagnes du Québec: 1792–1812," 36. Christian Dessureault's analysis of the post-mortem inventories for the seigneurie of Lac-des-Deux-Montagnes in the Montreal district between 1795 and 1824 reveals that the average peasant family owned two horses, though a few had three or four, and others none. Dessureault, "L'inventaire après décès et l'agriculture bas-canadienne," 132.

17 Beutler, "L'outillage agricole dans les inventaires paysans de la région de Montréal reflète-t-il une transformation de l'agriculture entre 1792 et 1835?," 122.

18 The ratio of farm inventories from the Montreal and Saint-Hyacinthe regions that included swing plows increased from 2 per cent in 1805–14 to 10 per cent in 1825–34. Dessureault and Dickinson, "Farm Implements and Husbandry in Colonial Quebec, 1740–1840," 116.

19 Bernier, *Le cheval canadien*, 98; Skeoch, "Developments in Plowing Technology in Nineteenth-Century Canada," 162,174; Dessureault and Dickinson, "Farm Implements and Husbandry in Colonial Quebec, 1740–1840," 116–17, 121. Dechêne claims that the Scotch plow was adapted more quickly in the Quebec district than in the heavier-soiled Montreal district. "Observations sur l'agriculture du Bas-Canada au début du XIXe siècle," 196. On this theme, see also Beutler, "L'outillage," 129.

20 Olmstead and Rhode, *Creating Abundance: Biological Innovation and American Agricultural Development*, 363. Until neck yokes were introduced by the British, Canadian oxen were fitted with a very simple yoke attached to their horns, though some were also fitted with leather collars. See Séguin, *La civilisation traditionelle*, 609–17; and Chamberlin, *Horse: How the Horse Has Shaped Civilizations*, 106–11.

21 Dessureault and Dickinson, "Farm Implements," 119.

22 Bouchard, "L'agriculture saguenayenne entre 1840 et 1950: l'évolution de la technologie," 361, 366, 370.

23 Skeoch, "Developments in Plowing Technology in Nineteenth-Century Canada," 160.

24 Johnston, *Notes on North America, Agricultural, Economical, and Social,* 2:101.

25 Oxen are not a separate category in this census. In 1850 oxen were only 25.8 percent of farm draft power in the United States, though they still predominated in New England and the upper Midwest. Olmstead and Rhode, *Creating Abundance,* 362–4.

26 Séguin, *La civilisation traditionelle,* 592.

27 Evans, *A Treatise on the Theory and Practice of Agriculture.* On Evans, see Hamelin and Roby, *Histoire Économique du Québec,* 186.

28 Johnston observed in Shepody Bay, New Brunswick, however, that teams with two oxen and one horse had "a boy to drive and a man to hold the plough." See Johnston, *Notes on North America,* 2:101.

29 Evans, *A Treatise on the Theory and Practice of Agriculture,* 118. On rural roads and marketing in Lower Canada, see Courville, Robert, and Séguin, *Atlas historique du Québec: le pays laurentien au XIXe siècle. Les morphologies de base,* 29–35.

30 This system posed the challenge of controlling the front horse, which was not restrained by the sleigh's shafts. Séguin, *La civilisation traditionelle,* 602–4.

31 Evans, *Treatise,* 146–7, 247–8. On the later attempt to force the habitants to modify their sleighs, see Kenny, "'Cahots' and Catcalls: An Episode of Popular Resistance in Lower Canada at the Outset of the Union."

32 Heriot, *Travels through the Canadas,* 120; Talbot, *Five Years' Residence in the Canadas,* 1:176. For other observations, see Bernier, *Le cheval canadien,* 89–92, 109; Bélanger, "Évolution du cheptel équin et de la culture équestre dans la vallée du Saint-Laurent sous l'influence britannique, 1760–1850," 35–7, 119–20; and Jones, "French-Canadian Agriculture in the St Lawrence Valley, 1817–1850," 120–2. For a critique of the British travel narratives as historical sources, see Dechêne, "Observations," 191.

33 McShane and Tarr, *Horse in the City,* 127.

34 Apps, *Horse-Drawn Days: A Century of Farming with Horses,* 26; Telleen, *The Draft Horse Primer: A Guide to the Care and Use of Work Horses and Mules,* 151–4. Olmstead and Rhode have calculated that, as of the mid-twentieth century, a horse required 30 per cent more grain than a dairy cow, but 25 per cent less roughage. Olmstead and Rhode, *Creating Abundance,* 266.

35 Henry, *Feeds and Feeding, A Hand-Book for the Student and Stockman*,
    138–9. L.H. Bailey's *Cyclopedia of American Agriculture*, published in 1908,
    states that oats yield 882 calories per pound, as compared with only 327 cal-
    ories from a pound of hay. McShane and Tarr, *Horse in the City*, 127.
36 McShane and Tarr, *Horse in the City*, 127, 136, 144.
37 On the assumption that 100,000 of the province's horses were fit for work,
    Evans calculated that the oats harvest was not half enough to feed them, and
    not one quarter enough if they were constantly worked. He assumed that
    only twenty-four bushels of oats were harvested per horse, however, and the
    total reported in the 1831 census actually represents thirty-three bushels per
    horse, using Evans's total of 100,000 horses. Evans, *Treatise*, 120–1.
38 Evans, *Treatise*, 118–19, 246; Bernier, *Le cheval canadien*, 95. The role
    horses played in French-Canadian sociability is reflected by the fact that
    there were more than twice as many pleasure carriages in rural Lower
    Canada as in rural Upper Canada in 1852 (150,833 and 68,043, respec-
    tively), though the average carriage in Upper Canada was valued consider-
    ably higher ($44.45 versus $25.01 for Lower Canada).
39 Bernier, *Le cheval canadien*, 98–9; Jones, "Old French-Canadian Horse," 147.
40 Jones, "Old French-Canadian Horse," 141–4, 150–2; Bernier, *Le cheval
    canadien*, 77, 81–3, 92–6, 119; Bélanger, "Évolution du cheptel," 41–3,
    48–53, 58–60.
41 The Upper Canada ratio is for horses aged three and older (McCalla,
    Planting the Province: The Economic History of Upper Canada, 106),
    while the 1831 census for Lower Canada reports an average of 1.4 horses
    of all ages per family (as defined by inhabited house).
42 Bélanger, "Évolution du cheptel," 43. Jones claims, on the other hand, that
    most early Upper Canadian horses were of French-Canadian stock, though
    increasingly interbred with inferior American imports. Jones, *History of
    Agriculture in Ontario, 1613–1880*, 144, 268–9.
43 Talbot, *Five Years' Residence*, 177.
44 On the Conestogas, see Greene, *Horses at Work*, 85–6. Because they were
    slower, the Shires were less popular than the Clydesdales. Jones, *History
    of Agriculture*, 146–8, 193–4, 267–8; Greene, *Horses at Work*, 106–7;
    MacEwan, *Heavy Horses: Highlights of Their History*, 14.
45 MacEwan, *Heavy Horses*, 3. A chief cause of intensive raiding for horses
    was that, during harsh winters, "nomadic hunters sometimes could not
    give their horses the rest, food, shelter, and protection from predators that
    they needed to survive." Binnema, *Common and Contested Ground: A
    Human and Environmental History of the Northwestern Plains*, 141–2.
46 MacEwan, *Heavy Horses*, 4.

47 In the United States threshing machines were powered by treadmills and sweeps until the end of the nineteenth century. McShane and Tarr, *Horse in the City*, 137. See also Jones, *History of Agriculture*, 208.

48 Bernier, *Le cheval canadien*, 100. For brief histories, see Apps, *Horse-Drawn Days*, 144–5.

49 See Fiege, *The Republic of Nature: An Environmental History of the United States*, 246–50.

50 Greene, *Horses at Work*, 45. On the link between railways and horses, see ibid., 77–8.

51 Greene, *Horses at Work*, 199.

52 Blouin, "La mécanisation de l'agriculture entre 1830 et 1890," 95–6. On the "lag" between invention of the reaper and its widespread adoption in Upper Canada / Ontario, see Pomfret, "The Mechanization of Reaping Nineteenth-Century Ontario: A Case Study of the Pace and Causes of the Diffusion of Embodied Technical Change," 87–93.

53 Blouin, "La mécanisation," 94, 96–7.

54 Hamelin and Roby, *Histoire Économique*, 201.

55 Blouin, "La mécanisation," 96, 98–9, 102–4; Armstrong, *Structure and Change: An Economic History of Quebec*, 162.

56 McShane and Tarr, *Horse in the City*, 132–3.

57 There were 36,874 reapers and mowers in Ontario in 1871, and only 5,149 in Quebec, 869 in New Brunswick in 1871, and none in Nova Scotia. Pomfret, "Mechanization of Reaping," 81–2; MacKinnon, "Agriculture and Rural Change in Nova Scotia, 1851–1951," 239.

58 Hamelin and Roby, *Histoire Économique*, 202–3. Steam-powered threshers drawn by horses from farm to farm made their first appearance in American wheat-growing areas during the 1870s and 1880s, but the Saguenay region did not even have horse-powered threshing machines until the late 1880s. Greene, *Horses at Work*, 194, 197; Apps, *Horse-Drawn Days*, 144–50; Bouchard, "L'agriculture saguenayenne," 366.

59 In 1871 Quebeckers would, nevertheless, report more light carriages than Ontarians, an average of 2.0 versus 1.2 per farmer (defined as occupant), a margin that was even more inexplicably wide for transport vehicles: 3.4 vs 1.7.

60 McCalla, *Planting the Province*, 268.

61 Ibid., 320; Jones, *History of Agriculture*, 228. Greene (*Horses at Work*, 135) is clearly mistaken in suggesting that relatively few horses were imported to the United States from Canada during the Civil War.

62 Jones, *History of Agriculture*, 274–5. According to McShane and Tarr, cornmeal was frequently substituted for oats in the United States because

it was cheaper, but the fact that it lacked niacin meant that horses were
wanting in the energy required for heavy urban work. Corn-fed horses
were also more prone to colic than those that were fed oats. *Horse in the
City*, 10, 145–6.

63 McCalla, *Planting the Province*, 222.

64 Greene adds that oxen "remained in wide use on New England farms into
the twentieth century." Greene, *Horses at Work*, 40–1.

65 See Rajala, *Clearcutting the Pacific Rain Forest: Production, Science, and
Regulation*, 12–14, 128–30; Hardy and Séguin, *Forêt et société en
Mauricie*, 103–5; Radforth, *Bushworkers and Bosses: Logging in Northern
Ontario, 1900–1980*, 57–62. Rajala (*Clearcutting*, 16) notes that
Percherons and Clydes were the favoured horses on the West Coast, and
that one logger paid as much as $500 each to replace his oxen with horses.
The use of farm horses in the woods during the winter helps to explain
why in 1861 the average number of bushels of oats harvested per horse in
New Brunswick and Lower Canada was 75.1 and 70.3, respectively, while
in Upper Canada and Nova Scotia it was only 56.2 and 47.2, respectively.
Oats, grown largely to feed horses, increased in production in the valley
of the St Lawrence and its major tributaries from an average of 8.4 bushels
per farmer in 1831 to 13.3 bushels in 1851, and 17.1 bushels in 1871.
Courville et al., *Atlas historique*, 53, table 2.

66 Apps, *Horse-Drawn Days*, 132–43. These machines were in turn eventually displaced in the prairies by the combine harvester, which cut, threshed,
and winnowed the grain in a single process. Because they were so heavy,
these combines required twenty to thirty horses to operate, and so were
not widely adopted on the prairies until suitable tractors became available
in the 1950s. Bellis, "Machines to Cut Grains"; John Herd Thompson, personal communication, 19 May 2013.

67 Lawr, "Development of Ontario Farming, 1870–1914," 240–1. See also
Apps, who notes that the mechanical hay loader appeared shortly after
the turn of the century. Apps, *Horse-Drawn Days*, 75, 80, 86, 88–9, 96,
118–19, 121.

68 The fact remains, however, that the national ratio of binders per farm was
still only 0.6 in 1931, when only the Prairie provinces reported an average
of close to one per farm.

69 Lawr, "Development of Ontario Farming," 242–3.

70 It was generally recommended that horses doing heavy work be fed approximately twice as much grain as horses doing light work (Telleen, *Draft Horse
Primer*, 162). The Department of Agriculture's advice was that a grain mixture of five parts whole oats and two parts bran be fed at the rate of one

pound per day for each 100 pounds of horse. (Archibald and Rothwell, *The Feeding of Horses*, 1–2). By this standard, the annual consumption of the oats portion alone by a mid-size horse weighing 1,200 pounds would be 3,132 pounds, equivalent to 97.9 bushels. See appendix 3, table A3.1, "Oats (Bu.) per Farm Horse," for a summary by province, 1851–1921.

71 MacEwan, *Heavy Horses*, 5.

72 Elofson, *Cowboys, Gentlemen and Cattle Thieves: Ranching on the Western Frontier*, 27–8, 48–9.

73 Breen, *The Canadian Prairie West and the Ranching Frontier, 1874–1924*, 131.

74 McEwan, "'Our Bugbear of War': The Development of Canadian Army Veterinary Practices in the Great War."

75 Elofson, *Cowboys, Gentlemen and Cattle Thieves*, 147.

76 Woolliams, *Cattle Ranch: The Story of the Douglas Lake Cattle Company*, 159.

77 Ibid., 67, 74–5; Sasges, "Colliers and Cowboys: Imagining the Industrialization of British Columbia's Nicola Valley," 17.

78 Rajala, *Clearcutting*, 16.

79 Quoted in ibid., 13.

80 Canada, *Census Reports*, 1931.

81 Greene, *Horses at Work*, 111.

82 MacEwan, *Heavy Horses*, 82, 84.

83 Quoted in ibid., 51. Percherons were first imported to the United States in large numbers in the 1880s. Belgians began to replace them in the 1890s because – despite being slower and relatively short-lived – they were better feeders. See McShane and Tarr, *Horse in the City*, 10–12; Telleen, *Draft Horse Primer*, 28, 30; Greene, *Horses at Work*, 104–6.

84 For details, see MacEwan, *Heavy Horses*, 22–6, 53.

85 See MacEwan, *Heavy Horses*, 55–6.

86 Tarr and McShane, *Horse in the City*, 11.

87 Contrast the farms of the United States, where in 1920 there were 70,600 registered purebred Percherons, but only 5,600 Shires, 5,000 Belgians, and 4,200 Clydesdales. Olmstead and Rhode, *Creating Abundance*, 369.

88 Ibid., 366.

89 The price for a mule in Victoria in 1861 was $225. The merchant Felice Valle packed 200 kilograms on each mule, but only 100 kilograms on each horse. Bowen, *Whoever Gives Us Bread: The Story of Italians in British Columbia*, 27, 30.

90 Derry, *Horses in Society: A Story of Animal Breeding and Marketing, 1800–1920*, 79–80; Greene, *Horses at Work*, 109–10; Jones, "Old French-Canadian Horse," 145–6.

91 Greene, *Horses at Work,* 103, 109–10.

92 Derry, *Horses in Society,* 81–2.

93 Barnard, *Manuel d'agriculture: le livre des Cercles Agricoles,* 266–7, 271–2.

94 A source published in 1944 claimed that it cost 20 per cent more to feed, raise, and maintain an 1,800-pound horse than a 1,500-pound one. Cited in Telleen, *Draft Horse Primer,* 37–8.

95 Derry, *Horses in Society,* 82.

96 Ibid., 91–6.

97 On this theme, see Greene, *Horses at Work,* 15–20.

98 Archibald, *Preparing Farm Horses for Summer Work,* 1–2.

99 Greene, *Horses at Work,* 222–3. On the attempts of engineers to improve horse efficiency, see ibid., 213–22.

100 Henry, *Feeds and Feeding,* 253.

101 Telleen, *Draft Horse Primer,* 336–41; Henry, *Feeds and Feeding,* 251–3. The dynamometer was invented in France in 1821. McShane and Tarr, *Horse in the City,* 3.

102 Apps, *Horse-Drawn Days,* 13.

103 Ibid., 60.

104 Greene, *Horses at Work,* 21. Humans, too, had an efficiency of approximately 20 per cent, as measured by the value of fuel returned as external work. Henry, *Feeds and Feeding,* 88–9. For Cornell University's table, published in the early forties, on the number of acres a good 3,000-pound team of horses could be expected to plow, harrow, drill, plant, cultivate, mow, or rake in a day, see Telleen, *Draft Horse Primer,* 261.

105 McShane and Tarr, *Horse in the City,* 4.

106 In 1913, when local communities were given the authority to allow cars to be driven on Mondays, Tuesdays, and Thursdays, "leaving market days and Sundays worry-free for horse owners," only thirteen of ninety-one did so. All roads were finally declared fully open to automobiles in 1919. MacEachern, "No Island Is an Island: A History of Tourism on Prince Edward Island, 1870–1939," 89–98.

107 Ankli, Helsberg, and Thompson, "The Adoption of the Gasoline Tractor in Western Canada," 12–13, 21.

108 Ibid., 15–16.

109 Carlson, *Studies in Horse Breeding.*

110 Greene, *Horses at Work,* 270–1.

111 Ankli, Helsberg, and Thompson, "Adoption of the Gasoline Tractor," 17–19, 33.

112 Armstrong, *Structure and Change,* 213.

113 Reporting horse ownership were 606,294 of the country's 667,032 occupiers, a number that drops to 622,976 if we exclude those who occupied ten acres or less.

114 See Urquhart and Buckley, *Historical Statistics,* 353, 367.

115 McShane and Tarr, *Horse in the City,* 128–9.

116 Quoted in Derry, *Horses in Society,* 97.

117 Corporation des agronomes, *Au Service de l'Agriculture* (Quebec, 1941), 96–7, quoted in Bernier, *Le cheval canadien,* 152.

118 Armstrong, *Structure and Change,* 213.

119 Ankli, Helsberg, and Thompson, "Adoption of the Gasoline Tractor," 10, 33, 35. See also Lew, "The Diffusion of Tractors on the Canadian Prairies: The Threshold Model and the Problem of Uncertainty.'

120 Courteau, "Horse Power: A Practical Suggestion That Would Transform the Way We Live."

121 See Olmstead and Rhode, *Creating Abundance,* 373–4.

122 Ankli, Helsberg, and Thompson, "Adoption of the Gasoline Tractor," 10, 33, 35.

123 MacEwan, *Heavy Horses,* 123.

124 Woolliams, *Cattle Ranch,* 246.

125 Baron, "L'éloge de *la Grise:* Le cheval et la culture populaire au Québec (1850–1960)," 146.

126 MacEwan, *Heavy Horses,* 7. According to Telleen (*Draft Horse Primer,* 11), during the late thirties and forties local machinery dealers took horses in "on trade," then sold them for meat, thereby eliminating the competition.

127 Ankli, Helsberg, and Thompson, "Adoption of the Gasoline Tractor," 35.

128 Telleen, *Draft Horse Primer,* 11. This book devotes considerable attention to the conversion of machinery designed for tractors to the use of horses.

129 Silversides, *Broadaxe to Flying Shear: The Mechanization of Forest Harvesting East of the Rockies,* 12–13; Rajala, "The Forest Industry in Eastern Canada: An Overview." 134–5; Rajala, *Clearcutting,* chap. 1.

130 In 1947 the Canadian Pulp and Paper Association published the second edition of a time-and-motion study in the interest of improving "efficiency of technique" in skidding with horses. Walker and Stevens, *Pulpwood Skidding with Horses: Efficiency of Technique.*

131 Rajala, "Forest Industry," 143–5; Silversides, *Broadaxe to Flying Shear,* 27–9. The average load a horse could skid when working for an extended period of time was about 14 cubic feet of logs, or about 800 pounds, though this was increased by a factor of 10 or more when hauling sleigh loads during the winter. Silversides, *Broadaxe to Flying Shear,* 10–11.

132 Quoted in Radforth, *Bush Workers*, 206–7.
133 The decline was relatively more marked in Canada than in the United
     States, where the number of horses dropped from six million to three mil-
     lion between 1950 and 1960. Greene, *Horses at Work*, 278. Olmstead
     and Rhode (*Creating Abundance*, 375) calculate that tractors accounted
     for 11 per cent of US farm horsepower capacity in 1920, 40 per cent in
     1930, 64 per cent in 1940, 88 per cent in 1950, and 97 per cent in 1960.
134 Ontario had an average of 1.3 tractors per farm in 1961, British Columbia
     had 0.9, and Quebec and the Maritimes each had 0.7. To arrive at these
     ratios I divided the number of farms as reported in the Canada census
     report of 1961 into the number of tractors as recorded for 1960, in
     Urquhart and Buckley, *Historical Statistics*, 381.
135 This controversial treatment still uses approximately 2,000 mares in
     remote areas of North Dakota and Canada. Telleen, *Draft Horse Primer*,
     346–9.
136 See Harper, *Changing Works*, 54–5.
137 Olmstead and Rhode, *Creating Abundance*, 377–8, 383.
138 Clark (*Three Centuries*, 180, 220) notes that on Prince Edward Island
     it was the smaller subsistence farmers, as well as the Highland Scots and
     Acadians, who were most likely to have draft horses in 1951. British
     Columbia and Alberta stood out in 1961, with 5.3 and 3.8 horses per
     farm, respectively, on those farms that reported horses, but most were
     clearly being raised for the market in riding horses.
139 Greene, *Horses at Work*, 198.
140 Harper, *Changing Works: Visions of a Lost Agriculture*, 49–50.
141 Telleen, *Draft Horse Primer*, 3, 7.
142 Bell, *The Little Book of Horse Poop*, 9, 14.
143 Wrigley, *Continuity*, 71–2. According to recent studies, the modern farm
     in the United States requires sixteen calories of energy to produce one
     calorie of energy from grain and seventy calories to produce a single
     calorie from meat. Harper, *Changing Works*, 32. Soil that has not been
     compacted can be plowed and seeded earlier, and it allows plant roots
     to penetrate deeper, thereby enabling crops to better survive periods of
     drought. Ibid., 56–8.
144 MacEwan, *Heavy Horses*, 8. See also Telleen, *Draft Horse Primer*, 262;
     and Kendell, "Economics of Farming with Horses."
145 Telleen, *Draft Horse Primer*, 349–51.

4

Horse Power in the Modern City

Joanna Dean and Lucas Wilson

Muscles were the original source of power: the muscles of humans, horses, oxen, dogs, and a range of other animals hauled goods and turned machinery in the pre-modern world. Animal energies continued to power the early industrial city, providing a flexible mid-size motor force alongside the emergent forces of steam and electricity. This chapter looks at the harnessing of equine energy in the Canadian city in the nineteenth and early twentieth century. We build on the work of F.M.L. Thompson, Joel Tarr, Clay McShane, and Ann Greene. Like them, we argue that the urban workhorse was not a holdover from the pre-industrial era, but a modern "industrialized organism," the potential of its massive body developed through selective breeding programs and amplified and directed with modern technology.[1] We depart from these authors, however, in emphasizing the animality of the horse and focusing upon the fact that *horsepower* was a cognitive form of power. Like humans, animals resist, and it is that resistance, and the consciousness and intelligence behind it, that distinguishes animal power from all other forms of power.

HORSES IN HISTORY

The working relationship between human and horse is an ancient one: horses were domesticated by 4000 BCE and humans were riding horses by 1675 BCE.[2] The relationship brought the horse back from the edge of extinction and revolutionized human society. Technology has always been central to human control of equine energies. The bit, the saddle, the stirrup, and the chariot made the horse a formidable partner on the battlefield. Elaborate harnesses, horse collars, horseshoes, and spoked wheels channelled the horse's energies for agricultural uses.[3] Mechanical devices put the power of the horse to early industrial purposes: hoists multiplied animal energy through the physics of ropes over wheels, and horse sweeps and treadmills directed the forward movement of the horse into rotary power. The technological innovations of the eighteenth century and the advent of steam power only increased the utility of the horse, and through the nineteenth century horses continued to serve as an intermediate form of power between human muscles and steam engines, as increasing numbers of horses were brought into the city to haul goods and people to and from railway stations.[4]

Demand for horsepower was so great in the late nineteenth century that the number of urban horses grew faster than the urban human population. F.M.L. Thompson calculates that the numbers of town horses in Great Britain increased from 350,000 in 1830, to over 500,000 in the 1870s, and 1.2 million in the early 1900s; the ratio of urban horse to urban human increased from 1:30 to roughly 1:20 between 1830 and 1900. In the United States, horses urbanized 50 per cent faster than humans, and by 1900, as Joel Tarr and Clay McShane show, the ratio of urban horses to humans ranged from 1:7.4 in Kansas City to 1:22.9 in Chicago to 1:26.4 in New York City.[5] Margaret Derry argues that the world population of horses peaked between 1910 and 1920 at 110 million horses. This amount was double the horse population of a century earlier, and four times the 27 million horses in 1720, before the industrial age.[6]

HORSES IN CANADIAN URBAN HISTORY

The main source of Canadian data is the census, which from 1871 to 1911 tracked populations of adult horses and horses under the age of three. Echoing the trends observed in the United States and Great Britain, Canadian urban horse populations increased dramatically in the late nineteenth century, as the population of other urban domestic animals declined.[7] Horse populations frequently grew faster than human populations.

Toronto illustrates the broader trends in Canada. Between 1871 and 1891, the city's adult horse population increased from 1,874 to 7,269. This growth partly reflected geographical expansion (Toronto annexed Yorkville, Brockton, and Riverdale over this period), but it also reflected a greater density of urban horses. The ratio of horse to human in Toronto increased from 1:28.4 in 1871 to 1:24.5 in 1891. The urban horse population dropped precipitously after the adoption of the electric streetcar in the 1890s. The ratio of horse to human fell to 1:62.3 in 1901 before recovering to 1:51.3 in 1911. The pattern varied slightly in other Canadian cities. The horse population grew more quickly than the human population in Halifax and Ottawa, but grew more slowly in Montreal, Quebec City, and St John. The ratio of horses to humans in 1891 for Canadian cities was the following: Ottawa 1:17.8; Halifax 1:23.0; Quebec City 1:23.1; Toronto 1:24.5; Montreal 1:32.1; St John 1:41.3.

Municipal assessment rolls also attempted to track horse populations in some Canadian cities. A comparison of the two data sets in Toronto, however, suggests that the municipal assessors had more difficulty in accurately capturing the number of horses. Municipal assessment rolls consistently recorded approximately half as many horses as the census. Amongst other things, this discrepancy likely reflected the fact that municipal taxes were levied on horses in Toronto. Even the census data, however, are likely to have under-counted the total number of working horses in Canadian cities, as they failed to account for horses stabled on farms and suburban lands just outside municipal borders. Urban workhorses tended to be geldings, and the numbers also fail to reflect the large populations of mares kept in rural areas to reproduce the urban workhorse population.

As these numbers show, the Canadian city was dominated by humans and horses in the late nineteenth century. Both laboured: muscle power combined with the emerging technology of the modern era to produce new synergies as important as early steam and electric power. Although our emphasis here is on the horse, human power paralleled horsepower, sometimes in partnership with the larger animal. Human power is about one-tenth that of the horse, and the two bodies often laboured in parallel, with humans taking the smaller burdens, and the horse the larger.[8] Men and boys pushed small handcarts and wheelbarrows; horses pulled larger carts and wagons. Women treadled sewing machines; horses walked the treadmill. Teams of men powered hoists to lift goods, until they couldn't, and then the horse was brought in to lift the heavier goods. Firemen pulled fire engines, until the introduction of the heavy steam pump necessitated a team of horses. Men pulled saws, but a horse-powered engine could pull a saw faster and longer. The heavy use of horse power in North America, as compared to Europe, or Asia, may have reflected the shortage of human muscle power on this continent, as well as the availability of cheap hay and oats.

The importance of the horse was brought home by the impact of the 1871–72 epizootic of equine influenza. The disease started north of Toronto and moved across North America, following transportation links, disabling 80 to 90 per cent of the working horses for a week, or longer, and killing an estimated 1–2 per cent of the horse population.[9] Entire cities were brought to a standstill as the disease passed through, demonstrating the close links between human power and horse power. "The illness of nearly 3,000 horses has caused between 5,000 and 10,000 men to be idle," said an observer from San Francisco. One set of muscles had to be replaced by another: "Our grocery men, provision dealers, etc are delivering their goods by means of hand-carts, wheel barrows, and wagons drawn by man-power," reported observers in Pottsville, Pennsylvania.[10] Reports also shed light on the conditions under which most horses lived. "And just in proportion to the squalor, the filth, the impurity, and the absence of a proper hygiene, so does the infection prove more severe and fatal," one veterinarian reported. "So it is in the close, unventilated, and undrained, or underground stables of cities, with air loaded to suffocation with the products of respiration and putrefaction. In these the mortality proves far in excess of that of the horses in the better-appointed stables, or in the country."[11]

In the sections below we look at different uses of horse power in the Canadian city. We look first at animal-powered machines: the sweeps, treadmills, and hoists that translated the power of the horse into rotary or lifting power. We examine the critical role of the urban horse in transportation, hauling goods and people. Then we look at the tensions caused by the growing number of urban horses, as cities struggled to deal with animal aspects of the horse, such as manure, and the unpredictable behaviour of a cognitive source of power.

MECHANIZING MUSCLE

Horses have powered grain mills, threshing machines, and water-lifting devices since Roman times. Animals are not easily turned into machines: the irregular energy of the animal body has to be converted into the steady power needed for mechanical purposes. The traditional sweep (referred to, in its various permutations, as a whim, horse-whim, gin, or, simply, a horsepower) redirected the forward movement of the walking horse into rotary power.[12] The horse walked in a circle turning a long bar connected to a central shaft, and the rotation of the shaft could be employed in a variety of ways. Rope coiling around the shaft, in the capstan or windlass, could be used to raise heavy material, such as water from a well or loads from a mine shaft.[13] Agricola described such a sweep being used to lift material from a 1556 mine. Sweeps were commonly used to lift water and also turned mill wheels: it was a horse-powered grain mill that James Watt used to establish the horsepower unit in

the late eighteenth century.[14] Watt calculated that a horse could turn a
millwheel 144 times an hour (twelve feet per minute), with a force of
180 pounds; he averaged this out to 33,000 pound-feet/minute, or one
horsepower. Some have argued that Watt was overly optimistic and suggest that the more realistic measure of a horse's energy is 0.8 horsepower;
others argue that his calculations accurately reflect the power of a good
strong draft horse.[15]

There is some debate about how long the industrial use of horsepower
persisted in the face of steam. In the early industrial period, teams of
horses turned sweeps to power cotton mills, breweries, foundries, and
corn mills, but their use declined in the face of cheaper and tractable
sources of power. As Karl Marx observed, "Of all the great motive forces
handed down from the period of manufacture, horse-power is the worst,
partly because a horse has a head of his own, partly because he is costly
and the extent to which he can be used in factories is very limited."[16] If
some industrial uses for horses were abandoned over the course of the
nineteenth century, the animals persisted as cheap and flexible sources of
power suitable for situations where intermittent and relatively low levels
of power (under ten horsepower) were needed. Jennifer Tann described
horses as "the bridge between human muscle and steam power for men
of moderate or larger capital" in Britain between 1780 and 1880,[17] and
Joel Tarr argued for the United States, "Although the evidence is limited,
the existing sources suggest that horsepower machines were widely used
for much of the nineteenth century in both urban and rural settings." Tarr
suggests that historians have neglected horsepower because they viewed
it as a traditional and largely agricultural power source, and because in
urban areas it was used primarily in small firms, businesses, and craft
shops.[18] His evidence is strongest for the early nineteenth century, when
horses were frequently used in the textile industry, and when there were
occasional references to horse-powered treadmills in small workshops. In
the second half of the nineteenth century, it appears that such uses were
diminishing, and by the turn of the twentieth century, the only definitive
evidence for the use of horsepower outside of agriculture was in the construction and transportation industries.

The evidence is similar for Canada. Horses were widely used in agriculture, transportation, and construction well into the twentieth century,
but the evidence for their employment elsewhere is limited. Gunter Gad
has noted that as late as 1880 steam power in Toronto was restricted to
larger industries, such as printers, and sash-makers, and planing mills.[19]
Census returns for 1871 show that most urban workshops and factories

relied upon hand or horse power; only 17 per cent used inanimate energy in the form of waterwheels and steam engines. This was especially true of smaller shops: 63 per cent of workshops with fewer than six workers used only animate power, as compared to 0.6 per cent of factories with over fifty-one workers. In Toronto, as in Montreal, about 44 per cent of establishments used animate power exclusively. It is not, however, clear whether much of this animate power was horsepower. Humans powered a range of industrial machinery, from the treadle sewing machine to the hand-operated lift, and it is possible it was human power that ran most of these early industries. An 1874 legal case reveals how common it was for men to put their muscles to work: a factory had contracted for an iron-hoisting machine, capable of raising 2,000 pounds, and the evidence provided in court indicated that this was to be powered with four, or possibly five, "good men."[20]

Such hoists could be powered by horses. The American Hoist and Derrick Company advertised a series of massive cast iron sweeps, referred to as "horsepowers" in their 1907–08 and 1914 catalogues (see figure 4.1). These were massive industrial machines. The catalogue's engravings emphasize the sleek brute force of cast iron; the horse is curiously absent. Whereas traditional wooden sweeps had provided low speeds (two rpm) and high torque, the gearing of the new horsepowers provided up to 200 rpm. The company offered a range of sizes, from the basic Single Speed Horsepower No. 4, to the Contractor's Horsepower No. 2, for general contractors, quarries, and bridge-building; the Double Drum Horsepower No. 1, with two speeds for heavier and lighter loads, for "stone contractors, builders, miners and all heavy and light hoisting"; and the "duplex geared" Quarrying Horsepower No. 3, for the heaviest hoisting. The company claimed that no. 3, "the most powerful machine ever made," could lift three tons at eighteen feet per minute when operated by a single 1,600 pound horse working "moderately," making four-and-a-half turns around a twenty-four-foot circle each minute. A 1923 photograph shows a horse-powered hoist loading logs in Alberta.[21] Those who needed more power than the horse could provide turned to steam engines.

Horse-powered sweeps were also commonly used in brick-making to power the "pug mill" or brick machine that kneaded the clay and pressed it into a brick mould.[22] An Ontario company, H.C. Baird, Son and Co. Ltd, sold both horse-powered and steam-powered brick-makers or "Quakers" in the 1910s (see figure 4.2). The horse-powered machine was adequate for many smaller operations, manufacturing 8,000–10,000 bricks per day, but steam-powered machines ran twice as fast, and the company

Contractors' Horsepower No. 2
Duplex Geared

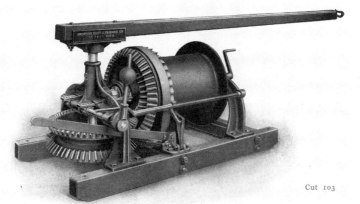

Cut 103

For General Contract Work

THIS machine is provided with our duplex motion, whereby light loads can be hoisted quickly and heavy loads slowly; the change from quick to slow being instantly made by the movement of a lever from left to right—throwing together the system of gears which causes the drum to make either one or two revolutions to one turn of the horse. The amount of time saved by this device is very noticeable in loading teams and laying wall where the stone run large and small. **Save time of men and teams and you save money.** Slack rope may be taken up with the hand crank at the end of the machine without starting the horse. This machine is designed for contractors, bridge builders, etc. Its use in quarries is extensive and it has proven itself **convenient, strong and durable**—the iron work alone weighing 500 pounds more than any other machine of its dimensions.

Power and Range of Machine with One Horse

Results obtained with a horse weighing 1,600 pounds, working moderately, making six turns per minute.

Single line, no block . . . . 1½ tons at 30 feet per minute.
Two lines, or one block . . . . 3 tons at 15 feet per minute.
Four lines, or two blocks . . . . 6 tons at 7½ feet per minute

The quick motion handles **one-half** above loads at **double** these speeds.

Length of machine	6 feet.	Weight of machine	1,700 pounds.
Height of machine	3 feet.	Horse travels in a circle of	18 feet diameter.
Width of machine	3 feet.	Drum	18 inches diameter by 20 inches long.
Price . . . . $		Telegraph Code . . . . FACET	

Miner's Whim

This attachment enables the power to be controlled by levers outside of the ring in which the horse travels, enabling the operator to watch proceedings down the shaft, and also tend bucket at the mouth. It can be attached to No. 2 or No. 4 horse powers.

Cut 448

Price of attachment, $ . . . .
Telegraph Code . . . . FATES

Figure 4.1   Horsepowers were still sold to power hoists in the early twentieth century.

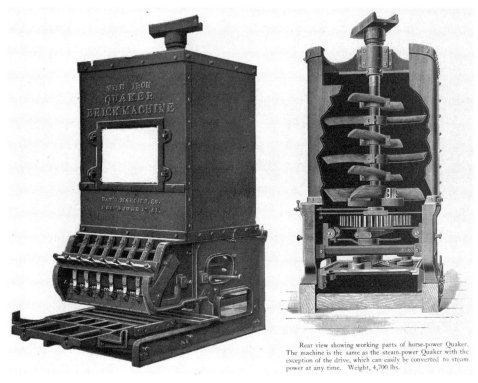

Rear view showing working parts of horse-power Quaker. The machine is the same as the steam-power Quaker with the exception of the drive, which can easily be converted to steam power at any time. Weight, 4,700 lbs.

Figure 4.2   The Quaker Brickmachine was powered by a horse turning a wooden sweep at the top of the machine.

anticipated that brick-makers might want to upgrade, pointing out, "The machine is the steam-power Quaker with the exception of the drive, which can easily be converted."[23] A photograph in the catalogue recorded the delivery of three steam-powered Quakers and two horse-powered Quakers to a Quebec town, where one assumes they were to be used simultaneously. Rothwell Machine Company of Hamilton, Ontario, also sold a two-horse machine, the Genuine Henry Martin Horse Power Brick Machine, which, they stressed, "Can be changed to steam power if desired."[24]

Portable sweeps were widely used by farmers to operate a large range of equipment. They transmitted power via a tumble rod or drive shaft from the central capstan to machinery; horses had to step over the tumble rod on each rotation. They powered such a wide range of machinery (the catalogue produced by E.W. Ross Company advertised a grinder, table saw, swivel carrier, corn husker, cob crusher, silage cutter) that it is likely that they were also used for similar purposes in the small workshops in urban and peri-urban areas. An 1898 court case referred to a "horse-power" that appears to have been an agricultural sweep adapted for urban uses. It was described as "consisting of a large wheel about five feet in diameter, with the necessary cog wheels, tumbling rod, etc. adapted to run the elevators for elevating the grain and feed to the warehouse."[25] This kind of adaptation was probably not unusual; this device appeared in the historical record only because a five-year-old child was injured when the horse and machine was left operating unattended.

In the late nineteenth century, treadmills began to replace the sweep or horsepower. The treadmill had been in existence for centuries, but like the sweep it took new modern form in the nineteenth century. Improved treads and cast iron gearing increased the speed. Horses were not the only animals put to work on treadmills: oxen and mules powered large treadmills, smaller dog and goat treadmills powered butter churns, corn shellers, and cream separators, and in the early nineteenth century human slaves and prisoners were also put to work on treadmills (although with humans the power output was less important than the punishing drudgery and shaming).[26]

The Moody Agricultural Machinery Company, based in Terrebonne, Quebec, encouraged farmers to replace their sweep with the treadmill. They pointed out that the treadmill was a more efficient use of horse power than the sweep: the sweep forced the horse to turn, and power was lost in the angle of the turn, also in the multiple gears needed to convert the slow speed of the sweep to the needs of modern machines. The horse walked in a straight line on a treadmill, and the power was transferred more directly to a large gear, so that a treadmill started at 30 rpm compared to 2.5 rpm in a simple sweep. Moody's claimed that a four-horse sweep could be replaced by a two-horse treadmill. Farmers were often reluctant to make the switch, because they were concerned about the impact upon their horses. As Moody's explained, "In some parts of the country where the modern Tread Powers are not well introduced, the Sweep Power is still in use to a very large extent, and the farmers are slow to discard them for a Tread Power, as they are afraid the latter is a horse killer." In response Moody offered to put any potential purchasers in touch with long list of persons who had used tread powers over the last sixty-five years of manufacture.[27]

There is little Canadian evidence for the use of treadmills in urban areas, but given their ubiquity in rural areas (and in museums today) it seems likely that they would have been adapted for urban uses. We do know that treadmills powered small ferries in Toronto and Montreal. Crisman and Cohn described the improved capture of horse power over the nineteenth century as ferries transitioned from a whim, to a turntable, and then a treadmill mechanism.[28] Toronto's "horse boat" operated for a number of years taking day trippers to the Toronto Islands (see figure 4.3). As one resident recalled,

> She was by no means a very large vessel, being only sixty feet in length by twenty three feet wide, and had what are now known as side wheels. These paddles were set in motion by two horses who trod on a circular table set flush with the deck in its centre. This table as it revolved worked upon rollers, which, being connected with the shaft, set the paddles in motion. The horses were stationary; the table on which they trod was furnished with ridges of wood radiating like spokes from the centre, which the horses caught with their feet, thus setting the table in motion. For some time the boat was worked with only two horses, but after about two years an alteration was effected in the arrangements, and in the vessel as well. Instead of two horses, five were introduced, and they walked round and round the deck, exactly as horses do when employed in working a threshing machine, and the vessel was set in motion precisely as such a machine is.[29]

Horse ferries competed with early steam-powered ferries and were often preferred for short haul trips because of their economy and reliability. It was only in 1850 that Toronto's five-horse ferry was replaced with a twenty-five-horse steam engine. Two horse ferries operated in Montreal in the 1860s and 1870s, and they lingered in remote locations until the early twentieth century.[30]

All of the catalogues emphasize the interchangeability between animal (human and horse) and steam power. The Ross Company sold small cutters that could be operated by men, and larger ones, like the Ross #16 Little Giant Cutter, that could be operated by a three-horse treadmill, four-horse sweep, or a steam engine up to ten or twelve horse power.[31] The brickmaking Quakers could be converted from horsepower to steam power. The hoists sold by the *American Hoist and Derrick Company Catalogue* could be powered by men, horses, or steam engines. This catalogue also depicts a horse-powered elevator, in which the horse pulls a rope that runs through pulley system to raise a construction elevator. The adjacent page of the catalogue shows a steam engine performing the same function as the horse, and the text reads, "The horse elevator can be readily worked by steam

Figure 4.3   A horse ferry operated between Toronto and its islands for many years.

– practically no changes being necessary."[32] Men, horses, and steam provided variable amounts of power and could be substituted for one another, depending upon the demands of the situation. It appears that the horse body was essential for mid-size power until the emergence of the internal combustion engine, and the small electrical engine.

The horse himself (most workhorses were gelded male horses) is curiously absent from the pages of most catalogues described here, his presence suggested only by the open bar of the sweep or the empty treads of the treadmill; he appears only occasionally in the text, for example when copy writers suggest that the newly designed treads on the Moody treadmills are less likely to catch the horses' feet. Yet his intelligent participation in the production of the power is assumed. As Joseph Glynn noted in his 1854 *Rudimentary Treatise on the Construction of Cranes and Machinery for Raising Heavy Bodies, for the Erection of Buildings, and for Hoisting Goods,* "Horses regularly engaged on such work display great sagacity and obedience to a word or sign, to hoist, lower or to stop."[33] When the five-year-old was injured by a horse-powered warehouse hoist in 1898, the court record shows that the horse had been left to work independently.[34] In another case, a man had left a horse to run a treadmill, while he worked on the attached turnip cutter in another room. Horses were frequently directed by children. Brick-makers were to be operated by a boy and a man; the boy presumably directed the horse. A series of 1914 photographs of an "overthrow stacker" in the Glenbow Museum show a small boy leading a horse away from a massive haystack; the horse is pulling a rope that hoists the hay from a wagon up a ramp to the top of the haystack, and adult men are occupied with the more arduous tasks of loading the hay.[35]

Animals capable of intelligent cooperation were also capable of resistance. We know of this mainly because of Humane Society accounts that describe the brutal treatment of recalcitrant horses. Recent experiments with the use of improved horse powers for developing countries are also revealing. Peter Lowe reported in *Animal Powered Systems* that a tired unattended donkey produced only 44 watts of power, whereas a whipped team of oxen produced 500 watts, at least in the short term. Terry Thomas's prototype horsepower, described in *Animal Traction for Agricultural Development*, did not account for the resistance of the ox: he reported ruefully, "Unfortunately the lack of radial constraint upon an angry ox resulted in damage to the prototype which is yet to be repaired."[36]

MOVING GOODS

The vast majority of urban horses were used to transport people and goods. The demand for delivery and haulage grew exponentially in the nineteenth-century city with the arrival of the railway. Horse carts and wagons provided the essential flexible linkage between the fixed lines of the railway and the multiple points of delivery. A horse parade organized by the Society for the Prevention of Cruelty to Animals in Toronto in 1904 illustrates the range of goods delivered. Five hundred horses took part, and prizes were awarded for "commercial horses" used by businesses in a broad range of categories: cartage company, brewery company, coal company, departmental, dry goods and clothing stores, hatters and furriers, bakers and confectioners, milk companies, laundry companies, grocers, florists, butchers, and, finally, "all trades not enumerated above."[37]

Probably the best-known delivery service was that of Toronto's T. Eaton department store. The company started delivery in 1870 with a pony and small wagon that made two circuits of the city on weekdays and three on Saturday. By the late nineteenth century a bird's-eye map of their Toronto offices showed a large multi-storey stable shadowing the department store. Their distinctive red, white, and blue delivery wagons and bay horses were still closely associated with the store; the competition, Simpsons, used grey teams.[38] At the other end of the scale, teams of draft horses were used to move entire buildings: photographs show Toronto's Queen's Wharf lighthouse being moved by horses in 1929.[39]

Fire horses were the elite of city horses, famed for their strength, speed, and calm under pressure (see figure 4.4). Fires in Toronto were initially

Figure 4.4   Manure and snow made city streets difficult, even for the Montreal
Fire Department's matched team. Note the hand-pulled sleigh at left.

fought with bucket brigades: human muscles working in unison to bring
water to the fire. The first pumps were hand powered and were drawn to
the scene by the firemen themselves. In 1861, when Toronto's hand pumps
were sold off to smaller towns and heavy steam engines were purchased,
carter's horses were enlisted to draw the new engines. In 1891 the City of
Toronto upgraded, and purchased two new steam engines, twenty-eight
massive city-owned horses to pull them, and a crew of full-time firemen.
By 1895 thirty-six horses were needed to pull engines, hose carts, chemical engines, and hook-and-ladder trucks.[40] Most fire halls had special
harnesses suspended from the ceiling that could be dropped directly onto
a horses' backs, allowing the team to leave the station within minutes of
the alarm. One enterprising Toronto fireman devised an automated system that released the horse from his stall when the alarm went off; he was
trained to move immediately into position for the harness to drop. The
bravery of both men and horses in the face of raging fires is emphasized
in early accounts of fire fighting. Photographs show fire horses standing
near fires, untied, and much was made of their calmness. But there are
also accounts of horses failing to live up to this legend, shying at explosions and running over pedestrians and firemen.[41] Horses were favoured

by fire departments for many years over motor engines because, unlike early motors, they could be relied upon to start, and Toronto's last fire horses, Mickey and Prince, were retired only in 1931.

Horses of different sizes and breeds were used for the different forms of transportation, but it was the large draught horse who epitomized equine power. Heavy horses nearly doubled in size during the nineteenth century to meet the demand for heavy hauling in the city. As J.I. Little explains in this volume, Canadian farmers preferred light, versatile horses, but, in response to the demands for larger urban horses, farmers imported heavy-bodied stallions, such as the Scottish Clydesdale, the English Shire, and the French Percheron, to breed with local mares. Improved feeding of young horses also produced stronger bodies, and the rations of oats given to working horses also improved the performance of the animal. Over the nineteenth century, the body of the urban draft horse reached an apex of size and power: Ann Greene says that between 1860 and 1880 the average size of the draft horse increased from 900–1,100 pounds to 1,800–2,000 pounds in the United States.[42] Leah Grandy notes that horses in St John were noticeably heavier and stockier in photographs taken after 1901.[43]

The power of the horse's body was amplified by the efficiencies of the wheel on a hard surface. It has been estimated that a packhorse can carry one-eighth of a ton, and a horse pulling a cart on a soft surface can pull five-eighths of a ton. But a horse can pull two tons on a macadam surface and eight to ten tons on hard stone tramway or iron rails.[44] Nineteenth-century engineers debated the merits of cobblestones, wooden blocks, granite blocks, brick, planking, macadam, and eventually asphalt roadways, and much money was expended on the development of the hardest, longest-wearing, least-slippery, most hoof-friendly road surface.[45] Macadam surfaces especially were destroyed by the pressure of narrow wheels, so bylaws regulated the size of cart and carriage wheels to limit wear and tear on these expensive surfaces.[46] In 1890, Toronto's wagons had to have wheels either three or four inches in width, depending upon the circumference, so as to spread the load over a greater area.[47] Horseshoes protected the horse's hooves from the impact of hard uneven surfaces, but the noise of iron on stone drove many urban residents to distraction.

Photographs of city streets suggest that few met the standard suggested in the literature on road surfacing (see figure 4.5). Side streets, especially in the spring and fall, were rutted, potholed quagmires. When Samuel Massey reinvigorated the Humane Society in the late 1880s in Montreal, he described the bad condition of Montreal streets and the difficulties in pulling heavy loads uphill: "For these reasons the life of a carters horse is

Figure 4.5   Horses and delivery wagon stuck in mud. Toronto, 1914.

generally a life of great hardship. I allude here more especially to those who cart coal and wood (over 600) and material of that kind, and to the horses employed on the shipping wharves. They are more heavily loaded and abused than any other horses in the city."[48]

Transportation in the winter was both eased and encumbered by snow. Fresh snow and ice provided the fastest surface for a horse and sleigh, so the winter was the preferred season for hauling, and habitants in Quebec were known for driving their fast Canadian horses along the frozen rivers and lakes. But in the city, as one Ottawa resident noted, accumulated manure could cause problems:

The condition of the main streets are such that sleighing is almost impossible – while from the depth of snow on every country and side road wheeling is on them equally impossible. Even in mid-winter, after a heavy thaw followed by a hard frost, the streets of Ottawa are covered with a coating of frozen manure, which makes them almost impassable for those heavy loads of wood or produce so easily then brought in from the country. For this there is at present no remedy but a fresh fall of snow. But at this season there may be weeks when at least half of every load which reaches the suburbs should be thrown off, when country people leave their vehicles outside and walk in, to save themselves and horses from becoming objects of comment and commiseration while creeping along the gutter and hunting up every detached piece of clean snow or ice ...

There is under the accumulation of filth in the streets a foot or more of solid snow and ice which if it can be got at would maintain the streets in passable order for a long time ... Can nothing be done? A few horses and scrapers at the proper time could rake the manure to the sides, from where it could be carted (and worth the cartage) and I am confident there is nothing in which a small outlay, scarcely worth considering, would produce such important economical results.[49]

The solution to uneven road surfaces and manure-covered roads was, ironically, more horse power. Horses were used to grade and clean the road surface, pulling the machines that graded and rolled the street, and pulling the sweepers that removed manure and the sprinklers that dampened the dust. Steamrollers were introduced at the end of the century, but as late as 1896, *The Illustrated Catalogue of Champion Road Machinery* argued that the horse roller was better: "In the judgment of many of our best engineers and practical road makers" the steamroller was inferior because it was too heavy, "causing it to push the material instead of compacting it" and tilting the bricks in brick streets out of place; because "it cannot mount a steep grade of coarse broken stone" (presumably because the wheels slipped where hooves could grip), and because it was " in a measure disastrous to life and property, frightening horses on our streets and highways."[50] The cyclist John Kirk would have agreed. He was permanently injured when a horse shied at the sudden appearance of a steamroller on Yonge Street in Toronto and knocked him from his bicycle. He sued and in 1904 won his case against the steamroller's operator. Moss CJO, for the majority of the Ontario Court of Appeal, found that steamrollers were "calculated to frighten horses of ordinary courage and steadiness" and held that the operator should have warned passersby of his approach.[51]

TRANSPORTING PEOPLE

It was mainly human muscles that moved people about the nineteenth-century city; only a very small elite could afford to hire a hackney cab or own a private carriage, and cities were small enough that most residents simply walked from place to place. Toronto, for example, with a population of 80,000 residents in 1882, was clustered along six kilometres of lakeshore, extending three kilometres back to Bloor Street.[52] Walking seems uncomplicated and pre-modern. But there were technological innovations and regulatory frameworks to make pedestrian mobility efficient. Extensive networks of planked and then granolithic sidewalks lifted the pedestrian out of the ruts and mud of the roadway. In Toronto, bylaws regulated efficient movement along these sidewalks from as early as 1834. All handcarts were banned from the sidewalks, and at the turn of the century bylaws mandated that pedestrians should move on the right-hand side. Even preaching was not allowed to impede free movement along these early thoroughfares.[53]

Toronto's first form of local mass transit was a horse-powered omnibus service, introduced in 1849, shortly after a similar service had been initiated in Montreal. The service was fairly rudimentary and slowed by poor road conditions but proved very popular. The early six-passenger omnibuses were replaced with ten-passenger carriages in 1850. Horse-drawn streetcars running along rail lines were introduced in 1861. Although far more efficient, streetcars had their difficulties. A reporter for the *Globe* described Toronto's inaugural run: "The first car, with the band, had only proceeded as far as the line which divides Toronto from Yorkville, when it ran off the track and came to a sudden stop. The passengers inside alighted and assisted in placing the vehicle on the rails again, and the driver received the order 'go-ahead.' This occurred several times with all the cars, but the passengers treated the delay as a joke, and the crowd were always ready to give a 'shove' or a 'lift' to keep moving."[54]

Streetcars were relatively slow, generally operating between four and six miles per hour, which was not much faster than human walking.[55] Heavy snow caused trouble, as the rails had to be clear for the streetcar to function. Ottawa's omnibuses had to be brought out when streetcars failed in the spring mud and slush. Snowplows were eventually developed to clear the streetcar lines, but they left dangerous gullies for the sleighs operating on the snowy streets.[56] Streetcars were expensive to operate: each required five to seven horses, working five-hour shifts. Horses were purchased for $125 to $200 in the 1880s, and they could be worked for only four to five years before they were sold at about 75 per cent of this value for less-demanding forms of cartage.[57] But it was difficult to replace the horse. Cable cars, powered by stationary engines, were faster and cheaper to operate, and functioned better than horses on hills, but were inefficient over longer distances and were dangerous because they could not reduce speed on corners.[58] They were practical only in cities with steep hills. So, despite the difficulties, the horse prevailed as a power source for streetcars for over thirty years.

In the late nineteenth century it appeared as if the bicycle might replace the horse. The same hard surfaces that allowed horses to pull tons of goods made it possible for humans to cycle, especially once gearing and rubber tires were developed. Proponents of the "steel horse" argued that bicycles "have all the advantages and none of the disadvantages involved in keeping a horse," and were also more humane:

> Its first cost paid, it requires no further expenditure except for occasional repairs. It does not have to be fed like a horse, and no one needs to be hired to take care of it … In the case of the horse, all men in whom the very human love of animals is not wanting will rejoice that at last a substitute for the horse has been found as a means of locomotion in the cities. The horse is capable of many things, but nature never intended he should be at the tender mercies of so cruel a taskmaster as the average street cab driver.[59]

The bicycle craze, however, had a short life. Darcy Ingram has argued for Montreal that the bicycle became a sport rather than a serious means of transportation.[60] It may also be that the presence of so many horses on the streets deterred cyclists. Even though the city of Toronto created bicycle lanes and regulated the place of the bicycle on the street, horse manure made the streets dirty and slippery, and as cyclist John Kirk found, horses physically dominated the streets.

It was the electric streetcar, rather than the bicycle, that eventually replaced the horse-drawn streetcar. The electric streetcar (invented by John Joseph Wright of Toronto) ran along electrical lines. It was faster, at ten miles per hour, cheaper, and simpler to operate. The electric streetcar was one of the most quickly accepted innovations in history. Toronto converted its entire streetcar line between 1891 and 1894. In 1891 1,500 horses powered the streetcar, and most were gone by 1893, with the last 2 retired in 1894.[61] In 1890, 70 per cent of the American streetcars were powered by horse or mule, but in 1902 97 per cent were electric. Toronto continues to operate electric streetcars.

The replacement of the horse-drawn carriage and cart by the motorcar and motor truck was almost as sudden. There were a few early experiments with electric powered vehicles. Toronto's Simpsons purchased an electric power truck to supplement their grey delivery horses in 1898, and Parker Dye Works bought an electric truck in 1899. But it was the internal combustion engine that swept the city after the First World War, especially for passenger traffic and for high-value shipping where speed was important. When the proprietor of Hendry's cartage returned from the First World War he purchased a truck, breaking with sixty years of experience with horses.[62] He was not alone. Between 1914 and 1925 the number of passenger motor vehicles travelling along Yonge Street increased from 229 to 3,339 per day and the numbers of commercial motor vehicles increased from 11 to 272, while the number of horses diminished from 87 to 19. In citing these figures, the Toronto and District Roads Commission noted that the trend would have been greater if the records had been made in 1910, and concluded, "One noticeable feature brought out by this tabulation is the constant dwindling, almost to the vanishing point, of horse drawn traffic on the main roads, pointing to the conclusion that the main roads of the future must be designed primarily for motor traffic."[63] Bulky low-value loads continued to be hauled by animal power for some time. As Gijs Mom has argued, business practices that had evolved with the horse, such as waiting for customers to try on goods and sampling the beer with the tavern owner, made it difficult to fully realize the economies of the motor vehicle. Horses lingered in the Canadian city through the 1930s and 1940s, making the heavy deliveries with milk carts, beer kegs, and coal wagons.[64]

THE HAZARDS OF HORSEPOWER

Urban streets in the pre-automobile era have been romanticized as a common area, even a playground for children. Graham Hill Grahame recalled of his childhood in New York and Toronto, "In those golden days of youth our pleasures were simple and many. We had the streets to play on, and the leisurely horse-drawn traffic allowed us greater freedom and confronted us with fewer perils than does the motorized highway of today. The gravest danger we encountered on the streets was the occasional runaway horse – and a couple of husky steeds hauling a heavy swaying dray was a peril indeed. But runaways didn't often happen."[65]

But his account is tinged with nostalgia. Contemporaries were less sanguine about the dangers. Horses are flight animals, predisposed to shy and run from any danger, and it is only because of extensive "breaking" and training, the technology of blinders, harnesses, and coercive devices such as check reins, the exhaustion of overwork, and perhaps the reassuring presence of so many other horses on the nineteenth-century street, that traffic proceeded as calmly as it did. Numerous examples of horses shying at unusual noises and sounds can be found, and they often led to serious injury.[66] A letter to the editor from "A Walker" in Ottawa in 1867 provides a contemporary assessment of the dangers of horse drawn traffic, more realistic than Grahame's rosy memories.

> Those of your readers who may ever have been so unfortunate as to pass down Sussex Street about five p.m. – when the evening train is due, and an hour when the street is thronged with passengers – cannot be ignorant of the whooping and yelling, and lashing of horses, with which on the first sound of the whistle, the carters dash into their vehicles, and totally regardless of the poor pedestrians who may be crossing George Street at the moment, tear away down to the station like demons just let loose.
>
> George Street is wide, and a person, particularly a lady, cannot gain the sidewalk in a moment ... Were the horses heads turned up the street, pedestrians would at least, have time to reach a place of safety before the rush comes, and ladies would suffer fewer alarms from the "shying and starting" of horses at other times, on snow falling from the neighbouring roofs, or on the bugle surrounding the barracks &c.[67]

Eric Morris notes that most analysis of the dangers of the car emphasizes the early years, when people and infrastructure were not adapted to the speeds or nature of motor cars. He argues that mortality rates were higher with the horse, pointing to the fact that 200 persons were killed by horses and horse-drawn vehicles in 1900 in New York City, whereas only 344 died in 2003, and concluding that the per capita mortality was 75 per cent higher with horses.[68] Horsepower was particularly dangerous for the drivers, who had to harness, control, and discipline the animals. Teamsters were the profession third most likely to be killed at work.[69]

Toronto bylaws repeatedly stressed the necessity of keeping horses under control. In 1834 horses were prohibited from "running, galloping or going immoderately" on city streets; citizens were empowered to "stop any horse, mare or gelding from running at large," and the mayor and alderman were expected to cause the horse "and the carriage (if any) to which the same may be attached" to be detailed and kept at the owner's expense.[70] Subsequent bylaws repeated the prohibitions. Stallions were a particular problem: in 1868 owners of loose stallions were fined five dollars, while other horses were fined only fifty cents.[71]

Courts frequently were called upon to pass judgment upon horse behaviour. A judge in a 1914 case cautioned against making municipalities liable for the havoc created by urban horses. He warned against "deciding that if a horse becoming restive or unmanageable runs or backs into a ditch or down a declivity or embankment on a highway, the municipality is therefore accountable for the consequences, or that they are bound to fence or guard all such places against the possibility of a vicious, baulky, or runaway horse, running or backing over the highway at such points."[72] The case of the cyclist, the horse, and the steamroller, described above, also captures a legal system struggling to assign fault for errant horse behaviour. One judge found that horses and their drivers should be expected to be familiar with a steamroller:

The roller was lawfully on the highway and a proper and necessary machine to use in the care and keeping in repair of the city streets. Unless progress is to be stayed, drivers of vehicles must adapt themselves to changed conditions caused by improvements in machinery and modes of locomotion ... Upon the admitted evidence it has been in use in Toronto for nine years and upwards, and horses have become accustomed to seeing it on the public streets. The plaintiff to recover must prove that it was so calculated to frighten horses of ordinary steadiness as to be a public nuisance on the highway.

The majority of the court, however, found that this horse was not unusual: "It was shown that at other times and on other occasions horses had been frightened by and had shied at the roller when in motion, and it must have been obvious to every one who had to do with it that it could not be used ... unless some precautions were taken." Similarly, a 1906 case describes a horse that shied at a flapping road roller: "Evidence was given at the trial to shew that the horse drove well and quietly, and was not apt to take fright unnecessarily."[73] Courts of the age were cognizant of the intelligence, nervousness, and wilfulness of this form of power.

MANURE

Manure disposal was one of the most intractable problems posed by the heavy use of urban horses. As one memoir noted, "On a warm summer day the heavy smell of manure on the streets passed unnoticed by Torontonians."[74] Estimates vary, but a horse produced about five tons of droppings in a year, along roadways, when waiting in traffic, and in their stables. Manure seeped into groundwater, and when dry would be pulverized underfoot to fill the city with a fine dust.

In the early years, horse manure was one form of manure among many, and compared to pig, chicken, and cow manure, or the night soil removed from outhouses, horse manure was relatively innocuous. It was also a valuable commodity, and stables could sell it to market gardens. In 1835 a select committee appointed in Toronto to enquire and report upon "fit and proper places" for the disposal of manure, offal, and rubbish, suggested, among other places, the creation of a "Dung heap" near the Old Parliament House, to be disposed by public auction.[75] But as cities grew, other animals were regulated off the streets, and horse numbers increased. As alternative sources of nitrogenous fertilizer appeared, manure became a costly waste disposal problem, especially in the spring. A 1913 public health paper described attempts of leading North American cities to manage the problem. Toronto reported that it had contracts with four companies for the removal of manure, at the stable owners' cost, at least once a week. The Department of Health required manure to be kept in fly-proof bins, which were to be waterproof, concrete, and connected to the sewer system if they were underground.[76] Winnipeg, Edmonton, and other Canadian cities reported that, as there was no demand for manure as a fertilizer, it was incinerated, "but how successfully and what cost we have not the figures at present to state." Cities dealt with manure left on the streets by sweeping and watering the streets.[77]

Manure also fostered disease. The tetanus vaccines routinely administered today are in part a legacy of an environment shared with horses. Humans and horses are both susceptible to tetanus. Horses shed it in their manure, and if the tetanus bacterium infects a deep wound, the horrors of lockjaw ensue.[78] Manure also provided an ideal breeding ground for flies. Nigel Morgan has argued that rates of infant mortality were levelling off until the 1870s in Britain, when they rose again, coincident with rising numbers of urban horses from 1880 to 1900. He points out that the main causes of infant death were enteric diseases spread by flies that bred in horse manure and argues that the incidence increased in houses with privies close to stables.[79] Patricia Thornton and Sherry Olsen tested his argument with a statistical analysis of infant mortality in Montreal. They agree that the hatching of flies at the end of May, and more rapid growth of bacteria, appear to have contributed to the rise of infant deaths in the summer, but they found that immediate proximity of carters' horses did not alter the statistics on a block-by-block basis. The fact that flies travel upwards of twenty kilometres suggests, however, that the broader argument, tying infant mortality to the general population of urban horses, might hold.[80] Certainly public health officials understood this to be the case. A 1913 public health paper noted that "the dangers of the house and stable fly as carriers of disease are well recognised and received." Over time, increasing restrictions were placed on stables. An 1890 Toronto by-law stipulated that livery stables were to allow no more than two wagon loads of manure to accumulate.[81] After 1904, Toronto did not allow stables to be built in areas where more than one-third of the buildings were residential, and in 1905 the Municipal Amendment Act created further restrictions on delivery horse stables.[82]

CONCLUSION: THE BEASTLY HORSE

In *The Horse in the City*, Joel Tarr and Clay McShane make frequent reference to the machine-like qualities of the horse; their subtitle is *Living Machines in the Nineteenth Century*.[83] Their emphasis was important for cutting through nostalgia around the horse, and they rely heavily on nineteenth-century engineers who reduced the animal to a kind of technology with their calculated comparisons of the energy provided by electricity, steam, and animal muscle. But the animal keeps re-emerging in the "Living Machine"; horses keep shying, misbehaving, and resisting. Horses, within the limited means available to them, force humans to work at their pace and respond to their will.

If we turn from the engineers who parsed horsepower to the people who lived and worked with horses, the horse emerges as another intelligent, ornery, and recalcitrant being. The judges who commented on the horse's ability to learn about steamrollers were acknowledging the horse's intelligence. The carters who whipped their horses, and the Humane Society volunteers who restrained the carters, were both (albeit differently) aware of the horse's consciousness and sensitivity to pain. City by-laws provide a sense of the horse's volition and agency, especially when dealing with the more purely animal aspects of horse's existence. A 1933 *Toronto Telegram* article about the deaths of four horses in a stable fire reveals a common sympathy with the animal as another sentient creature. Emphasis was placed on the stable manager's assurances that the horses who died in the fire were suffocated before they burned. He was thinking of the animals' terror and pain, and the reporter who further insisted on this point in an article clearly thought that his readers would share these concerns.[84]

The horse, like the human, was a troublesome source of power. It resisted, and it called upon our sympathies. The great advantage of steam, electricity, and gasoline was that these new forms of energy took the horse out of horsepower.

NOTES

1 Thompson, "Nineteenth Century Horse Sense"; McShane and Tarr, *The Horse in the City: Living Machines in the Nineteenth Century*; Greene, *Horses at Work: Harnessing Power in Industrial America*; Grandy, "The Era of the Urban Horse: Saint John, New Brunswick, 1871–1901"; Derry, *Horses in Society: A Story of Animal Breeding and Marketing Culture, 1800–1920*. For the term *industrialized organism*, see Schrepfer and Scranton, *Industrialising Organisms: Introducing Evolutionary History*. For a discussion of the question of animals as technology, see Envirotech, "Are Animals Technology?"

2 Olsen, "Introduction." Although metal bits have been dated to 1675 BCE, it is possible that horses were ridden with leather bits, which leave no trace, at an earlier date.

3 Langdon, "The Work Horse as Technological Innovation"; and J.I. Little, chapter 3, this volume.

4 As Derek Fraser pointed out, "Far from displacing horses, railways created new and expanding demands for horse labour" (*History of Modern Leeds*, 137).

5 Thompson, "Horses and Hay," 166. McShane and Tarr, *Horse in the City*, 16.

6 Derry, *Horses in Society*, 47. Derry cites Barclay, *The Role of the Horse in Human Culture*, 339.

7 In Toronto, for example, there were 1,102 dairy cows in 1861, 500 in 1891, and 29 in 1911. Kheraj, "Living and Working with Animals in Nineteenth Century Toronto," 126. Donkeys and mules were widely used in the southern states, but there is little evidence of their use in Canadian cities. Working oxen were frequently employed on Canadian farms but were generally not used in cities (the 1891 census, for example, records only twenty-four working oxen in Toronto). We would like to acknowledge the assistance of Sean Kheraj with the analysis of the census data.

8 A healthy, fit horse can produce 1.0 horsepower all day, whereas normal humans can produce 0.1 horsepower briefly, and 1.2 horsepower in short bursts, and athletes can produce 2.5 horsepower. At the 1935 Iowa Fall Fair, peak powers for a horse over a few seconds were measured as high as

14.9 horsepower. Stevenson and Wassersburg, "Horsepower from a Horse." See also Avallone, Baumeister, and Sadegh, *Marks' Standard Handbook for Mechanical Engineers.*

9 Law, "Influenza in Horses."

10 See Judson, "History and Course of the Epizootic among Horses upon the North American Continent in 1872–1873." See also the maps in Board of Health, "Report on the Origin and Progress of the Epizootic among Horses in 1872, with a Table of Mortality in New York. Illustrated with Maps."

11 Law, "Influenza in Horses," 227.

12 "The name of horse power has become technical," noted the *Journal of the Franklin Institute* in 1833, "and is applied to any apparatus by means of which a horse is made to exert his power in propelling machinery," 36, cited in McShane and Tarr, "The Horse as Technology: The City Animal as Cyborg."

13 Tarr, "A Note on the Horse as an Urban Power Source," 434. See, for example, the sketch depicting a horse whim in Lethbridge Alberta, ca 1890, from *Western World*, by Acton Burrows, Glenbow Museum, NA 430-16. There are many photographs of horse sweeps powering agricultural machines. See, for example, at the Glenbow Museum: NA 1328-64450 (coiling cable); NA 3046-10 (grinding grain); NA 1095-2 (8 horse sweep threshing grain); NA 2685-36 8 (pre-1912 eight-horse windlass grinding grain).

14 For a history of animal-powered machines, see Kenneth Major, *Animal Powered Machines*; "The Pre-Industrial Sources of Power: Muscle Power"; and "Animal-Powered Machinery in the Medieval Period."

15 Many others made similar calculations. Hills and Pacey describe the calculations made by designers like Thomas Savery and Marten Trenwald, and the rough calculations of mill owners who thought in terms of the power of "6 strong horses." See Hills and Pacey, "The Measurement of Power in Early Steam-Driven Textile Mills." Although some have argued that Watts was overly optimistic, and that a horse pulls only about 0.8 horsepower, Stevenson and Wassersburg argue that the rate of 1 horsepower could be maintained by healthy draft horses. Stevenson and Wassersburg, "Horsepower from a Horse," 364.

16 Cited by Hribal, "Animals Are Part of the Working Class Reviewed," 18.

17 Tann, "Horse Power 1780–1880."

18 Tarr, "Note on the Horse," 437.

19 Gad, "Location Patterns of Manufacturing: Toronto in the Early 1880s, Urban History," 113.

20 *Hamilton v Myles*, [1874] OJ No. 141, 24 UCCP 309 (CA).

21 See "Log jammer in operation, McDonald Lumber company, Whitecourt, Alberta, 1923," photograph, Glenbow Archives, NA 3219-31.

22 Census returns showing a high percentage of animate power in Yorkville, north of Toronto, in 1871, can be explained by the presence of numerous brick-makers, and their horse-powered pug mills. Bloomfield and Bloomfield, "'Our Prosperity Rests upon Manufactures': Industry in the Central Canadian Urban System, 1871," 87.

23 Baird, Son, *Clay-Working Machinery and Supplies Catalogue*. See also Pursell, "Parallelograms of Perfect Form: Some Early Brick-Making Machines."

24 Rothwell Machine Company, *Catalogue of the Latest Improved Brick Machinery and Brick Makers Supplies*. Tarr describes a Massachusetts company, Nourse, Mason and Co., that also sold machines adaptable to horse, steam, or water power. Tarr, "Note on the Horse," 438.

25 *Smith v Hayes*, [1898] OJ No. 96, 29 OR 283 (H Ct J). This took place in Ingersoll, Ontario.

26 Pierce, "Dog and Horse Power"; Shayt, "Stairway to Redemption: America's Encounter with the British Prison Treadmill."

27 Moody Labour Saving Agricultural Implements, *Autumn Catalogue*, 42. E.W. Ross Company, *Catalogue*, similarly suggested that the treadmill generally required half the number of horses needed for the sweep. For modern calculations, see Roosenberg, "Animal-Driven Shaft Power Revisited." Roosenberg explains that the ratio between the surface area of the belt and the surface area of the pinion gear that it turned was such that the first shaft in a treadmill turned much more quickly than the first gear in a horse sweep.

28 *Crisman and Cohn, When Horses Walked on Water: Horse-Powered Ferries in Nineteenth-Century America.*

29 Robertson, *Robertson's Landmarks of Toronto*, 2:763.

30 William Notman, "Horse boat, Huntington Copper Mine, Bolton, QC, 1867," photograph, McCord Museum N-0000.94.53; Alexander Henderson, "Horseboat and S.S. 'Parisian,' Montreal Harbour, QC, about 1870," photograph, McCord Museum, MP-1980.47.70.

31 E.W. Ross Company, *Catalogue*.

32 American Hoist and Derrick, *American Hoist and Derrick Catalogue*, 112.

33 Joseph Glynn, *Rudimentary Treatise on the Construction of Cranes and Machinery for Raising Heavy Bodies, for the Erection of Buildings, and for Hoisting Goods*. Cited in Tarr, "Notes on the Horse," 439.

34 *Smith v Hayes*, [1898] OJ No. 96, 29 OR 283 (H Ct J).

35 "Building hay stack on farm, Edmonton area, Alberta, c 1914," photograph, Glenbow NA 1328-64832. Kenneth Major includes an engraving of a very small boy with a whip perched in a platform above the central "safety gear" in a four-horse engine, probably to emphasize the safety of the gearing. Major, *Animal Powered Machines*, 25. Jason Hribal describes Jeremy Bentham's observations on the potential of the child/horse team: "An old blind horse, an ox, perhaps even an ass, will turn a wheel, a little boy will serve for driving, the keep of both together will not exceed the keep of one man, certainly not equal that of two." Hribal, "Animals Are Part of the Working Class," 30.

36 Thomas, "Animal Power Production, Mechanisms for Linking Animals to Machines"; Lowe, *Animal Powered Systems*.

37 "Open-Air Horse Parade," *Globe and Mail*, 2 July 1904.

38 Eaton's delivery trucks were introduced in Winnipeg in 1912. Blanchard, *Winnipeg: Diary of a City*, 119; Macdonald, "The Scribe," 43, 94–5; Hayes, *Historical Atlas of Toronto*, map 163.

39 "Moving Queens Wharf Lighthouse, 1929," in Filey, *A Toronto Album: Glimpses of a City That Was*, 106. *See also* "Moving a House, Huxley Area, Alberta, 1915," photograph, Glenbow NA-1520-2.

40 J. Ross Robertson, *Landmarks of Toronto*, 606. For a detailed description of the history of Toronto's fire brigades, see chaps 196, 197.

41 Kirkpatrick, *Their Last Alarm: Honouring Ontario's Firefighters*, chap. 1.

42 Greene, *Horses at Work*, 174.

43 Grandy, "Era of the Urban Horse," 23.

44 Numbers approximate, cited from Skempton (1954) in Evans, "Roads, Railways and Canals: Technical Choices in 19th-Century Britain." The most efficient mode surface was water, but the canal barge, which carried fifty tons, was not suitable for most urban traffic.

45 Holley, "Blacktop: How Asphalt Paving Came to Urban United States."

46 See, for example, Thomas Young's discussion of the merits of Macadam's methods in "Road Making," *Encyclopedia Britannica*, 4th ed., 1824; cited in Evans, "Roads, Railways and Canals," 5.

47 City of Toronto Bylaw 2464, 1890. The law did not apply to farm produce or lumber coming from a distance. See also Bylaw 467, 1868. See Evans, "Roads Railways and Canals," 22, on the advantages of a two-wheel cart versus a four-wheel wagon.

48 Massey, *Dumb Animals: A Plea for Man's Dumb Friends, Being the Substance of an Address Delivered in Salem Church by Rev. Samuel Massey*.

49 "Outsider," letter to the editor, *Ottawa Citizen*, 25 March 1867.

50 Ellis Keystone Industrial Works, *The Ellis Keystone Industrial Works Catalogue*.

51 *Kirk v Toronto (City)*, [1904] OJ No 249, 8 OLR 730, at para 28.

52 Gad, "Location Patterns of Manufacturing Toronto," n70.

53 The 1834 Nuisance Act banned dumping of material on sidewalk "so as to incommode or obstruct the free passage or use thereof" and forbade drivers to "leave any horse, cart or waggon over any such sidewalk," section 16, bylaw 4, 1834.

54 "Opening of the Street Railway: The Dejeuner, Concert and Ball," *Globe*, 11 September 1861. Derailment continued to be a common source of trouble for the horse-powered streetcars. See, for example, "The Street Railway," *Globe*, 15 April 1880.

55 City of Toronto Bylaw 2467 (1861) restricted their speed to six miles per hour.

56 For an amusing account of a sleigh falling into the gully created by a streetcar snow plow, see Duff, *Toronto Then and Now*, 85.

57 Hilton, "Transport Technology and the Urban Pattern."

58 On the problems of horse power and early designs for steam cars, see White, "Steam in the Streets: The Grice and Long Dummy."

59 "The Horse and the Bicycle," *Scientific American* 73, no. 3 (1895): 43; McKee, "The Horse and the Motor," *Lippincott's Monthly Magazine*, 381. Both cited in Furness, *One Less Car: Bicycling and the Politics of Automobility*, 232.

60 Ingram, "'We Are No Longer Freaks': The Cyclists' Rights Movement in Montreal."

61 Fleming, "The Trolley Takes Command: 1892–1894." For the invention of the streetcar, see the entry under his name in the *Dictionary of Canadian Biography*; and Black, *Canadian Scientists and Inventors: Biographies of People Who Shaped Our World*.

62 Duff, *Toronto Then and Now*, 85.

63 Toronto, *Toronto and York Roads Commission: A General Report of Work Completed*. The statistics are taken from traffic census reports prepared by the Department of Public Highways, Ontario. Although the report does not specify, it is likely these figures refer to daily traffic.

64 See Mom and Kirsch, "Technologies in Tension: Horses, Electric Trucks and the Motorization of American Cities, 1900–1925." See Barker, "The Delayed Decline of the Horse in the Twentieth Century," for the British experience. For the reference to coal horses, see "Four Horses Die in Fire at Stables," *Toronto Telegram*, 3 December 1933. City of Toronto Bylaw 15215 (1939) authorized the conversion of the Street Cleaning Department stable to a garage for motor vehicles. The City of Ottawa purchased their own horses for scavenging in 1920, built stables in 1928 (Bylaw 6448), and then in 1946 sold the horse stalls and equipment

(Bylaw 478). Photographs at Library and Archives Canada show delivery vehicles late into the twentieth century, possibly because they were remarkable at that date. For example, see "Loaded horse drawn delivery sleighs set out to deliver the *Montreal Star* throughout the city," photograph R11284-47-8 (1930s), Montreal Star fonds, Library and Archives Canada.

65 Grahame, *Short Days Ago*, 56–7.

66 See, for example, the incidents cited in Kheraj, "Living and Working with Animals," 122.

67 "A Walker," *Ottawa Citizen*, 20 February 1867.

68 For a different views on the safety of early streets, see Morton, "Street Rivals: Jaywalking and the Invention of the Motor Age Street." "Before the American city could be physically reconstructed to accommodate automobiles, its streets had to be socially reconstructed as places where cars belong. Until then, streets were regarded as public spaces, where practices that endangered or obstructed others (including pedestrians) were disreputable." For a critique of this romantic view of the street, see Epperson, "Fighting Traffic." See also Baldwin, *Domesticating the Street: Reform of Public Space, 1850–1930*. Photographs of empty streets can be misleading, because the long exposure times of early films blurred any moving objects into invisibility. Morris, "From Horse Power to Horsepower."

69 See Kruse, "The Best Dressed Workers in New York City: Liveried Coachmen of the Gilded Age," 6, cited in Grandy, "Era of the Urban Horse," 66.

70 Bylaw 4 (1834).

71 City of Toronto Bylaw 474 (1868). Entire (i.e., uncastrated, male) horses were banned from running at large in any part of the city at this time, whereas horses generally were banned only from specific parts.

72 *Little v Smith*, [1914] OJ No 160. See also *Halifax Electric Tramway v Inglis*, [1900] SCJ No 7, 30 SCR 256 for descriptions of horses with a mind of their own.

73 *Poulin v City of Ottawa*, [1916] OJ No 718, 9 OWN 454.

74 Grahame, *Short Days Ago*. See Tarr, "The Horse: Polluter of the City."

75 Minutes, Toronto City Council, 12 May 1835.

76 Hall, "Disposal of Manure."

77 Anders, "The Dust Menace and Municipal Diseases," 276.

78 See Dean, "Species at Risk: C.tetani, the Horse and the Human."

79 Morgan, "Infant Mortality, Flies and Horses in Later-Nineteenth Century Towns: A Case Study of Preston."

80 Thornton and Olsen, "Mortality in Late Nineteenth Century Montreal: Geographic Pathways of contagion."

81 City of Toronto Bylaw 2477, Section 45 (1890).

82 City of Toronto Bylaw 4408, Section 49 (1904); City of Toronto Bylaw
    4571 (1905).
83 McShane and Tarr, *Horse in the City.*
84 "Four Horses Die in Fire at Stables," *Toronto Telegram,* 13 December
    1933.

5

Hewers of Wood:
A History of Wood Energy in Canada[1]

Joshua MacFadyen

When the weather is bad [the settler] may employ his time within doors ...
amusing himself by the fire, which can always be made a warm and cheerful
one, from the profusion of fuel that the poorest person has continually at
command.

John Howison, *Sketches of Upper Canada*, 1821

INTRODUCTION

Energy historians can usually identify the precise time when modern
energy carriers – fossil fuels and hydroelectricity – overtook biomass and
other traditional carriers as the largest energy source in industrializing
countries. In England and Wales, coal already represented over three-
quarters of energy consumption in 1800, but fossil fuels did not make up
the majority of energy use in the United States until 1884, and in Russia
not until the early 1930s.[2] New research shows that Canada experienced
this transition around 1906, significantly later than its American neighbour, and the consumption of firewood remained a critical source of
energy for heating and cooking well into the twentieth century. This delay
was partly because Canada adopted coal later and produced firewood
longer than the United States. Canadian coal consumption quintupled in
the first two decades of the century and quickly dwarfed traditional
energy carriers, wood included. However, this chapter shows that the
woodpile continued to grow steadily in the early twentieth century, as

well. Energy from wood rose from around 400 petajoules in the late nineteenth century to over 480 in the 1930s. As the Great Depression temporarily interrupted coal's growth, wood represented a large share of Canada's total energy consumption.[3] In 1941, household wood consumption accounted for over 490 petajoules and likely reached a much higher apex during the energy-starved war. "Peak wood" occurred in the middle of the fossil fuel century.

This chapter presents an overview of this process, providing explanations for the persistence of wood energy and new evidence on its rates of consumption in Canada. Canadians were hewers of wood, and hewers of far more wood than historians have imagined. Firewood was displaced more slowly in Canada than in other nations for many reasons. One is demand. Canadians live in a cold climate, and as their economy and population grew at unprecedented rates in the late nineteenth century, they needed energy to support them. Most energy histories have focused on how modern energy carriers responded to this demand, particularly in critical industries such as steel manufacturing. But in reality it was not yet an age of iron. Most goods and processes were still made from and powered by biomass. Most urban and virtually all rural families heated their homes and cooked their food with wood. Railway and steamboat lines traversed Canada's vast distances powered mainly by firewood.

For several reasons, Canadian historians emphasized modern energy carriers and assumed that these technologies displaced biomass. Histories of new technologies often sell better than stories of things-in-use,[4] and national historians seeking to understand Canada's role in the second industrial revolution may have steered away from interpretations that presented Canadians as hewers of wood and drawers of water. Railways and steamships, those pinnacles of progress and instruments of expansion, converted many of their locomotives to coal along trunk lines in the 1870s, and perhaps historians assumed that the rest of Canadian engines would simply follow suit. But why would nineteenth-century Canadians shift from wood to coal, as many eventually did, when they lived in one of the world's largest forests? One answer lies in the geography and dynamics of energy flows. When Europeans arrived in Canada they were literally surrounded by a perennial supply of carbon, building materials, and habitat for game and other nutrients. Yet as John Weaver describes, Canada and all of the "Little Europes" became interested primarily in one thing: apportioning their new land and removing most of the native flora, fauna, and people so the soil could provide nutrients for a handful of domesticated plants and animals.[5] As agricultural clearing depleted the supplies of solar carbon and the expanding infrastructure reduced the cost of transportation between cities and other single points, it eventually became more cost effective, as Andrew Watson shows in this volume, to move the concentrated supplies of fossil fuels, channelling carbon from single points in Pennsylvania or Lethbridge to consumers in southern Ontario and the Prairies. These mineral concentrations were North America's version of what Rolf Sieferle calls "the subterranean forest."[6] The incoming streams of mineral carbon followed the course of Canada's new transportation infrastructure, and David Walker shows how railways were critical to the creation of coal markets in southern Ontario, the first part of Canada to be crisscrossed with a network of railways and canals.[7]

Looking beyond southern Ontario's railways, however, we see that most of the rest of Canadians, including most people in the boreal and Acadian forests, remained well connected to the fuel supplies produced by Canadian forests. Historians are increasingly interested in the stability of these traditional energy systems. Whereas most wrote about the quickness of the transition, Richard Unger and John Thistle have recently argued that, relative to other industrializing countries, Canada was slow to adopt modern energy carriers.[8] Indeed, the new estimates introduced in this chapter suggest that those who were not connected by the new, modern networks to fossil fuels continued to burn biomass until at least the mid-twentieth century. Canada's transition to fossil fuels does not mean it abandoned its firstborn – biomass. Rather, it adopted new fuels to meet new energy demands.

The second reason for this slow transition was supply. Throughout most of their history, most Canadians lived in or not far from the forest and continued to use products of the forest every day for living, building, heating, and cooking. Canada contains one of the largest boreal forests in the world, fringed by mixed wood and Acadian forest hardwoods to its south, and connected by water transportation. Forests supplied the fuel for virtually all of North America's home heating and steam energy before 1850, and in the relatively sparsely populated dominion the wood supply was large indeed. The human presence created dramatic disturbances in woodland ecosystems, especially through farming, but woodlots were also valued and maintained as a workplace and an investment. Rural Canadians continued using woodlots well past the period commonly understood as the transition to fossil fuels, and although some scarcities emerged in heavily deforested regions, most woodlots were resilient and productive elements of the rural landscape.

Thus woodlots are both a carbon store and a flow, and in Canada they provided consistent and predictable energy supplies for households, industries, and transportation. As we will see, single point "consumers" such as cities and industries imported firewood long distances over land and even longer over water. Like all renewable energy flows, forests are limited by the ability of *photosynthesis* to convert solar energy to carbon. Densely populated peri-urban areas and rural settlements began to supplement wood with other fuels when woodlots were depleted, but for most of the pre-war period this was restricted to small areas in southwestern Ontario and the Montreal plain. Many Canadians were distributed consumers, living in low densities and close proximity to these renewable energy sources. As Sandwell points out in her introduction, it was not until 1941 that more than half (51 per cent) of people lived in communities larger than 1,000 people.[9] In some places wood was free, although families had to consider the opportunity cost of dedicating labour to its production. On private land, farmers maintained their woodlots on steep, rocky, or wet terrain, places that were not suitable for agriculture or other clearing. So firewood was a relatively abundant and renewable resource harvested from woodlots.

For most of Canadian history, in all areas outside of the southern prairies and the Far North, firewood was a ubiquitous part of everyday life. Indeed it was firewood's ubiquity that can, paradoxically, help to explain its limited appearance in the historical record. It was so much a part of everyday life that most people never thought to record their experiences with this form of energy. More surprisingly, experts rarely discussed firewood production either, even though, as forest historian Michael Williams argues, they discussed almost every other aspect of farming and farm life. "Fuel wood was an unexpected by-product, an unasked for increment that was not enumerated," according to Williams. "There is no other commodity of such importance about which we are so ignorant."[10] In other words, we know surprisingly little about wood energy systems in Canada, at least in part because they comprised a component of the largely unrecorded subsistence activities hiding in the shadows of both the diurnal rounds of household activities and their more economically visible forest sector siblings, logging and lumber.[11] On closer examination, however, the wood energy trade actually intermingled with markets, engaged more Canadians, and represented a much larger share of the economy than the logging industry. And, along with clearing, firewood harvesting accounted for more deforestation and biomass extraction than any other land use practice.

From the vantage point of the staples framework of Canadian economic history, subsistence and internal trade of forest products took a back seat to unprocessed exports. This chapter explores the activities of farmers away from their fields, and finds that what many historians have seen as insignificant distractions from farming were often central to survival in a mixed economy.[12] These everyday exchanges were performed by most rural Canadians with access to a woodlot. As Douglas McCalla has argued, if non-timber products of the forest had been recorded as exports were, we would have a much different sense of the scale of this sector.[13]

WHAT IS BIOMASS ENERGY?

Biomass energy is the useful energy derived from plants and animals when they are burned. *Combustion* causes the solar energy stored by a plant or animal to be released as heat. This is a renewable energy, and it is limited by the efficiency of photosynthesis and the space of land available for the conversion. Biomass energy has included everything from dried straw to animal carcasses and dried bison dung. The most readily accessible form of biomass energy for most Canadians and First Nations has been wood, and wood has been the primary heating and cooking fuel for First Nations south of the Arctic for millennia.

Domestic wood energy is defined in this chapter as any heat, light, or cooking energy produced by harvested wood that has been dried to less than 25 per cent moisture content. The cord is the standard measurement for wood, equalling 128 cubic feet. It was usually measured in units 4 feet wide, 4 feet high, and 8 feet long. Deforestation or clearing is the long-term removal of a forest so that the land may be used for other purposes such as agriculture, roads, or urban development. Harvesting is the selective or total removal of trees from a forest for sale or use, and it is usually followed by successive regrowth or plantations. Harvesting land for firewood always disturbs the forest, but in Canadian woodland ecosystems it rarely removes it completely.

Biomass includes all organic sources of useable carbon, and although the biosphere is remarkably prolific, most of the useable carbon is stored in one main form – trees. Trees are nature's solar energy batteries; they conduct part of the hydraulic cycle through transpiration, but their main function is in the carbon cycle, where they act as both energy converters and storage containers. Trees may have been abundant in Canada, but converting solar energy to useable energy was labour intensive, and like any other fuel, wood was a commodity that required processing and transportation to markets. Farm families devoted enormous effort to producing energy from the forest; roughly the same amount of energy was expended in its production as in its consumption in stoves. The opportunity cost of this labour was potentially quite high, but in most of the period discussed here, firewood production was flexible enough to fit with cycles of rural work and family labour.

All tree species produce the same amount of energy per unit of mass; hardwood was preferred because it produced more heat for the same amount of work as softwood. Douglas Sobey has shown that beech was widely considered the most preferred species of hardwood for fuel in the eastern Acadian forests, but overharvesting and disease in the hardwood uplands caused the destruction of this once dominant species.[14] Since cordwood is a unit of volume, not weight, the rate of consumption was much higher in areas that burned softwood for fuel. People could burn both, and Sir Andrew Macphail recalled burning softwood primarily in the summer for cooking and cool nights.[15] Softwoods are often easier to light and they burn out more quickly, so they might have been preferred for year-round cooking in kitchen stoves.

Wood energy was an incidental good and a by-product of clearing at the very early stages of settlement, but most cleared wood was burned on site, and the vast proportion of firewood was produced from woodlots as opposed to new clearings.[16] Firewood was used for all forms of heating requirements, from homes to locomotives and industrial engines, but the most important use was in domestic space heating and cooking. An important secondary use for firewood in the early nineteenth century was *potash*, an alkali export produced by burning, leaching, boiling, and roasting hardwood ashes. This industry was more than a by-product of clearing, and its persistence into the post-pioneering period suggests that potash was both an important form of wood energy and a regular part of the early Canadian economy.[17]

Another form was in the burgeoning transportation sector. Early railroads helped forge an urban wood supply chain, but they also hurt it by consuming fuel themselves and starting forest fires almost everywhere they went.[18] Similarly riparian forests were among the first to disappear, as the result of coastal settlement and the demand for cordwood along steamboat lines.[19] Railway companies like the Great Western initially consumed as much firewood as a small city, but locomotives were among the first to make the switch to coal. At first the shift was less in response to scarcity and more for convenience; coal is a denser and more easily loaded solid fuel than wood, so loading times decreased and distances between fuelling stations increased. But eventually railways found they were committed to travel between the most densely populated landscapes, and this meant travelling at the greatest distance from woodlots and paying the highest price for firewood. Retrofitting engines for coal burning was a modest expense by comparison.

As the twentieth century progressed, the consumption of firewood for energy decreased, first in railways and urban industries, and then in steamboats, and eventually in Canadian homes. As we will see, the decline in household consumption did not begin until mid-century, and at that point it dropped dramatically. By the 1960s most Canadians had traded their axes for oil furnaces. In 1973, when the Organization of the Petroleum Exporting Countries (OPEC) raised oil prices, a resurgence in wood energy and other biofuels began to develop in Canada and the United States.[20] Governments and other agencies began to experiment with biomass energy initiatives in burning wood pellets, sawdust, and garbage.[21] A slight corresponding recovery in firewood production is evident in the data, but this form of Canadian biomass energy never regained its pre-war lustre (or at least has not yet).

FINDING, PROCESSING, AND BURNING WOOD FOR HEATING

Producing the large amounts of fuel needed for heating buildings required a never-ending cycle of cutting, trimming, blocking, hauling, chopping, splitting, and stacking. Michael Williams's reading of American farm diaries shows that farmers spent 14–17 per cent of their working days on these activities, and in Upper Canada J. David Wood estimates that as much as 25 per cent of farm work was for this purpose. Although the initial work of clearing a farm might diminish relatively quickly, the time Canadian farmers spent in the woods did not. In many cases, farmers and their horses would expend as much energy harvesting wood, as the fuel burned in their homes.[22] As it was in the logging industry, winter was the prime season for cutting firewood, mainly because more labour was available from farms and other seasonal work and because snowy fields and frozen rivers facilitated the transport of wood by horse and sleigh.

The technology of harvesting and processing firewood did not change much before the introduction of the portable chainsaw in the 1950s, and even then chainsaws likely had more impact on the lumber industry.[23] Human muscle and certain basic tools were indispensable for producing wood energy. As thousands of these hewers of wood turned to the forest in December, they undoubtedly brought the same tools with them that their fathers and grandfathers had used. The axe was important, but they accomplished much more with steel wedges and crosscut saws. Trees could be felled and blocked much more quickly with saws, and the wedges were used to keep the tree from jamming the blade. Sawing was best done with two or more, but when the help of children and neighbours was not available, farmers went alone. "It was marvelous what one man with an axe and saw could do," George Cairns stated in his memoirs of growing up in Stanchel, Prince Edward Island.[24] Cant hooks provided leverage for spinning logs during cutting, and they were used to roll larger logs onto wood sleighs. As J.I. Little notes in his chapter in this collection, another critical factor in firewood processing was the horse. As soon as cleared fields began to separate houses and forests in the first few years of pioneering, a horse or ox was essential for transporting wood to homes and markets. Sleighs could hold up to half a cord of wood, so the average farmer made dozens of trips to the woodlot each year, just to haul the season's harvest home.

Burning and insulation technologies also remained surprisingly consistent, considering the long ripple effect that even small improvements would have on labour costs. The open hearth was a prehistoric heating and cooking technology with as low as 10 per cent heating efficiency.[25] It was said that New England homes burned forty cords per year, and manors with several hearths could consume seventy cords of wood per winter.[26] The hearth remained in wide use in the eighteenth and even nineteenth centuries, although Benjamin Franklin's circulating stove was a dramatic improvement in heating efficiency and started coming into use quickly after it was designed in 1741.[27] In fact, stoves were used extensively in mid-eighteenth-century Quebec, where they were manufactured at the Forges du Saint-Maurice in Trois Rivières and installed in a third of the homes in some rural districts. Almost all Montrealers cooked and heated with wood stoves. which were easier than hearths to install and possible to move in summer.[28] Stoves also travelled well; in 1764 surveyor Samuel Holland purchased one in Quebec for use while charting the Maritime provinces, and his crew used it both on land and aboard the *Canceaux*.[29] English Canadians were much slower to abandon hearths, and wood stoves were gradually adopted in Upper Canada only in the 1830s.[30]

Improving a house's energy efficiency requires increasing thermal insulation and controlling air infiltration, but again changes in these technologies took centuries. Inuit igluit were the most energy-efficient dwellings in Canada until the late twentieth century. Other aboriginal dwellings and European log and timber frame houses were uninsulated by comparison. In the 1830s, the modern technique of balloon framing created stud walls covered with sheathing, plaster, and building paper. These created moderately controlled air pockets and provided adequate barriers against temperatures as low as -40°C.[31] However, internal heaters struggled to keep rooms warm at those temperatures, and heating in cold spells often caused fires to break out in chimneys.[32] The walls of Prairie houses were typically wrapped with extra layers of building paper for protection against the wind, and Atlantic Canadian houses were commonly "banked" or surrounded with piles of straw, seaweed, and snow to reduce draft.[33] In addition to keeping wind out, snow and other banking materials provided thermal insulation to keep warmth in, but these banks had to be replaced every winter. Thermal insulation for wall and ceiling cavities eventually became common with the development of rudimentary cellulose fibre insulation in the 1920s and fibreglass insulation in the 1950s.[34] Prior to this period, fibre of any kind was expensive, and nineteenth-century Canadians preferred to use it for clothing, cordage, and newspapers rather than stuff it in their walls. A variety of substitutions were available, from straw and sawdust to grass and seaweed, but as natural fibres contracted moisture from interior condensation and exterior leakage, they lost their insulating value. Still, there were other ways to stay warm. Although they were otherwise uninsulated, houses were also usually small and contained many small rooms that could be closed and heated as needed. Canadians conducted some government research on insulation in the interwar period, but in the age of abundant energy and extensive forests there was little incentive to focus on the energy efficiency of home heating and cooking.[35] Even as late as 1976, a Tennessee Valley Authority conservation program struggled to get a handful of participating homes to insulate their attics to R-19.[36] The energy efficiency programs of the 1970s and 1980s helped to increase building standards, and by 1982 the Canadian "R-2000 Standard" set the minimum insulation value to R-40. It is now common to insulate attics up to R-60.

For those with the capital to buy it or the tools, labour, and woodlots to produce it, most Canadians found abundant fuel supplies from biomass. Contemporary observers sometimes discussed the amount of forest a farmer required for fuel, fencing, and lumber, and these estimates ranged from ninety-five down to "a few" acres of woodlot.[37] In reality, the size of a farm's woodlot often depended on more practical factors such as the resources available for clearing, including labour, the size and quality of the farm property, and the pressure placed on its field-and-forest products by its proximity to urban markets.

However, there were alternatives, and southwestern Ontario's dense rail network quickly began to supply affordable coal to rural areas in the late nineteenth century. Indeed, Ontario's railroads and the Maritime provinces' schooner and steamboat network gave some rural Canadians access to fossil fuels at reasonable prices. In southern Ontario, where many woodlots were under strain and where wealth was created by the successful agricultural and manufacturing sectors, coal was an affordable option and most people took it. Still, in most of Canada's woodland agro-ecological systems, woodlots produced a renewable crop of fuel and the pluri-occupational and seasonal nature of farming allowed most rural people to consume and often produce their domestic energy from wood fuel. It required enormous labour and substantial capital and infrastructure, but it was still less costly than coal.

CHANGING ENVIRONMENTS OF SCARCITY AND ABUNDANCE

Coal had played a determining role in Britain's industrialization from the early nineteenth century, but recent research confirms that wood, not coal, played a dominant role in Canada until the early years of the twentieth. Whereas massive and easily available forest reserves provided a cheap and abundant source of energy, one used in increasing amounts in many regions as populations rose in the twentieth century, deforestation was characterizing other regions from the early nineteenth century onward.

Historians have documented Canadians' considerable ambivalence, and at times downright hostility, towards the country's abundant forests.[38] But eliminating a perennial commodity with multiple by-products in order to simply wage a war against the forest would not have made sense to farmers. Canadian regions burned wood at different rates and in different ways. Most landscapes were shaped in some way by the need for fuel, but again the experiences were varied. Until recently, energy historians argued that all modern economies replaced a subsistence economy based on wood with an industrialized system based on fossil fuels.[39] A plate in the *Historical Atlas of Canada* depicts a 45 per cent drop in firewood production per capita between 1871 and 1891, arguing that the best wood supplies in peri-urban counties were exhausted. However, many other counties in Ontario, the Eastern Townships in Quebec, and most of Nova Scotia and Cape Breton (despite being the home of Canada's most accessible coal belt), experienced an *increase* in per capita wood energy production. Per capita data mask the scale of firewood production during a time of rapid population growth. As areas like southwestern Ontario and the Montreal plain experienced urban growth and extensive deforestation, the per capita production dropped significantly. But new harvesting in the Canadian Shield, in the northern Prairies, and throughout British Columbia compensated for the decline in more densely settled areas. Quebec remained a monolith throughout the entire period, and it became the largest firewood producer in Canada, surpassing Ontario in my estimates by 1891. Ruth Sandwell's data show that in 1951, over 75 per cent of the homes in most of the Eastern Townships and all districts along the Lower St Lawrence (with the exception of Quebec City) were heated with wood (see plate 5.1).

Despite its relatively consistent consumption, the most common narrative in the history of firewood is scarcity, perhaps because these shortages precipitated a major episode of change.[40] According to energy historian Vaclav Smil, wood shortages in England "led to widespread combustion of coal long before the commonly cited take-off of the industrial revolution in the eighteenth century."[41] In a sort of macabre way, extensive deforestation is the hallmark of industrialization. According to Kenneth Pomeranz, deforestation usually reached 70–96 per cent of the landmass of any region that was making a sharp transition to fossil fuels, and industrialization. Denmark's forest was only 4 per cent of its landmass in 1800, and the reset of "'insular and peninsular Europe' – Italy, Spain, the Low Countries, and Britain – was down to 5–10 per cent forested by 1850." These ecological constraints were similar in East Asia, where the most densely populated regions of China had reached similar levels by the 1930s.[42] In North America, Michael Williams identified firewood shortages in the northeast states as early as the early eighteenth century, and the residents of Canadian cities including Montreal, Toronto, and even Charlottetown complained of periods of scarcity, or at least high prices, in the firewood supply chains.[43]

Canada had a much smaller population and more recent resettlement than Europe, and its deforestation was far less extreme. However, the areas around urban centres, places like the Montreal plain and lakefront regions in southwestern Ontario were harvested to the point of near-total clearing. By the end of the nineteenth century, pockets of the country's woodlands began to resemble the rest of the industrializing world. Many forests in the most densely populated areas of Quebec were two-thirds cleared in the twentieth century and are getting smaller still. The Montérégie region, south and east of Montreal, was only one-third forested in 1984 and has decreased to 26 per cent in the twenty-first century. Important agricultural areas within this region such as the municipality of Maskoutains, surrounding Saint-Hyacinthe, are now only 15 per cent forest.[44]

A similar pattern emerged in southern Ontario, home to over 90 per cent of the province's and a quarter of the country's population. This region comprises the southern edge of the Canadian Shield and the densely populated area bounded by Lake Huron on the west, Lake Erie and Lake Ontario on the south, and the upper Saint Lawrence River on the east. The ecosystems here are a mix of deciduous and coniferous trees, comprising 11 per cent treed bogs or fen and only 15 per cent forests. Thus its forest cover resembles places like the Lower St Lawrence Valley or central Prince Edward Island in extent.[45] And like the Montreal and Saint-Hyacinthe hinterlands, Ontario's southwestern agricultural heartland is the most severely deforested, especially in the Toronto-London-Windsor corridor. This southernmost part of Ontario is a deciduous Carolinian ecosystem with 7 per cent treed bog or fen and only 10 per cent forest remaining.[46]

Essex and Kent Counties were important firewood markets in the nineteenth century, and with only 4 per cent of the landmass currently in forest or wetland, the marks of clearing are more evident in this rural landscape than perhaps any other in Canada.[47] In the 1860s, four small lake ports – Amherstburgh, Wallaceburgh, Sarnia, and Windsor – accounted for over three-quarters of Upper Canada's firewood exports, and despite its distance from Canadian cities, this region continued as a firewood-producing area for several decades.[48] Admittedly, exports were not a large proportion of the total production, and if only because of their proximity to the border these towns reveal a fuel trade that would have been typical of many lakefront towns. In mid-nineteenth-century Toronto, for instance, fuel merchants brought cordwood to port on schooners from up and down the lakefront and on grain ships returning from upstate New York.[49] The Essex and Kent wood trade was also typical of the stress placed on riparian forests in a rapidly growing province. The relatively late clearing of farmland in these southwestern counties meant that its hardwood forests were being cleared just as urban fuel markets were expanding along the Great Lakes. This timing, combined with the region's excellent agricultural land and access to lake ports, led it to become one of the most extensively disturbed areas of the province. In 1871, the two counties were half covered in forest, but in only twenty years the forest was reduced to 22 per cent and by 1911 it was down to 16 per cent. The situation was most severe along the lakes and Thames River. Three-quarters of Dover Township, a lakefront township south of Walpole Island, was in forest and wetlands in 1871, but by 1891 farmers had cleared all but a third of the forest. Dover is now one of the most deforested townships in Canada. Apart from the St Clair National Wildlife Area and a few wetlands along the shore of Lake Huron, only a single twenty-acre woodlot remained in 2014.

These ribbons of deforested land raise questions about the geography of firewood production. How did deforestation influence firewood consumption, and vice versa? If Canadian woodlots were disappearing in the late nineteenth and early twentieth centuries, then why, and where, did firewood consumption continue to increase? Scarcity was actually less important than the story of stability and management over the long run, and thus many Canadian farmers must have recognized the limits and specialized products of their woodlots. It is possible to compare the production of firewood with the amount of unimproved woodland that remained on these farms.[50] The estimates vary, but it is feasible to harvest anywhere from 0.75 to 1.25 cords of wood from a hectare without depleting it. Houses consuming between 12 and 30 cords annually could have clear-cut anywhere from 0.5 to 1.5 hectares every year. In PEI, George Cairns recalled clear-cutting an acre (0.4 hectare) every year "when we were getting the wood out, just by enlarging the same patch from year to year."[51] After a few years, suckers growing from the stumps of hardwood species would repopulate the patch and fill in the opening.

In a handful of counties along Lake Ontario and the fluvial section of the lower St Lawrence (the Montreal plain), farmers were harvesting close to that amount in 1871 (see plate 5.2a). Of course farms in these areas were also still clearing small amounts, too, so we cannot assume that all production came from sustainable yield harvesting. The four districts that stood out as the highest producers per hectare of forest were Napierville, Beauharnois, and Soulanges just upriver from Montreal, and the Îles de la Madeleine in the Gulf of St Lawrence. By 1931, the farms in Soulanges and other districts on the south shore of the Montreal plain were producing negligible amounts of firewood, and they had cleared 85–91 per cent of the forest from their farms. This record corresponds with testimony in J.H. Morgan's report to the Canadian Forestry Commission, which in 1882 warned that there is a "large district of good agricultural land, south of Montreal, where the scarcity of firewood, which is a matter of life and death in a climate like ours, has compelled many a farmer" to leave.[52] Montreal and its neighbours were the first to feel a significant shortage of firewood in the late nineteenth century. Woodlots in southwestern Ontario and the Eastern Townships seemed ready to follow in 1931. In other Canadian districts the production of firewood on farms remained at steady and sustainable levels (see plate 5.2b).

Most rural Canadian homes were not part of a late nineteenth-century transition to fossil fuels – a factor that kept Canadian firewood consumption figures high, for, as noted above, Canada remained a rural and small-town country until the mid-twentieth century.[53] Wood remained the most common source of fuel for heating in virtually all districts in British Columbia, Quebec, and the Maritimes in 1951 (see plate 5.1). The northern fringes of agricultural settlement in the prairies and Ontario were also heated mostly by wood at mid-century. Only southern Ontario, the treeless districts of the prairies, and the coal-rich districts of Newfoundland, the southern Gulf of St Lawrence, and the southern Rocky Mountains were home to more dwellings heated by coal. Outside of the grasslands, there were only a few rural areas where people found alternatives to wood heat. Coal was the most common heating fuel in southern Ontario, the Rockies, and parts of the Maritimes, whereas several districts in the Montreal plain were adopting oil fuels. In addition to being influenced by geography, people were more likely to heat their dwellings with a fuel other than wood if they rented or if they were classified as urban.

The new consumption estimates presented in this chapter also confirm what was previously evident only in farm production data: the total amount of wood energy consumption dropped significantly for the first time in 1951. In Ontario, the percentage of farm dwellings heated principally with wood dropped from 75 to 50 per cent in just ten years after 1941. Even in 1951, there were still large firewood supply chains to Quebec cities, fuelling almost five times as many homes as urban Ontario. Interestingly, the overall Quebec consumption stayed constant, but on-farm production dropped by one million cords, and the amount sold off-farm dropped in half that decade.

The decline in farm firewood production and the even sharper drop in the number of farms reporting products of the forest (from 403,000 to 235,000) by 1951, indicates both a postwar shift in consumer preferences and the decline of mixed farming and agro-forestry in Canada. Many Canadians were now able to purchase oil and gas furnaces, and they increasingly used them, as well as hydroelectricity, for heating and cooking. Natural gas furnaces first became available in many parts of the country in the early 1960s. At the same time, farms were decreasing in number, increasing in size, and turning away from mixed products like agro-forestry. For many farms, fuel became part of the new off-farm inputs on which large capitalized farms now relied.[54]

Quebec's firewood consumption and market is a story that needs to be told; on the one hand, its high consumption rates, famously warm houses, and active farm firewood markets set it apart in Canadian energy history; on the other (and with a level of generalization appropriate to this chapter) is the province's expanding firewood consumption and active wood trade that helped define energy history and rural life in the Canadian Shield and Acadian Forest. By far the largest group of homes heated principally with wood was in Quebec, but northern Ontario, New Brunswick, and Nova Scotia shared some similarities.[55] Most significant is the number of *non-farm* rural families who heated their dwellings with wood. Of all rural dwellings heated principally with wood in Ontario, Quebec, New Brunswick, and Nova Scotia, non-farm dwellings represented 40, 45, 51, and 56 per cent, respectively. More than a story about Quebec, firewood in the mid-twentieth century was about prolonged biomass consumption in the rural (including both farms and non-farm dwellings) Canadian Shield and Acadian Forest regions.

THE BUSINESS OF BIOMASS

Firewood was one of Canada's most important "manufactured" products, but due to its low barriers to entry (almost anyone could cut wood), local trade, and exemption from most taxes and regulations it had no trade association or lobby group. As we saw earlier, it attracted only the occasional comment from promoters, experts, and statisticians, and the amounts recorded in the Census of Agriculture were probably met and surpassed each year by commercial harvesters. We cannot know, however, because the Canadian government did not make commercial firewood production and sales part of the census, and although the Canadian Forest Service provided some estimates, it never systematically tracked the activities of commercial firewood merchants. This industry was sometimes referred to as the cordwood business to differentiate it from small-scale firewood production. In 1871 the US census reported seventy "kindling-wood" industries, located mostly in New York and a few other large cities.[56]

Firewood processing required extensive work. As we saw above, farmers spent a great deal of time dealing with firewood for their own home consumption, and commercial operations took no less; considerable time and energy were needed to chop, section, dry, relocate, and reduce a great beech tree to small pieces of carbon. What solar energy and photosynthesis created in 150 years, men and women dry for months, process for days, and burn in just hours. Just as fossil fuels are plants reduced to pure carbon over millennia, trees require seasoning or drying or else they are difficult and unpleasant to burn. This represented an outlay of capital for wood merchants.[57] Farmers were better able to plan fuel needs and wood supplies years in advance, and so organizing this fuel chain on the farm or in local markets was sometimes more feasible. Even here, however, the supply of labour was a critical factor that could not always be predicted.

Farmers also participated in the trade to some degree, especially in the early years of settlement. During the clearing phase in Essex and Kent Counties, timber and cordwood were the main sources of revenue for early settlers. Not all trees are suitable for firewood or lumber, so even with the large market from this part of Kent, Dresden area farmers apparently burned a large portion of cleared forest and sold it as potash. There was an ashery on the riverbank in Dresden, and a local boatbuilding business supplied the cordwood export business directly.[58] Joshua Wright operated a coal and wood business on the riverbank, and up until 1900 most of the fuel used in Dresden was wood from the surrounding district.[59]

The commercial sale of firewood was shaped by significant variations in the price of firewood across the country. Shortages around Montreal and Toronto began to drive up the price of wood in the mid-nineteenth century. By 1867, the promoter of a new railroad called for public subsidies on the basis that his line could help break the cordwood monopoly held by Toronto's Northern Railroad. Shipped in from the Collingwood area, the Northern charged $8.00 per cord in Toronto, and railway boosters claimed that a fair market value was $4.50.[60] The Toronto, Grey & Bruce and the Toronto and Nipissing light gauge railways were chartered the next year mainly to provide lumber and firewood to the city.[61] In the 1870s, as Canada began its second wave of railroad expansion, much of it was done, and partly subsidized, to bring firewood to urban markets.

In the early 1870s the price of good hardwood per cord ranged from $2.50 at Goderich where wood was plentiful to $8.00 in Hamilton where it was not.[62] The influence of woodlot scarcity and increased demand near older settlements was visible even in the centre of this part of southwestern Ontario. *The Woodstock Review* reported that firewood in western Ontario towns like Woodstock and Ingersoll ranged from $2.50 to $3.50 per cord, whereas towns farther east like Paris ranged from $4.00 to $5.00 per cord.[63] The average price of firewood in Ontario also fluctuated over the 1870s, peaking sharply at $5.37 in 1873 and then dropping to $3.71 the following year. By comparison, the price of coal remained more constant over the same period, and still much higher than wood.[64]

By the end of the nineteenth century, the gap between the prices of coal and wood was closing in the larger cities. In 1892, the average retail price

Table 5.1   Retail prices of firewood and coal in Canadian cities, 1892

Page	City	Coal lowest, $/ton	Coal highest $/ton	Firewood lowest (green), $/crd	Firewood highest (dry), $/crd
7-8	Quebec			3.50	5.00
7-20	Montreal	6.00	6.00	5.50	5.75
7-32	Halifax			2.50	3.50
7-39	St John			6.00	8.00
7-42	Sherbrooke			3.00	3.50
7-49	Ottawa			2.50	5.00
7-53	Kingston			4.00	5.00
7-58	Toronto	6.00	6.00	5.50	5.50
7-67	London			4.50	5.50
7-82	Hamilton	4.00	5.25	4.00	5.50
7-103	Port Arthur			2.75	3.00
7-115	Winnipeg			3.50	5.50
7-120	Brandon			3.50	4.50
7-126	Regina			4.50	4.50
7-130	Calgary	6.00		3.50	3.50
7-135	Vancouver	7.00	7.00	4.00	5.00
7-143	Victoria	8.00	8.00	4.00	4.50

Source: "List of Retail Prices of the Ordinary Articles of Food and Raiment Required by the Working Classes," Canada, *Sessional Papers of the Dominion of Canada*, vol. 5, Second Session of the Seventh Parliament, Session 1892, no 7.

of firewood ranged from $3.93 for green wood to $4.87 for a dry cord, but in the nation's two largest urban markets, Toronto and Montreal, the prices ranged from $5.50 to $5.75. One ton of coal had a heat value equivalent to that of a cord of dry wood, so with coal available for $6.00 in both metropoles the wood market must have been significantly displaced in the early 1890s. However, residents in most other Canadian cities could still get wood for substantially less than the price of coal. In Sherbrooke and Quebec City, green wood could still be bought for $3.00 and $3.50 per cord, respectively, and the average price in Ontario towns outside of Toronto was $3.75. In Prairie cities, the average price was $3.55 for green wood, and in Vancouver and Victoria, green wood was $4.00 per cord, roughly half the price of coal in those cities (see table 5.1).

The census began to assign a value to firewood only in 1901, and began to differentiate wood sold off the farm only in 1921. The value of firewood and fence posts made up over three-quarters of the value of forest products from farms in 1911. "The greater portion," according to census officials, was "no doubt employed for home consumption and the values represent what the farmer considered these commodities worth to him rather than their market price."[65] But later censuses would show that this was not quite right. By mapping the percentage of wood sold in census districts in 1931, we see areas where farmers were in the fuel supply business, and other areas where they were not (see plate 5.3).

The census revealed a striking aspect of the interwar energy regime when it began to differentiate between wood used on the farm and products of the forest sold to others. By the early twentieth century, when data are first available, farmers sold less than a quarter of their woodpile. In 1931, only 19 per cent of total production entered the market, although this ranged from 25 per cent in Quebec to 16 per cent in Ontario and even less in most of the Maritime districts outside of the St John River Valley. Quebec farmers chopped and sold over 1 million cords for neighbouring stoves in addition to another 2.3 million for their own heat. Plate 5.3 demonstrates some clear patterns. Farmers in interurban areas like Waterloo and Wellington Counties, where harvesting firewood put the highest stress on Ontario woodlots, were also the least likely to sell their fuel. By contrast, farms in the lower St Lawrence Valley and Upper St John River Valley traded firewood almost as much as they consumed it.

In Prince Edward Island, we are able to see that the westernmost district, Prince County, sold fence posts (mainly cedar) and firewood (mainly softwood) to the other two districts in 1931. This area was the most forested and least populated on the island, and its wet swampy cedar forests produced excellent fence posts. Energy consumption varied across the province and over time, according to factors like the quantity and quality of the forest, the number and types of stoves used, and the availability of fossil fuels. Households in the centre of the province had better access to hardwood, and residents in West Prince and Kings Counties tended to burn more softwood.

The commercial cordwood trade is sometimes visible in other sources, such as business directories, newspapers, and atlases. Cordwood prices and products usually ranged according to quantity and quality: from softwood to hardwood, from green wood to dry, and from split blocks to uncut lengths. Hardwood was worth about twice as much as softwood, and drying the wood added about 25 per cent to its value.

Part of the urban and small-town fuel footprint is visible in late nineteenth-century fire insurance plans. Piles of cordwood ranging in height from six to twelve feet were common sights in many mill yards. Certain industries were attracted to cordwood over coal, especially if cleanliness was important to the mill product. The William Hay & Company flour mill in Ailsa Craig, Middlesex County, contained three woodpiles, and the largest held over 400 cords, possibly for multiple businesses (see plate 5.4).[66] The Crediton Flour Mill, owned by H. Switzer, was also flanked by three piles of cordwood, ranging from four to eight feet high.[67] Blacksmiths and foundries needed a wider variety of fuels, and H.C. Baird's Eagle Foundry in Parkhill had a large cordwood pile, a small one-storey coal shed, and a one-storey "Coke & Sand" building. Large wood- and metal-fabrication companies like the Ontario Car Company in London powered their kilns, smithies, and machine shops with "wood & shavings" stored in a massive cordwood pile beside the company's rail siding in the 1880s.[68] Even in urban and small-town industry, the Canadian energy landscape was very mixed.

The rail yards and fuel depots in any medium- or large-size town would have held a mix of biomass and fossil fuel carbon stores, and in almost every location the *Western Ontario Gazetteer and Directory, 1898–99,* contained dealers in "coal & wood."[69] Four businesses concentrated along the railway through Stratford, Ontario, serve as examples of these fuel merchants. The plan of Duncan & Gray's coal and wood yard specifies an area of "Cordwood Piled," in addition to a "motor Wood Cutting" building and a coal shed, along the railway; across the tracks the Lennox Coal Company owned four coal sheds, but still had two piles of slabs; the E. Burdette & Son coal and wood company consisted of only one large coal shed; and the C. Schneider & Son coal and wood yard contained three buildings, one for coal and two smaller buildings for "motor wood chopping" and wood storage.[70]

Coal and wood dealers usually carried both forms of carbon, although the primary market in many southwestern Ontario towns must have been coal. Dealers could focus on one fuel or the other with some small modifications, and in 1918, the Federal Commission of Conservation noted that many former wood dealers had converted their operations to coal because of wartime shortages.[71] The *Western Ontario Gazetteer and Directory* identified only two businesses that supplied cordwood specifically. They were located in neighbouring towns, Wallaceburg and Dresden, in the area of Kent County known for its large-scale cordwood production.[72] They were the remnants of the large export operation once located on the province's westernmost lake ports of Amherstburgh, Wallaceburgh, Sarnia, and Windsor.

Since it is clear that the firewood data recorded in the Census of Agriculture – the basis for most estimates of firewood consumption and biomass energy in Canada – are incomplete, it is important to look to other sources for determining the most likely amount of firewood produced and consumed in this critical period between 1850 and 1950.

RE-ESTIMATING WOOD ENERGY CONSUMPTION

The best-known histories of energy speak in terms of clear transitions, overlapping trajectories of wood fuel followed by coal and hydroelectric power, and finally oil, gas, and nuclear.[73] Even David Nye's captivating corrective to determinist views of energy is framed as "an overlapping series," focusing more on the emergence of new technologies than the persistence of old systems.[74] But biomass is not a life-cycle energy that peaks and then gradually disappears; it should be considered a baseline that remains relatively stable under certain conditions, especially in rural areas, as other fuels rise and fall.[75] One recent Canadian treatment of this subject, a *Historical Atlas of Canada* plate titled "From Firewood to Coal," draws its data directly from F.R. Steward's 1978 tabulations, which are in turn based on reported farm production in the census.[76] It suggests that the consumption of wood climaxed in 1881 and began a steady descent as coal's overlapping trajectory took over in the late 1890s.

In figure 5.1 I present a new estimate of the consumption of wood energy in Canadian homes by applying relatively conservative consumption rates to the proportion of dwellings that were heated principally with wood. In some years this is known, based on census reporting (1941, 1951) or on lack of alternatives (1871), and in others it is estimated by interpolating rates for urban and rural dwellings in each province between 1941 and 1871. Using Ruth Sandwell's population[77] and fuel[78] databases from the printed census, as well as the Canadian Century Research Infrastructure data tables,[79] I was able to estimate the total consumption of firewood in Canadian dwellings.

The Census of Agriculture's data on firewood production are problematic because the number of Canadian households claiming to burn mainly wood for heating and cooking (first recorded in 1941) does not line up with the wood production data. In parts of the country, the homes burning firewood would have had access to no more than three or four cords per house if they each received an equal share of the fuel reported by farms in the census. A fuel supply that small would not have sustained even the thriftiest of households into the new year. Yet this is still the amount used in many Canadian economic histories.[80] It is possible to provide a new estimate of annual consumption between 1871 and 1951 by determining appropriate consumption rates for each province and

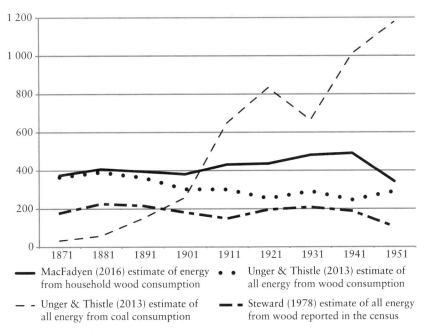

Figure 5.1 Wood and coal energy consumption estimates, Canada, 1871–1951 (petajoules)

Sources: Steward, "Energy Consumption in Canada since Confederation"; Unger and Thistle, *Energy Consumption in Canada in the 19th and 20th Centuries*; and the author.

period as well as the reported or approximate number of rural and urban dwellings that were heated principally with wood.

There are several alternative methods of estimating firewood consumption per Canadian household using the 1941 and 1951 censuses.[81] The most accurate way to measure household energy consumption in Canada is by dwelling. Occupied dwellings required a relatively consistent amount of energy for heat, regardless of the number of occupants.[82] Per capita rates do not take into consideration the fact that the average number of people per dwelling dropped steadily over time. Between 1881 and 1921, for instance, the number of occupants in Canadian dwellings dropped from 5.76 to 4.97.[83] Furthermore, the most accurate way to measure consumption of wood energy in Canadian homes is to determine the number of dwellings relying principally on wood heat each year and multiply that by a reliable rate of consumption for that year. The printed census provides ways of getting at both variables: the number of homes burning wood is available in the Census of Households and Families for 1941 and a reliable rate of consumption is available from the Census of Agriculture.

The number of homes burning wood is available for farm, non-farm rural, and urban areas starting in 1941, when the Census of Housing and Families began to provide data on heating and fuel use in Canadian homes based on a 20 per cent sample.[84] We know that the percentage of homes burning wood as the primary source of heat dropped from nearly 100 per cent in 1871 to 46 per cent in 1941, varying widely by province and between urban and rural areas, as consumers found access to alternative (fossil) fuels. Extrapolating backward from the mid-twentieth century to 1871, when virtually all Canadian dwellings were heated by firewood, it is possible to estimate the number of urban and rural dwellings heated primarily by wood in Canada in each census year.

I formed a dynamic estimate of the number of cords used per dwelling in each decade, starting with the very precise data on reporting farmers in 1931, 1941, and 1951. Products of the forest were becoming less important on Ontario farms in this period, and they were reported by only 53 per cent of farms in the Atlantic provinces. Considering that only these farms were reporting firewood production to the census, the farms produced almost sixteen cords (450 gigajoules) per farm for on-farm consumption in 1941, and almost fourteen cords per farm in 1951. It seems that many Canadian farmers were not active in their woodlots, and others may have systematically avoided recording or reporting their forest products. Still, the figures representing those who did report firewood and other forest products offer a precise record of the rates of firewood consumption in Canadian farmhouses. In Ontario and Quebec, the two provinces that produced over half of the nation's firewood, reporting farmers consumed nineteen and twenty-four cords per farm, respectively, in 1931. These are large amounts, and the downward trajectory in 1941 and 1951 suggests that woodpiles were even larger in the nineteenth and early twentieth centuries when stoves and building materials were even less efficient (see table 5.2).

These ratios typically declined throughout the twentieth century as more efficient house-building and wood-burning technologies developed. The rate of consumption per farm dwelling in the 1931 to 1951 period declined from eighteen to fourteen cords per farm, based on the higher estimate of farms reporting products of the forest. In 1871 the number of farms reporting is not known, but even the rate per all occupiers of land was twenty-four cords per occupant. Thus I estimated that consumption rates decreased by at least one cord per dwelling in the six decades leading up to 1931. This places the total rate of consumption in 1871 at four cords per person, a familiar estimate to economic historians of the United

Table 5.2 Census farm production and household consumption of firewood by province, 1871–1951

Cords (thousands)	1871	1881	1891	1901	1911	1921	1931	1941	1951
Alberta									
Production on census farms	–	–	–	–	78	353	566	661	255
Consumption per dwelling (rural)	–	–	–	–	16.5	15.5	14.5	13.6	12.6
Consumption estimate (rural)	–	–	–	–	633	835	826	856	498
Consumption estimate (urban)	–	–	–	–	142	119	43	48	12
British Columbia									
Production on census farms	–	160	157	174	136	174	181	199	59
Consumption per dwelling (rural)	–	21.9	19.9	17.9	15.9	13.9	11.9	11.8	7.8
Consumption estimate (rural)	–	170	232	298	502	800	774	1,006	325
Consumption estimate (urban)	–	26	89	175	286	286	381	467	461
Manitoba									
Production on census farms	–	637	275	253	157	401	418	531	279
Consumption per dwelling (rural)	–	18.2	18.2	17.2	16.2	15.2	14.2	16.2	13.3
Consumption estimate (rural)	–	188	382	556	668	888	890	1,170	566
Consumption estimate (urban)	–	24	83	112	233	223	214	271	35
New Brunswick									
Production on census farms	546	782	616	495	430	427	426	418	224
Consumption per dwelling (rural)	22.5	21.0	19.5	18.0	16.5	15.0	13.5	12.3	9.7
Consumption estimate (rural)	872	864	878	768	678	686	640	683	530
Consumption estimate (urban)	78	128	95	128	131	126	124	128	111

Table 5.2 Census farm production and household consumption of firewood by province, 1871–1951 *(Continued)*

Cords *(thousands)*	1871	1881	1891	1901	1911	1921	1931	1941	1951
Newfoundland									
Production on census farms	–	–	–	–	–	–	–	–	12
Consumption per dwelling (rural)	–	–	–	–	–	–	–	–	9.6
Consumption estimate (rural)	–	–	–	–	–	–	–	–	261
Consumption estimate (urban)	–	–	–	–	–	–	–	–	23
Nova Scotia									
Production on census farms	526	3,639	704	536	462	469	403	372	167
Consumption per dwelling (rural)	18.2	17.2	16.2	15.2	14.2	13.2	12.2	12.4	9.4
Consumption estimate (rural)	993	1,008	950	806	693	661	569	663	458
Consumption estimate (urban)	64	107	107	144	150	110	62	74	43
Ontario									
Production on census farms	4,519	5,435	5,184	4,031	2,584	2,856	2,307	1,884	785
Consumption per dwelling (rural)	27.8	26.3	24.8	23.3	21.8	20.3	18.8	15.0	13.5
Consumption estimate (rural)	5,901	5,905	5,142	4,575	3,890	3,634	3,420	3,155	1,905
Consumption estimate (urban)	1,180	1,631	1,875	1,697	1,716	1,366	786	729	263
Prince Edward Island									
Production on census farms	138	220	161	121	124	131	101	128	76
Consumption per dwelling (rural)	18.3	17.3	16.3	15.3	14.3	13.3	12.3	12.4	12.2
Consumption estimate (rural)	254	262	246	223	196	168	149	158	126
Consumption estimate (urban)	17	22	19	15	11	8	2	2	0

Quebec

Production on census farms	3,122	82	3,380	3,070	2,876	3,303	3,065	3,374	1,859
Consumption per dwelling (rural)	33.4	31.9	30.4	28.9	27.4	25.9	24.4	20.3	16.2
Consumption estimate (rural)	4,656	4,933	4,885	4,934	4,635	4,381	4,070	4,071	3,558
Consumption estimate (urban)	981	1,231	1,431	1,658	1,904	2,015	2,416	2,460	1,625

Saskatchewan

Production on census farms	–	–	–	–	54	416	621	812	414
Consumption per dwelling (rural)	–	–	–	18.1	17.1	16.1	15	13.9	13.7
Consumption estimate (rural)	–	–	–	219	1,113	1,292	1,239	1,177	749
Consumption estimate (urban)	–	–	–	23	201	217	197	201	30

Canada (from national totals)

Production on census farms	8,851	10,955	10,477	8,680	6,903	8,529	8,087	8,378	4,129
Consumption estimate (rural)	11,243	11,738	10,917	10,218	11,126	11,638	13,021	13,220	9,292
Consumption estimate (urban)	1,949	2,648	3,002	3,182	4,078	3,764	3,954	4,113	2,783
MacFadyen wood energy estimates	13,192	14,386	13,919	13,399	15,204	15,402	16,975	17,333	12,074

Note: Years marked with a "–" had no data available. National total estimate is produced from the reported household data for the nation. The national total is therefore slightly different from the sum of the more granular estimates of rural and urban consumption in all provinces.

Sources: CCRI-IRCS, "Census of Canada, Contextual Data, Geography: Digitized Published Tables"; Canada, Census of Canada, 1931, table 5; Canada, Census of Canada, 1941, table 10; Canada, Census of Canada, 1951, tables 14, 25.

States and perhaps too conservative for Canada's longer, colder winters, according to Marvin McInnis.[85] For urban consumption estimates I reduced the interpolated consumption rates by 25 per cent and multiplied it by the proportion of urban dwellings burning wood. With these conservative estimates, my presumption that at least 3 per cent of dwellings were not heated with wood, and our omission of wood energy for transportation and industry, the revised estimate for the first few years in table 5.2 is undoubtedly lower than actual production. These data show domestic consumption alone, ignoring the smaller amounts used in factory, railway, and steamship engines, as well as mills heating a wide variety of products from wool and whiskey to maple sugar.

The new estimate based on interpolating these rates and the number of homes burning wood in each province starting in 1871 suggests that wood was not simply the first of a series of energy sources that rose with settlement and then fell as woodlots became scarce. This was the case in certain subregions, but on the whole, the wood-energy industry was larger, longer, and more stable than the data from census farms suggest. Wood for home heating was a baseline energy source that proved remarkably resilient, even as fossil fuels appeared and satisfied new energy markets (see figure 5.1).

Taking into consideration data on the number of dwellings burning primarily wood for heat (1941 and 1951), the new estimate for national firewood consumption is much higher and more constant than the decennial amounts reported by farmers and reprinted in most statistical and historical texts. In the late nineteenth century, the new estimate adds about 50 per cent to the decennial census figures. From 1911 to 1941 I augment the census data by about 120 per cent, and in 1951 I find that the rate of consumption was over three times higher than the production recorded by census farmers. The new estimate shows that the production of firewood increased into the interwar period, experienced relative stability, and declined only after the Second World War. New harvesting in the Prairie provinces, British Columbia, and the Canadian Shield actually compensated for the gradual decline in southern Ontario and the Maritimes.

CONCLUSION

The transition from biomass to fossil fuels is widely considered the critical stage in industrialization, and the literature on Canadian energy suggests that industrialization was accompanied by a rapid transition "from firewood to coal." However, recent research is confirming that the shift to modern energy carriers was limited to energy for transportation, urban industry, and some coastal and riparian communities before the mid-twentieth century. The enormous amounts of wood harvested annually for Canadian stoves and engines continued steadily throughout the period of industrialization, although much new growth was made possible by additional supplies of fossil fuels. In certain places harvesting wood fuel for urban and industrial expansion contributed to extremely deforested landscapes and local energy scarcities, but the transition to fossil fuel was slow elsewhere because biomass energy systems were stable and well managed. As a winter activity, chopping firewood did not interfere with time-sensitive summer work, and it contributed to other important inputs in the rural wood economy such as potash, stave, and maple sugar production. Wood was a valuable commodity that farmers could sell locally and in cities using systems that were already in place for home fuel.

Wood disappears from the historical literature for a variety of reasons, but in reality, the story is less about old energies dying out and more about new energy supplies being born. Many rural homes and virtually all farms in Canada outside of the Prairie grasslands and southwestern Ontario continued to burn wood into the middle of the twentieth century. Firewood did occupy a declining share of total energy consumption after 1871, but the census data for wood energy remain surprisingly constant and, as I show in this chapter, it grew in absolute terms. This increase was particularly true in rural Canada. Firewood could not satisfy the new energy requirements of industrial and urban growth, yet on most farm woodlots consumption rates hovered within the range of sustainable yield. After the Second World War, the number of Canadian homes burning firewood dropped precipitously, and by 1971 it was no longer included in the census of households as an enumeration question. The brief resurgence in wood energy since the 1973 OPEC oil crisis has led to new industries in wood pellet and other biofuels, promising alternatives to fossil fuels, but these forms of wood energy are still very small by comparison to the firewood consumed in the early twentieth century.

NOTES

1 The author would like to acknowledge support for this research from the Social Sciences and Humanities Research Council (Postdoctoral Fellowship) and the Network in Canadian History and Environment, as well as valuable input from Christopher Jones, Douglas McCalla, Alan MacEachern, Eric Sager, and Ruth Sandwell.

2 Warde, *Energy Consumption in England and Wales, 1560–2000*; Smil, *Energy Transitions: History, Requirements, Prospects,* 100.

3 Unger and Thistle, *Energy Consumption in Canada in the 19th and 20th Centuries: A Statistical Outline,* 124, table 1.

4 Edgerton, *The Shock of the Old: Technology and Global History since 1900,* ix.

5 Weaver, *The Great Land Rush and the Making of the Modern World, 1650–190*; Crosby, *Ecological Imperialism: The Biological Expansion of Europe, 900–1900.*

6 Sieferle, *The Subterranean Forest: Energy Systems and the Industrial Revolution.*

7 Walker, "Transportation of Coal into Southern Ontario, 1871–1921," 15–30. For a detailed exploration of the importance of transportation to the mineral energy regime in the United States, see Jones, *Routes of Power: Energy and Modern America.*

8 Unger and Thistle, *Energy Consumption in Canada,* 51.

9 Sandwell, "Notes toward a History of Rural Canada, 1870–1940," 37–8.

10 Williams, *Americans and Their Forests: A Historical Geography,* 13.

11 On the importance of the subsistence economy in industrializing Canada, see Murton, Bavington, and Dokis, *Subsistence under Capitalism: Historical and Contemporary Perspectives.*

12 Many twentieth-century historians saw work in the forest as a distraction from more productive farming. See, for example, Clark, *Three Centuries and the Island: A Historical Geography of Settlement and Agriculture in Prince Edward Island, Canada,* 150.

13 McCalla, *Planting the Province: The Economic History of Upper Canada, 1784–1870,* 61, 230–1.

14 Sobey, *Early Descriptions of the Forests of Prince Edward Island: II. The British and Post-Confederation Periods (1758–c. 1900). Part A: The Analyses,* 77; Sobey, "The Most Beautiful of Woods: Prince Edward Island and the Beech Tree."

15 Macphail, *The Master's Wife.*

16 Lorain, *Nature and Reason Harmonized in the Practice of Husbandry,* 333.

17 McCalla, *Planting the Province,* 22–3; Unger and Thistle, *Energy Consumption in Canada,* 32–4.

18 Pyne, *Awful Splendor: A Fire History of Canada,* 177, 196; Williams, *Americans and Their Forests.*

19 Merchant, *American Environmental History: An Introduction,* 69.

20 Mendell and Lang, *Wood for Bioenergy: Forests as a Resource for Biomass and Biofuels.*

21 MacEachern, *The Institute of Man and Resources: An Environmental Fable.*

22 Williams, *Americans and Their Forests,* 134; Wood, *Making Ontario,*
13–14, 87. Vickers shows that in earlier settler societies like New England,
"clearing forests and enclosing fields alone consumed close to one-quarter
of the working year." Vickers, *Farmers and Fishermen: Two Centuries of
Work in Essex County, Massachusetts, 1630–1850,* 49; Unger and Thistle,
Energy Consumption in Canada.

23 McNeill, *Something New under the Sun: An Environmental History of the
Twentieth-Century World.*

24 Cairns, "Memories of My Life and Times," 136.

25 Reynolds and Pierson, "Fuel Wood Used in the United States, 1630–1930,"
1–20, 3–4.

26 Cronon, *Changes in the Land: Indians, Colonists and the Ecology of New
England;* Williams, *Deforesting the Earth: From Prehistory to Global
Crisis;* Brewer, *From Fireplace to Cookstove: Technology and the Domestic
Ideal in America,* 28.

27 Brewer, *From Fireplace to Cookstove,* 31–2.

28 Unger and Thistle, *Energy Consumption in Canada,* 29–30.

29 Hornsby, *Surveyors of Empire: Samuel Holland. K/W/F/ Des Barres, and
the Making of the Atlantic Neptune,* 50–1.

30 Unger and Thistle, *Energy Consumption in Canada,* 30.

31 Hutcheon and Handegord, "Evolution of the Insulated Wood-Frame Wall
in Canada," 434–5.

32 "Cold Wave in America," *Courier Mail,* 12 February 1934.

33 See entries on "bank" and "banking" in Pratt, *Dictionary of Prince Edward
Island English,* 10–11; Mellin, *Tilting: House Launching, Slide Hauling,
Potato Trenching, and Other Tales from a Newfoundland Fishing Village,*
195–6.

34 Bomberg and Onysko, "Heat, Air and Moisture Control in Walls of
Canadian Houses: A Review of the Historic Basis for Current Practices."
For the development of suburban home insulation standards, see Rome,
*The Bulldozer in the Countryside: Suburban Sprawl and the Rise of
American Environmentalism,* 81–2.

35 See for example, Greig, "Wall Insulation."

36 Scanlan, Bayne, and Johnson, "Investigation of Attic-Insulation
Effectiveness by Using Actual Energy-Consumption Data," 503.

37 Sobey, *Early Descriptions of the Forests of Prince Edward Island,* 77.

38 Kelly, "The Changing Attitude of Farmers to Forest in 19th Century
Ontario," 64; Walder, "The Utilization of Wood as an Energy Resource
in Ontario," 52.

39 Walder, "Utilization of Wood," 44–6; Tillman, *Wood as an Energy Resource.*

40 Warde, "Fear of Wood Shortage and the Reality of the Woodland in Europe, c. 1450–1850"; Stoll, "Farm against Forest," 65; Jehlen, *American Incarnation: The Individual, the Nation, and the Continent,* 67; Olson, *The Depletion Myth: A History of Railroad Use of Timber,* 3.

41 MacFadyen, "Fuel Wood"; Smil, *Energy, Food, Environment: Realities, Myths, Options,* 25; Chandler, "Anthracite Coal and the Beginnings of the Industrial Revolution in the United States."

42 Pomeranz, *The Great Divergence: China, Europe and the Making of the Modern World Economy,* 222.

43 Williams, *Americans and Their Forest,* 79; Prudham, Gad, and Anderson, "Networks of Power: Toronto's Waterfront Energy Systems from 1840 to 1970," 180.

44 Ménard and Marceau, "Simulating the Impact of Forest Management Scenarios in an Agricultural Landscape of Southern Quebec, Canada, Using a Geographic Cellular Automata," 254.

45 Prince Edward Island's forest was reduced to 31 per cent in Queens and Kings (the province's two most densely populated counties) by 1900. MacFadyen and Glen, "Top-Down History: Delimiting Forests, Farms, and the Agricultural Census on Prince Edward Island Using Aerial Photography, c. 1900–2000," table 1.

46 Watkins, *The Forest Resources of Ontario, 2011: Geographic Profiles of Natural Resources,* 73, 85.

47 OMAFRA, Land Use Geographical Information Systems data.

48 Walder, "Utilization of Wood," 58–9, 76.

49 Prudham, Gad, and Anderson, "Networks of Power," 180.

50 We cannot use the size of census districts as a denominator because they often included massive tracts of Crown land and other forest that were not accessible to farmers in more settled areas. This is particularly true of Quebec districts along the north side of the St Lawrence. Even using colour to display a more representative figure like cords per hectare of woodlot can be misleading, so I have used a semi-transparent mask showing the edges of the ecumene in the early twentieth century.

51 Cairns, "Memories of My Life," 139.

52 "Forestry Commission: Summary of Preliminary Report of Mr J.H. Morgan," 6.

53 Sandwell, "Notes toward a History of Rural Canada, 1870–1940," 37–8.

54 Ibid.; and Sandwell, "Introduction," this volume.

55 Census of Canada, 1951, vol. 3, table 25.

56 United States Secretary of the Interior, *Ninth Census: The Statistics of the Wealth and Industry of the United States, 1870*, table VIII(b), The United States, by Industries, 1870, 396.

57 Many merchants found it easier to supply markets with coal than to judge the supply of seasoned wood one or two years in advance. Leavitt, Commission of Conservation Canada, "Wood Fuel to Relieve the Coal Shortage in Eastern Canada," 5.

58 Brandon, *A History of Dresden*, 11–12.

59 Ibid., 36–7.

60 Laidlaw, *Reports & Letters on Light Narrow Gauge Railways by Sir Charles Fox and Son*, 44–6; Toronto Grey and Bruce Railway, *Hand-book of Useful Information, respecting the Line, and the Country Tributary to It*.

61 McIlwraith and Muller, *North America: The Historical Geography of a Changing Continent*, 227.

62 Young, *Labor in Europe and America: A Special Report on the Rates of Wages, the Cost of Subsistence, and the Conditions of the Working Classes in Great Britain, France, Belgium, Germany, and other Countries of Europe, also in the United States and British America*, table IV, 834–9.

63 *Woodstock Review*, 16 October 1874.

64 Young, *Labor in Europe and America*, table IV, 834–9.

65 Census of Canada, 1911, vii.

66 Goad, *Fire Insurance Plan of Ailsa Craig, Ont.*

67 Goad, *Fire Insurance Plan of Crediton, Ont.*

68 Goad, *Fire Insurance Plan of London, Ont.*

69 *Western Ontario Gazetteer and Directory, 1898–99.*

70 Goad, Fire Insurance Plan of Stratford, Ont., sheet 15.

71 Leavitt, "Wood Fuel to Relieve the Coal Shortage," 5.

72 *Western Ontario Gazetteer and Directory, 1898–99.*

73 Williams, *Energy and the Making of Modern California*; Smil, *Energy, Food, Environment*, 24–5; Smil, *Made in the USA: The Rise and Retreat of American Manufacturing*, 22; Marchetti, "The Future of Natural Gas: A Darwinian Analysis"; Schurr and Netschert, *Energy in the American Economy, 1850–1975*; Cole, "The Mystery of Fuel Wood Marketing in the United States"; Dewhurst and Associates, *America's Needs and Resources: A New Survey*; Dwyer, "Wood and Coal: A Change of Fuel"; Steward, "Energy Consumption in Canada since Confederation," 242.

74 Nye, *Consuming Power: A Social History of American Energies*, 7.

75 See, for example, the long-run chart modified by Mendell and Lang, *Wood for Bioenergy*, 8.

76 Muise, Langhout, and Walder, "From Firewood to Coal: Fuelling the
    Nation to 1891."

77 *Census of Canada, 1931*, vol. 1, table 5; *Census of Canada, 1941*, vol. 1,
    table 10; and *Census of Canada, 1951*, vol. 1, table 14.

78 These data were graciously supplied by Ruth Sandwell, who received an
    SSHRC grant for compiling them from the printed censuses; fuel data from
    Census of Canada, 1951, vol. 3, table 25. See Sandwell, "Mapping Fuel
    Use in Canada: Exploring the Social History of Canadians' Great Fuel
    Transformation."

79 CCRI/IRCS, "Census of Canada, Contextual Data, Geography: Digitized
    Published Tables."

80 In his study of Gross National Product, M.C. Urquhart divided firewood
    and other forest products into farm and non-farm production in order to
    derive the value and consumption rates of each. He extrapolates these
    interwar ratios across the entire post-Confederation period and suggests
    that *rural* non-farm consumption was 11 cords per house in 1870 and only
    5 in 1931, a low and problematic figure. No doubt the nation's hewers of
    wood kept their homes a little warmer than those who bought fuel, but
    there is little evidence that non-producers endured the winter with half or
    two-thirds the amount of heating and cooking. Similarly, F.R. Steward's
    summary of energy consumption in Canada used early wood data from
    decennial records in the *Canada Year Book*, which were in turn drawn
    exclusively from production data in the *Census of Agriculture*. American
    energy scholars such as Schurr and Netschert (*Energy in the American
    Economy*, building on Reynolds and Pierson) relied on more plausible per
    capita consumption ratios. Reynolds and Pierson ("Fuel Wood Used in the
    United States") used regional consumption ratios ranging from 0.19 to
    4.9 cords per person (somehow calculated to account for industrial con-
    sumption, burning technologies, and the displacement of wood by fossil
    fuels!). The new estimates from Canadian data show that per capita house-
    hold firewood consumption in 1930, the end of the period examined by
    Reynolds and Pierson, was much higher in Canada (1.8 cords per capita)
    than in the United States (0.6 cords per capita), and the latter included
    industrial and other forms of energy. Urquhart, *Gross National Product,
    Canada, 1870–1926: The Derivation of the Estimates*, 191, 226; Steward,
    "Energy Consumption in Canada," 239; Reynolds and Pierson, "Fuel
    Wood Used in the United States," 17n3.

81 Scandinavian countries have better data on consumption, but little on pro-
    duction. Kunnas, "Fire and Fuels: CO_2 and SO_2 Emissions in the Finnish

Economy, 1800–2005"; Lindmark and Andersson, "Household Firewood Consumption in Sweden during the Nineteenth Century."

82   One variable in dwellings is of course the amount of space being heated. Another level of estimation would therefore include a determination of fuel consumption per occupied room of dwellings, multiplied by the number of occupied rooms in Canadian dwellings. The size of houses changed over time, but it also varied significantly from province to province. In 1941 the number of occupied rooms per farm dwelling increased from 6.0 in the Maritimes and 5.8 in Quebec to 6.6 in Ontario. It dropped to just over 4.0 rooms per farmhouse in British Columbia and the prairies.

83   Statistics Canada, *The Canada Year Book 1927/1928*, table "Dwellings and family households, by provinces, 1881 to 1921."

84   There were 206,015 dwellings in Canada with supplementary wood heat. 26,120 of those were in BC (17,000 of those homes in Lower Mainland [District 4] and Vancouver Island [District 5]).

85   Schurr and Netschert, *Energy in the American Economy*; Dewhurst and Associates, *America's Needs and Resources*; McCalla, *Planting the Province*; McInnis, "Wood, Wind, and Water," in *A Contrarian's View of Canadian Economic History*.

6

Wind Power: Sails, Mills, Pumps, and Turbines

Eric W. Sager

Wind energy is the kinetic energy of air in motion. Wind power is wind energy converted into useful forms of energy. Wind power has vast potential: the potentially extractable power available from wind is much more than present human power use from all other sources. While the theoretical potential is vast, there are serious limits to the application of this energy. Wind speed is critical, because wind power is proportional to the cube of wind speed: thus available power increases eightfold when wind speed doubles. Wind, however, is highly variable, subject to daily and seasonal variations in strength and to interactions with local topographical features. Time is self-evidently critical: bursts of short duration yield little power. Canada's geography also creates limits, and the potential for wind energy applications in Canada varies enormously among regions. Furthermore, even in regions of high potential there are technological limitations: the extraction, storage, and application of power from wind energy is extremely difficult.[1]

Wind energy can be transformed into many uses. In Canada, as elsewhere, there have been four main uses of wind energy. First, sails convert wind energy into the propulsion of waterborne vessels, on oceans, lakes, and rivers; this is the oldest use of wind energy in Canada, and economically the most significant. Second, windmills have been used to create mechanical power, either for grinding grain or for pumping or draining water (wind pumps). Third, windmills have been used in creating power in industry, as in sawmilling. The fourth and most recent application is the use of wind turbines to transform wind energy into electricity. Since transportation uses are very different from the other three uses, this chapter will deal first with transportation, and then with the other applications.

WIND ENERGY IN TRANSPORTATION

How and When Did Wind Energy Come into Use?

We do not know when wind energy was first used in waterborne transportation. Certainly the people who migrated to Australia around 50,000 years ago, and the ancient mariners who populated Pacific islands 12,000–15,000 years ago, possessed ocean-going craft and considerable navigation skills. How and when people from East Asia migrated to the Americas remains a subject of ongoing research; the idea that some ancient migrants came by sea has won increasing support in recent years. If some came by sea, it is certainly possible that their ocean-going vessels were propelled at least in part by sails. The canoes of some pre-contact Aboriginal peoples likely used small sails to assist in propulsion.[2] Nevertheless, sail-powered transportation is associated mainly with Europeans who came to the Americas. The application of wind energy to marine transportation peaked in the nineteenth century. The transition from sailing ships to vessels powered in whole or in part by coal or petroleum-based fuels was gradual and uneven.[3] Sail-powered vessels remained in Canadian export trades into the 1930s, and there were a few Canadian sailing ships carrying cargoes in the middle of the twentieth century.[4] Today there is a resurgence of interest in sail technology in ocean carriers, and manufacturers are developing sophisticated sails and kites ("free-flying sails") as auxiliary to traditional energy sources.[5] Although Canadians have a longstanding interest in sails for yachts, we have little involvement in the recent development of sails for ocean vessels.

Rethinking the Significance of Wind and Sail in Canada's Economic Growth

Wind energy can be measured once it is converted. Thus we can measure the kilowatt hours of electrical energy created by a wind turbine. There is no comparable measure of the power produced by wind applied to a sailing ship. And before the twentieth century, wind power could not be commodified. Precise measurement of energy produced and consumed is more feasible with the centralized stocks of energy (coal, oil, and electricity) in the mineral energy regime. Sail-powered vessels were part of the organic energy regime, which Jones defines as "a human society where energy is derived from flows of solar energy entering the ecosystem."[6]

Wind, the bulk movement of air, is ultimately the result of solar energy, which induces changes in atmospheric temperature and differences in atmospheric pressure. Human beings learned to use wind as an energy source, but like most energy sources of the solar energy regime, wind is fickle, dispersed, and not easily transported. The energy operation – in this case, a wind-propelled transportation machine – had to move to and with the site of energy abundance. In this energy regime the displacement of human muscle power is limited. Sailing vessels were, therefore, wind-energy conversion machines that depended on the application of manual labour and skill to the capture of a highly dispersed and intermittent energy source.

The total energy consumed by Canada's sailing ship fleet at its peak was likely to be very small (Unger and Thistle suggest "an all-but-insignificant two petajoules per year").[7] But any attempt to measure energy consumed by sail-powered transportation runs the risk, not only of imprecision, but also of anachronism. The commonplace units of measurement are derived from a subsequent mineral energy regime in which energy was centralized, commodified, and marketable. The implied comparison with later energy volumes risks underestimating the importance of wind transportation systems in the context of a very different energy regime. We have no choice but to observe the wind-powered transportation system, and then try to estimate the contribution of that system to economic development.

For most of the time that Europeans have spent in the northern half of North America, their presence here was impossible without sailing ships. Wind power was the necessary condition of European arrival, trade, and settlement for at least 900 years. Changes in sailing ship technology were enabling conditions of other changes. The arrival and settlement of the Norse on our eastern shores could not have happened without the development of the Viking *knorr*, a sail-powered cargo ship capable of North Atlantic voyages, carrying twenty-four tons of cargo and a crew of between twenty and thirty.[8] Changes in sailing ship technology were a condition of European expansion in the fifteenth and sixteenth centuries and the arrival of European fleets off our eastern coasts in those centuries: these changes included larger sails, celestial navigation, astronomical tables, astrolabes, and maritime cartography. The technical development and numerical growth of the Royal Navy in the eighteenth century was a necessary condition of the major geopolitical changes in North America in that century. The technology of wind power in sails punctuated the rhythm of European activity in the colonies.

If there is a measure of production and consumption, it must be found in the outcomes of wind-powered transportation in the economy. A sailing ship is a vehicle whose function is to produce and sell a change of location. The change of location adds value to a commodity by taking it to market. In the British North American colonies, and especially those on the Atlantic seaboard, wind-propelled ships were essential to the production and circulation of timber, fish, farm output, and other colonial products. These same ships were also essential to the movement of people, the arrival of immigrants, and the creation of a labour force. As is well known, there was a symbiotic relationship: the same ships that took wheat or timber to Britain provided cheap transportation of immigrants in the other direction.

The other major location for sailing ships was the Great Lakes, where schooners and other craft contributed significantly to settlement and trade, even after the arrival of railways.[9] The War of 1812 on the Great Lakes has been described as "a shipbuilder's war."[10] Sailing ships were adapted to the St Lawrence canal and river system, and remained in use until the end of the nineteenth century. Railways did not suddenly supplant canals and natural waterways as the main transportation system. Competition with waterborne transportation is basic to the political economy of railways in the third quarter of the nineteenth century.[11] There are no estimates of the contribution of lake shipping to economic development, but certainly lake schooners and sail-powered canallers were major contributors to the early export trade from Upper Canada, to trade with the United States for much of the nineteenth century, and to the movement of population between British North America and the United States.

In the historiography of Atlantic Canada, two older hypotheses put sailing ships at the centre of the region's economic development. First, sailing ships facilitated the export of raw materials, and in doing so were critical to the rise and stability of the staple-based economy.[12] Second, the rapid obsolescence of the wooden-hulled sailing ship in the late nineteenth century changed the distribution of economic and political power in Canada, by robbing the Maritime provinces of a major engine of relative prosperity and economic growth.[13] So long as these hypotheses were held to be true, Canadians had a simple and appealing story about the great historical impact of wind and sail.

The story, of course, can no longer be told in these ways. The staple framework, which held that economic development was a function of dependence on the export of wheat, timber, and other primary resources, has lost much of its explanatory appeal among historians of Canada.[14]

Staple dependence can hardly be the defining characteristic of colonies that also exported an increasingly sophisticated manufactured good – a sailing ship – in large quantities.[15] It remains true, however, that expanding fleets of sailing ships, built and often owned in the colonies, provided the transportation service for a range of staple exports. Smaller sailing vessels performed that same service for trade within and between the Maritime colonies and with the United States. This close integration of sailing ships with the economies of the Maritime provinces continued into the last decades of the nineteenth century. Sailing ships took the place of roads and, for some time, even of railways.

The undeniable importance of sailing ships for production and trade led to the second hypothesis: the loss of this shipping sector must have dealt a catastrophic blow to the entire region. Harold Innis insisted, "Nothing that has occurred in this Province [New Brunswick] has done more to create difficulty, has been a more serious blow to that section of the Dominion than the decay of the industry of shipbuilding."[16] Perhaps he and his co-author C.R. Fay meant to include ship-owning within "shipbuilding," although the two are not the same thing. If Fay and Innis were right, then wind and sail were indeed powerful engines of economic growth, and not merely icons of a golden age.

Writing in the 1960s, the American economist Peter McClelland offered a very different and much more thoroughly documented perspective on the age of wind and sail. McClelland argued that the shipbuilding industry was too unimportant and too meagre in its linkage effects to have been a major contributor to growth, and the decline of the industry had little to do with subsequent economic retardation. Shipbuilding and ship-owning may have diverted capital from more productive sectors. "To the extent that more attractive opportunities were available elsewhere, the growth of shipowning in the province was regrettable."[17]

McClelland's postulation of "more attractive opportunities" gestures to the key issue in both historical and future-oriented analyses – the issue of opportunity costs. One cost of investment in sailing ships was the cost of the opportunity foregone in some other sector that might have strong future returns. Today, the cost of every dollar invested in petroleum is the foregone opportunity of investment in another energy source. Whether for past or future investments, the cost calculations can never be definitive, since they always involve hypothetical conditions. We cannot know definitively what returns New Brunswick investors might have gained by larger investments in railways or woollen mills and smaller investments in ships.

Whether or not there were "more attractive opportunities," sailing ships were critical to solving the transportation problem faced by merchant exporters. Freight rates represented a cost to the exporter or importer of timber, fish, or wheat, and freight rates were the source of profit to the ship-owner. North Atlantic freight rates, always subject to extreme short-term fluctuations, trended upwards in the decades after 1815, and then the trend flattened between 1850 and 1875 (apart from a sharp peak in 1873). Throughout these decades, the volume of exports from British North America increased much more rapidly than did freight rates: by increasing the supply of shipping tonnage and improving its efficiency, the shipbuilding industry contributed directly to the stabilization of shipping costs in the long term (although it could not mitigate the short-term fluctuations).

The cost of ocean transport declined significantly across the nineteenth century, from between 5 and 10 pence per tonne-kilometre in the early 1800s to between 1.5 and 2 pence at the end of the century.[18] In the North Atlantic, where Canadian sailing ships were heavily deployed, freight rates declined steeply in the last three decades of the century, and the decline had an enormous impact on international trade. These broad changes cannot be attributed to a single cause – the arrival of iron, coal, and steam. The changes also owed much to the continuing relative efficiencies of the sailing ship as a wind energy conversion system. Fleets of sailing ships grew in Atlantic Canada and elsewhere well into the 1870s, and at the end of the century they were still substantial contributors to international ocean transportation. Sailing ships retained an advantage in the carriage of bulk cargoes where speed and predictability were not essential; their other advantage was that the energy source, although inconstant, was free. Coal was a very expensive fuel, until the triple expansion engines of the 1880s allowed a significant reduction in coal consumption per unit of power. Liquid petroleum fuel followed in the early twentieth century. The transition to the mineral energy regime was slow and intermittent, and it included several forms of hybrid that combined wind power with carbon-based fuels. The energy transition reflected and reinforced the new temporal regime of capitalist industry, in which speed, regularity, and scheduled delivery took priority. The transition to a mineral energy regime occurred in Atlantic Canada, in ocean transportation and on land, and it is important to avoid the older determinist assumption that the technologies of iron and steam (and their falling costs) were inevitably fatal to the shipbuilding and ship-owning industries in Atlantic Canada.[19]

What did these wind and sail industries contribute to gross provincial product? McClelland estimated total shipyard sales in 1870 to be slightly more than 3 per cent of provincial product for New Brunswick. But shipbuilding output for that year was below average for the third quarter of the century, and the value given by the census is certainly an underestimate. My own estimate is that shipbuilding output was around 5 per cent of GPP in the third quarter of the century. In addition, New Brunswick's ocean-going fleet of the early 1870s earned gross revenues of sixteen to twenty dollars per ton per year, or between 9 and 11 per cent of GPP. Taken together, the wind and sail industries were very substantial parts of the economy. As a point of comparison, Canada's wheat exports in the first decade of the twentieth century accounted for about 3 per cent of GNP. For several years Canada may have had the fourth-largest merchant marine in the world, in terms of total tonnage on registry. At no other time in our history did industries based on wind power contribute so substantially to our economy.[20]

What Changes Did the Wind and Sail Industries Bring About?

The question of change takes us back to the issues of growth and retardation that Peter McClelland addressed. Wooden ships by themselves did not retard the development of industrial capitalism. Indeed, the profits from ship-owning became investments in a range of other enterprises, including manufacturing. Nevertheless, McClelland's understanding of the dynamics of change comes closer to the truth than does the generalization of Fay and Innis. To the extent that the shipping industries sustained fishing, farming, lumbering, and mining, by providing low-cost transportation to both local and distant markets, "merchant shipowning worked toward the survival of independent commodity production and against a basic condition of the transition to industrial capitalism – creation of a larger capitalist labour market. Massive investment in sailing ships was therefore not simply a 'diversion' of capital from productive to unproductive sectors. It affected the entire structure of social relations in which the transition to industrial capitalism was occurring. As an experiment with capitalist class relations, shipowning was cautious and ambiguous."[21]

As Maritime capitalists invested more heavily in ships, investors in Upper Canada put much larger proportions of capital into iron foundries, railways, machine shops, and other industrial enterprises. It is probably no coincidence that relatively slow capitalist development in the Maritimes began in the third quarter of the nineteenth century, at the height of the age of sail.

The wind and sail era introduced profound changes in human capital: the stock of knowledge and skill accumulated in coastal populations. All forms of wind power evolve: windmills, wind pumps, wind turbines, and sailing ships. In Atlantic Canada in the nineteenth century the wind-energy transportation technology evolved quickly, resulting in improvements in productivity that contributed to the longevity of sailing ships. The average carrying capacity of the sailing ship increased in every decade. The three-masted square-rigged ship of the 1880s was four times larger than its predecessor of the 1820s. The colonial sailing ship, often accused of being a fragile death trap in the 1830s, became much more durable and long-lived, as construction methods changed and more iron was used.[22] Builders learned the qualities of hull design that yielded greater sailing speeds. Astonishing changes in sailing speeds (unknown to historians prior to the 1970s) were achieved, not just by the exceptional "clipper" ship, but by entire fleets of colonial and Canadian deep-sea ships. No less remarkable was the change in capital-labour ratios, due not only to increasing vessel size but also to the shedding of labour. The same vessel that required twenty-five or twenty-six sailors for an ocean voyage in the 1860s was being worked by sixteen sailors in the 1880s.[23] Canada was a major location for advances in marine wind-energy conversion.

In the nineteenth century a knowledge of wind and its many forms and uses became widely dispersed among settler populations. The understanding of wind was part of the fund of local knowledge among peoples in both coastal and inland communities. Women, the "skippers of the shore crew," used wind in the drying of salted fish.[24] Men and women in farm families knew what wind foretold of changes in weather; they knew the patterns of wind-fallen timber, the effect of wind on crops, and the volatile connections of wind and fire.[25] They learned from Aboriginal peoples, as did Catharine Parr Traill, that venison could be preserved by drying it in the wind.[26]

Among the men and women who went to sea in schooners, for trading or for fishing, wind was one element in a complex mental mapping of wind, currents, waves, tides, and the vessel itself, with its array of sails, masts, spars, blocks, ropes, rudder, and other gear. Shipmasters and deck-hands could distinguish different winds by their feel around neck and ears; they knew the difference between true and apparent wind (the latter being the sensation of wind from the forward motion of the vessel); they could see a wind before they felt it and knew its character, by its effect on the surface of the water. Stun breeze, close-reef breeze, liner, screecher, fairy squall, nor'easter: the many varieties of wind were part of the vocabulary of Newfoundlanders.[27] The fore and aft sails of the schooner were finely tuned to highly variable winds, allowing swift trimming of sails and tacking (bringing the vessel directly into the wind and across it, so that the sails take the wind from the other side). The schooner, with its fore and aft sails on two masts, was the preferred coastal vessel because of its ability to sail "close-hauled": the wind enters the sail diagonally and creates lift, so that the vessel moves *into* the wind. The aerodynamics of wind and its action on sails was not learned from manuals; it was a social knowledge acquired from observation and experience, through childhood to adulthood and passed across generations, among men and also among many women.[28]

As the ocean-going fleets expanded, the stock of knowledge changed and its distribution became more hierarchical. To the knowledge of the coastal mariner, the deep-sea shipmaster added the understanding of navigation, sail-handling, and ocean winds required for the master's certificate of competence. The preferred ocean-going vessels in eastern Canada were the two-masted brig, carrying square sail on each mast; the three-masted barque, with its square sails on the fore and main masts and fore and aft sail on its mizzen mast; and the three-masted ship, which carried square sails on all three masts. The techniques of sail-handling in such vessels filled chapters in seamen's manuals. The art of trimming sail to the wind was more complicated in the large square-rigger than in the schooner, if only because the number of parts to the machine was much greater. Even if masters had not learned that wind speed aloft is greater than at the water line, they knew that sails at different heights could be trimmed at different angles to the wind, to beneficial effect; "spiral bracing" maximized the collective force of sails upon yards and yards upon hull. Effective sail trimming meant the right balance between yards, sails, and wind; among other things, it required minimizing the effect of blanketing, when sails prevented wind from entering other sails before them. Such were the elementary principles in a vast stock of knowledge.[29]

Sail-handling and navigation were two sides of the same coin, and as ships from eastern Canada moved across oceans their masters gained a new level of competency and new tools in their application of wind energy. They needed charts of ocean routes and also of any approaches, bays, and anchorages that they were likely to encounter. To the sextant and other tools was added a deeper knowledge of nautical astronomy as well as the *Nautical Almanac,* which gave the information required for working out the vessel's position by celestial navigation. In 1847 the American navigator Matthew Fontaine Maury published *Wind and Current Charts of the North Atlantic,* the first of many works in which he showed sailors how to use ocean winds and currents to their advantage in order to shorten sailing times.[30] In one form or another, Maury's charts found their way into the hands of Canadian shipmasters and contributed to the success of their search for speed. The wind-powered vessel was also changing in the third quarter of the nineteenth century: it was built not only of wood but also of iron and copper; it acquired wire standing rigging and patent iron windlasses; it acquired taller masts and wider yards, and new kinds of sail. The application of wind energy had become an applied science that integrated the knowledge and innovations of manufacturers, nautical chart-makers, navigators, shipmasters, and deckhands.

General Observations

From the history of wind-assisted transportation we may draw no simple lessons for the present and future. Certainly we discover a warning against technological determinism, and against the easy extrapolation of a technology from one historical context to another. There is also a clear warning against the idea that investment patterns are the outcome of rational calculations of costs and rates of return on capital. Critical to the choices of Maritime capitalists was the flow of information into the Maritimes about investment opportunities in Canada as a whole.[31] Investment decisions cannot be seen merely, or even primarily, as acts of instrumental rationality and utility maximization. Furthermore, it is impossible to understand the application of wind energy to transportation as the result of market forces acting independently of state policies and actions. We are not observing a world of laissez-faire. The rise of the shipping sector in the nineteenth century occurred in the world of the British Navigation Acts and British tariffs.[32] The decline of the shipping sector occurred in the context of the Canadian national policies, which gave no direct protection to shipbuilding or ship-owning. The decline of the wind and sail industries meant the gradual waning (although not the disappearance) of a vast fund of local knowledge of wind energy, an accumulation of observation and application widely diffused in the culture of large segments of the population. Change in the utilization of energy sources is a product of culture and the social relations in which economic change is embedded.

FROM WINDMILLS TO WIND TURBINES

How and When Did Wind Energy Come into Use, Outside Shipping?

In the history of transportation, the transition from an organic energy regime to mineral energy is a story of displacement, as sails gradually gave way to coal and petroleum. In the history of wind-energy conversion on land, the transition tells a somewhat different story. In the organic energy regime on land, wind was always a minor aid to muscle power. As mineral energy systems expanded in the nineteenth and twentieth centuries, there was a displacement effect, at least initially: windmills and wind pumps were replaced by machines using other energy sources. Nevertheless, there was another effect, especially apparent by the late twentieth and early twenty-first centuries. Wind experienced a modest recovery as an adjunct to mineral energy systems, and despite the dispersed and intermittent character of wind, technological changes allowed wind to meet a small but increasing share of total energy demand. By the early twenty-first century it was even possible for advocates of wind energy to argue for its potential to displace a small but significant volume of power supplied from mineral energy sources.

Windmills were brought to the northern half of North America by Europeans, beginning in the seventeenth century. They were used mainly for milling, and from the beginning they competed with other milling techniques: hand-milling, horse-powered milling, and water-milling. In general, windmills appeared where there was both a good supply of wind and the absence of a useable flow of water, since water-milling was usually the preferred method.

According to Dominique Laperle, there were thirty-six windmills in New France in the first half of the eighteenth century.[33] The provision of a mill was a seigneurial responsibility, and habitants were required to use the seigneurial mill, at least for that portion of grain used for domestic consumption (the banality was extended in the 1790s to all grain). Windmills were easier to build than watermills, but windmilling had the reputation of producing an inferior flour because the source of power for grindstones was so variable. By the end of the French regime almost all settled seigneuries had either a windmill or a water-mill, and some had both. The milling banality was often critical to the profits of the seigneury, and Allan Greer argues that they were an important part of the "feudal burden."[34]

The use of windmills for mechanical power in industry was not widespread in Canada in the nineteenth and twentieth centuries. The availability of alternative sources (wood, coal, water power, hydroelectricity) helps to explain the scarcity of windmills. A comparison with California helps to confirm the point. In that state windmills were relatively widespread and contributed significantly to agriculture and early industrialization between the 1860s and early 1900s, only gradually being displaced by steam, gasoline, and electricity. The specific composition of energy sources in California helps to explain the popularity of windmills: prevailing Pacific breezes reached well inland; supplies of coal, wood fuel, and running water were either poor or not so readily available as they were in the eastern United States and Canada.[35] In Canada, by contrast, wind power contributed little to industry. The industrial census of 1871 reports only fifty-eight wind-powered establishments in Canada, and most of these (forty-three) were in Quebec, survivals of the older application of windmills in gristmilling. Although most of the nine Ontario windmills were located near lakes, where there was likely to be a good supply of wind, the breezy coastal provinces, Nova Scotia and New Brunswick, reported only six windmills. Of the fifty-eight windmills, almost half were flour or gristmills. Twelve were part of sawmill operations, and the other applications included a few cabinetry workshops, tanneries, carriage shops, a furniture factory, and a boat-building shop in Falmouth, Nova Scotia. Waterwheels and steam were far more important sources of power for manufacturing.[36]

Windmills were used in the Prairie provinces in the late nineteenth and early twentieth centuries, mainly for pumping water. So far as I know, there is no estimate of the number of wind pumps for any period in Canada. In Alberta between 1885 and 1911 there were at least fifty-two windmills (I assume that most or all were wind pumps), according to the ongoing Alberta Land Settlement Infrastructure Project.[37] "Waterpumping windmills dotted the U.S. landscape," says Robert Righter, and by some estimates there were six million of these pumps in the late nineteenth and early twentieth centuries. Wind pumps were particularly important wherever herds of livestock were not conveniently located near lakes or rivers.[38] There were wind-pump manufacturers in Canada in the late nineteenth and early twentieth centuries, but it is likely that most wind pumps were imported from the United States.[39]

Small wind machines were also used to generate electricity for farms prior to rural electrification.[40] One American company, the Jacobs Wind Electric Company, produced more than thirty thousand small wind-electric machines between 1927 and 1957, and many were sold in Canada.[41] As centralized electrical grids expanded into rural areas, they displaced the independent wind-electric generators and the co-operatives that operated such machines.[42] In the early twenty-first century this trend was being reversed: grid-connected wind energy systems appeared, and wind energy was increasingly a tradeable commodity as well as a means of energy independence for small farmers and other producers.[43] Stand-alone wind turbines have increased significantly in the last decade, and by 2010 market sales in Canada reached $20 million.[44] Wind turbines and wind farms are discussed at greater length below.

How Much Was Produced and Consumed? What Proportion Was Commodified?

It is impossible to quantify the production or commodification of wind power. Nevertheless, it is safe to say that in only one time and place did wind power dominate production: Red River in the middle decades of the nineteenth century. The grinding of grain by wind power superseded other milling methods and accounted for the major portion of mill output. The first functioning windmill appeared in Red River in 1825. By 1838 there were fourteen windmills, and by 1849 there were eighteen.[45] The winds were adequate for much of the year, but prolonged calms resulted in flour shortages in Red River. By 1856 there were nine operating watermills as well as eighteen windmills. The volume of production from each type of mill cannot be estimated, but the majority of flour produced in Red River came from windmilling. The relative importance of windmills can be attributed to the nature of local waterways: there were few fast-flowing streams draining into the Red and Assiniboine rivers; dams were required, and these were not easy to build and maintain; water mills could not be operated at all in winter, and in some summers the creeks dried up completely.[46]

Among the major uses of wind energy, the production of electricity is the most recent. The technology now has a history of over 126 years. It developed during the rise of the mineral energy regime, and both development and application were limited by competition from other forms of electricity generation. The first wind-powered electricity generation took place in the late 1880s (the innovators were James Blyth in Scotland and Charles Brush in the United States). Although commercial applications began in the 1890s, development was slow and localized, and there was no substantial transfer of wind-turbine electricity into regional or national electrical grids until the last half of the twentieth century.[47] In general, there was an inverse relationship between the price of fossil fuels and the pace of technological development in wind turbines. It was not until about 1990 that the cost of wind-generated electricity, in pennies per kilowatt-hour, converged with the cost of electricity generated by other sources.[48]

Canada was not a major contributor to wind turbine technology, and the application of the technology was slower in Canada than in northern Europe and the United States. The main reason for Canada's relative absence lies in the availability of energy from other sources, including hydro power. Canadians did undertake research on wind turbines, however, and it began in the public sector. In the 1960s the Brace Research Institute at McGill University designed and tested both vertical axis turbines (VAWTS) and horizontal axis turbines (HAWTS).[49] In 1966 the National Research Council began to experiment with wind power, and two NRC researchers, Raj Rangi and Peter South, are credited with the "rediscovery" of the Darrieus vertical axis wind turbine, originally invented by the French engineer Georges Darrieus in 1925.[50]

As in other countries, the rise in oil costs after 1973 led to increased interest in wind power. In 1974 the federal government established the Office on Energy Research and Development and a panel of senior public servants to co-ordinate research on all types of energy, including wind energy.[51] Hydro-Québec also took an interest in wind power through the Institut de recherche de l'Hydro-Québec, and in 1976 Hydro-Québec installed a large vertical axis turbine at Île-de-Madeleine. The turbine was manufactured by Dominion Aluminium Fabrication Ltd (DAF), which also installed five smaller vertical axis machines at other locations. Bristol Aerospace Ltd of Winnipeg also installed several vertical axis machines to power telecommunication systems in remote areas, including the Far North. By the 1980s DAF was exporting wind turbines to California, and Hydro-Québec constructed a few large megawatt turbines, which fed into the Hydro-Québec grid.[52]

This flurry of activity was not sustained. By the mid-1980s the global oil crisis had ended. Federal investment in research on alternative energy sources virtually ceased, and the wind-energy research program at the National Research Council ended. VAWTS, while having certain advantages, did not win support among major manufacturers. In the late twentieth century, Danish and other European engineers seemed to have confirmed the technical and economic superiority of horizontal axis machines, and the advantages of horizontal axis machines deployed in wind farms were being demonstrated in California in the 1980s. This

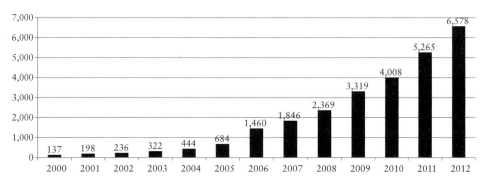

Figure 6.1   Canadian-installed wind energy capacity (megawatts), 2000–2012
Source: Canadian Wind Energy Association (CanWEA).

does not mean that Canadian research on VAWTs was a mistake. In recent years, long after the pioneering Canadian research on VAWTs had ended, the debate over the relative merits of HAWTs and VAWTs for wind farms resurfaced.[53] The story of wind energy research and development in Canada remained one of unfulfilled promise.

Renewed interest in wind-generated electricity, using turbines manufactured mainly elsewhere, occurred in the 1990s, and growth has been very rapid since then (see figure 6.1).[54] This renewed interest is strongly correlated (although not perfectly) with the rise in crude oil prices in world markets since 1998. The Cowley Ridge wind plant, Canada's first commercial wind farm, was constructed in southern Alberta in 1993–94.[55] Since then, all of Canada's provinces have some level of wind energy; production is highest in Ontario, Quebec, and Alberta.[56] Canada's annual production of 6,578 megawatts in 2012 meant that Canada was ninth among countries in total output.[57] Only 3 per cent of Canada's demand for electricity is currently met by wind power. By this measure, the world leader is Denmark, where about a quarter of electricity comes from wind power. The European Commission goal is that 20 per cent of all EU power (not just electricity) be met by renewable energy sources, including wind, by 2020. Despite our long coastlines, Canada has only one offshore wind farm, built off Cape Scott in British Columbia. In February 2011 the Ontario government placed a moratorium on all offshore projects, saying further study was needed on their impact on health and the environment.

For the advocates of wind power, an encouraging trend in the early twenty-first century is the growth of demand for small, stand-alone wind systems for domestic and local use. Small turbines, defined as turbines of

300 kilowatt capacity or less, provide power for battery charging, residential consumption (often to complement electricity from the grid), buildings (cottages, hunting lodges) and leisure craft vessels that the grid does not reach. Small turbines also provide power for farm equipment, commercial buildings, and equipment in remote locations (weather stations and telecommunication devices, for instance).[58] By 2013 one could buy a 600-watt twelve-volt HAWT generator for under $1,000, or a 15-watt twelve-volt VAWT for around $300. Canada has developed a small but relatively important manufacturing sector for mini-turbines, and a large proportion of production is exported.[59] A combination of environmental and geographic factors has allowed Canadian-based manufacturers to discover a market niche not fully occupied by producers of large wind energy systems.

General Observations

The Canadian Wind Energy Association (CanWEA), established in 1984, is the wind industry organization that leads the campaign for investment in wind energy in Canada. It is, as may be expected, boosterish and optimistic about the potential of wind and even its capacity to displace a portion of energy provided by the mineral energy regime. In 2008 CanWEA published *Wind Vision 2025*, a plan in which wind energy would reach a capacity of 55,000 megawatts by 2025, fulfilling 20 per cent of the country's energy needs. The strategy, said CanWEA, could create over 50,000 jobs and represent around $165 million annual revenue.[60]

If history offers pertinent lessons, it is unlikely that such a strategy will be widely accepted, let alone implemented in the near future. First, the application of wind energy in Canada has always been geographically dispersed and limited to specific, localized sites and applications, as it was in the era of sailing ships. This localized character continues in the era of wind turbines, even within major producing provinces such as Ontario. Such localism makes it difficult to create a national strategy with support in all regions.

Second, the age of wind transportation was characterized by an integral connection between the manufacture of machines (shipbuilding) and the application of the machines (ship-owning and ship operation). Merchant-ship-owners were rarely shipbuilders, but the personal and business connections between the two groups were close. In the absence of a large manufacturing capacity in turbines in Canada, the business sector lobbying for wind farms is small. Some attempts to bring branch plants to Canada for wind turbine production have succeeded, and in the early twenty-first century there was a small export trade in turbine blades, as well as the trade in mini-turbines.[61] Nevertheless, there have been disputes over regulations, complaints about the levels or inconsistency of government support, and laments about the failure of the federal government to renew its ecoEnergy for Renewable Power Program.[62]

Third, the presence of alternative energy sources, combined with the virtual impossibility of calculating opportunity costs, has always impeded the development of new energy sources and technologies, and it continues to do so. The presence of fast-moving water in rivers and creeks was an important limit to the building of windmills. Wind power became dominant in waterborne transportation when alternative energy sources were either non-existent or wholly inadequate. In North America, a major obstacle to further development of wind farms is the current and projected supply of energy from hydroelectric power and petroleum fuels, especially natural gas. The organic energy regime, as Christopher Jones argues, has never gone away, but our Faustian bargain with the mineral energy regime is not easily undone.

Finally, the biggest obstacle to the development of wind power, on the scale envisaged by CanWEA, is the dogma that market forces – the interaction of supply, demand, available resources, and cost factors – will determine outcomes, especially when left unfettered. Free-market environmentalism holds that private property rights, freed from state interference, would allow affected parties to seek redress from polluters, thereby imposing costs on environmental degradation and so reducing the undesirable "externalities." Over time, those sources of energy producing costly externalities will give way to sources, such as wind, that minimize external costs. The history of wind power offers no support for such ideas. If there was a moment when market forces acted independently of collective or judicial intervention to determine the choice between wind power and other energy sources, perhaps that moment occurred in Red River in the first half of the nineteenth century. Such moments were rare and highly localized.

The sailing ship industry grew in the context of imperial preference and protection. Wind turbine development in Germany, Denmark, California, and elsewhere occurred with substantial state support.[63] Even the modest beginnings to wind power research and turbine installation in Canada depended critically on agencies in the public sector or on government assistance.[64] The argument for increased public and private investment in wind energy rests not only on comparative cost analysis and projected rates of return. It also rests on a historical truth claim: markets may adjust supply to demand through the mechanism of price, but prices are not the same as costs. Prices do not reflect the full costs of production and distribution, including social and environmental costs.

More optimistically, the history of wind power tells us that changes in energy sources and applications have occurred, often over short spans of time, and that such changes can be implemented by a favourable conjunction of available resources and astute state-supported policies. Total energy consumption is always the sum of a mix of many sources. There have been attempts to estimate the relative contributions of many different energy sources in the mix, but the estimates of the contribution of wind power in these models are flimsy at best.[65] If there is any value to these models of energy consumption, it lies in the reminder that energy supply is always a composite, nationally and locally. Energy, like the composite sailing ship that used both wind energy and auxiliary power from steam engines, is always a hybrid. Even pure sailing ships depended on a mix of wind, human labour, and a multitude of mechanical devices, many of which were the products of steam or coal and human or animal energy. The wind turbine does not act alone, and its potential lies only in its capacity to work together with rather than replace other energy sources. The challenge for the energy-intensive mineral-regime economy is not to replace one source of energy with another. It is to apply political will, collective action, and local knowledge to the evolving mix of energy supply and consumption, to reduce our dependence on carbon-based energy.

NOTES

1 Introductory studies include Manwell, McGowan, and Rogers, *Wind Energy Explained*; Sathyajith, *Wind Energy: Fundamentals, Resources Analysis, and Economics*; Burton et al., *Wind Energy Handbook*; Musgrove, *Wind Power*. Righter focuses on electricity in *Wind Energy in America: A History*.

2 Among the peoples of the Pacific Northwest, simple split cedar sails may have been used in ocean-going canoes before the arrival of Europeans. Roberts and Shackleton, *The Canoe: A History of the Craft from Panama to the Arctic*, 104. In Nootka Sound in 1778 James Cook observed Native canoes and noted, "Sails are no part of their navigation." Ibid., 97.

3 The volume of sailing ship tonnage clearing Canadian ports increased between the 1870s and 1880s, and remained flat until the mid-1890s, before beginning a slow decline.

4 Sailing ships carried lumber across the Pacific in the 1920s and 1930s. See also MacDonald, *North Star of Herschel Island: The Last Canadian Arctic Fur Trading Ship.*

5 Geoghegan, "Designers Set Sail, Turning to Wind Power to Help Power Cargo Ships"; Graham, "Sailing into the Future of Global Trade"; B9 Energy Group, " – Flagships of the Future." See also the Integrated Power Technology Corporation, "Sail Manufacturers."

6 Jones, *Routes of Power: Energy and Modern America*, 15.

7 Unger and Thistle, *Energy Consumption in Canada in the 19th and 20th Centuries: A Statistical Outline*, 48.

8 Sawyer, *The Oxford Illustrated History of the Vikings*, 182.

9 Bamford, *Freshwater Heritage: A History of Sail on the Great Lakes 1670–1918.*

10 Ibid., 68.

11 Deloges and Gelly, *The Lachine Canal: Riding the Waves of Industrial and Urban Development, 1860–1950*, 72; Karamanski, *Schooner Passage: Sailing Ships and the Lake Michigan Frontier*, 183; Norrie, Owram, and Emery, *A History of the Canadian Economy*, 162–3.

12 The classic study from the staples perspective is McCusker and Menard, *The Economy of British North America, 1607–1789.* An early popular study emphasizing the integral role of sailing ships in the economies of fish and timber is Wallace, *Wooden Ships and Iron Men: The Story of the Square-Rigged Merchant Marine of British North America.*

13 Fay and Innis, "The Maritime Provinces," 663.

14 There is a cautious, qualified endorsement of the staple approach in Harris, *The Reluctant Land: Society, Space and Environment in Canada before Confederation*, 31, 335, 456. Recent studies of Harold Innis creatively reinterpret the intellectual context of Innis's writing on staples. See especially Bonnett, *Emergence and Empire: Innis, Complexity, and the Trajectory of History*; Buxton, *Harold Innis and the North: Appraisals and Contestations.*

15 By the mid-1820s no less than 80 per cent of the value of exports from Prince Edward Island was accounted for by sailing ships. Sager with Panting, *Maritime Capital: The Shipping Industry in Atlantic Canada, 1820–1914*, 14, 177. To preserve the staple model, the staple theorist has to argue that these machines themselves were staples, and thus robs the word of its meaning. Watkins refers to Quebec-built ships as "staple exports" in his "Staple Theory of Economic Growth," 68–9. More plausibly, Watkins treats transportation systems as backward linkages from the export trade. Export-base theory (although not always connected to staples) remains important in the economics of regional development. See also

Ommer, "The Decline of the Eastern Canadian Shipping Industry, 1880–95."

16 Fay and Innis, "Maritime Provinces," 663.

17 McClelland, "The New Brunswick Economy in the Nineteenth Century," 231.

18 Fouquet, *Heat, Power and Light: Revolutions in Energy Services*, 173–4.

19 On freight rates, see Sager and Panting, *Maritime Capital: The Shipping Industry in Atlantic Canada, 1820–1914*, 108–11; North, "Ocean Freight Rates and Economic Development, 1750–1913"; Harley, "Aspects of the Economics of Shipping, 1850–1913." Canadians, including those in Atlantic Canada, participated in the technological transition from wood and sail to iron and steam, and ocean-going merchant fleets expanded again, especially during and after the First World War and during the Second World War. The rejection of the older technological determinism is central to Sager and Panting, *Maritime Capital*.

20 This paragraph summarizes Sager and Panting, *Maritime Capital*, 176–84.

21 Ibid., 195.

22 Ibid., 54–70.

23 Sager, *Seafaring Labour: The Merchant Marine of Atlantic Canada 1820–1914*, 204.

24 Porter, "'She Was Skipper of the Shore Crew': Notes on the History of the Sexual Division of Labour in Newfoundland."

25 Catharine Parr Traill, *The Backwoods of Canada*, 193. Susanna Moodie often refers to wind and its effects, but see especially *Roughing It in the Bush*, 189, 487–93.

26 Traill, *Lost in the Backwoods: A Tale of the Canadian Forest*, 114.

27 Story, Kirwin, Widdowson, *Dictionary of Newfoundland English*.

28 Sager, *Seafaring Labour*, 58–65. The classic study of the history of sailing ship technology and ship-handling is by a Canadian: Harland, *Seamanship in the Age of Sail*.

29 Sager, *Seafaring Labour*, 84–6.

30 Hearn, *Tracks in the Sea: Matthew Fontaine Maury and the Mapping of the Oceans*.

31 This is the theme of Sager and Panting, "A Culture of Entrepot Growth," chapter 8 in *Maritime Capital*.

32 Protection for colonial shipbuilding and ship-owning did not end with the Navigation Acts: Sager and Panting, *Maritime Capital*, 70.

33 Laperle, *Le grain, le meule et les vents: Le Métier de meunier en Nouvelle-France*, 41–8. See also De Blois, "Les Moulins de Terrebonne (1720–1755) ou les hauts et les bas d'une entreprise seigneuriale."

34 Greer, *Peasant, Lord and Merchant: Rural Society in Three Quebec
    Parishes 1740–1840*, 129–33. See also Harris, *The Seigneurial System in
    Early Canada: A Geographical Study*, 72–5. Louise Deschene states that
    the ten or so larger, well-maintained commercially oriented grist mills
    (*minoteries*) that emerged in the second quarter of the eighteenth century
    to supply the nascent trade in wheat exports were watermills with two,
    three, or four wheels, located on rivers. The remaining seigneurial mills in
    the colony, many of which were windmills, provided indifferent service to
    the peasants who frequently complained to the intendant. Dechêne, *Le
    Partage des subsistances au Canada sous le Régime français*, 33, 35.

35 Williams, *Energy and the Making of Modern California*, 70–9.

36 The count of wind-power establishments is from the *Census of Industrial
    Establishments, 1871*, Canadian Industry in 1871 Project (CANIND71),
    University of Guelph, Ontario. I am indebted to Kris Inwood for his assis-
    tance in providing the data. According to the Bloomfields, 158 establish-
    ments in Canada in 1871 reported using water power with a force of
    100 horsepower or more; these accounted for 17.9 per cent of all industrial
    establishments. Bloomfield and Bloomfield, *Patterns of Canadian Industry
    in 1871: An Overview Based on the First Census of Canada*. Wind power
    was of so little importance that it does not appear in their summary tables.

37 Data as of March 2013, provided by Peter Baskerville.

38 Righter, *Wind Energy in America*, 25–6.

39 The tables of industrial establishments in the published census volumes
    do not have a separate category for windmills; the category is "pump and
    windmill factories," but many of these may have manufactured steam
    pumps. There were 305 such factories in 1891: *Census of Canada, 1891*,
    vol. 4, table 7. By 1911 this number had fallen to 29: *Census of Canada,
    1911*, vol. 3, table 1.

40 "When the Nor'wester, a Canadian journal, in 1937 offered to provide
    readers with instructions on how to build a small (6-volt) wind plant, cor-
    respondents swamped the editors with requests." Righter, *Wind Energy in
    America*, 100–1.

41 Ibid., 92, 94.

42 Ibid., 113–17, 123–5.

43 Natural Resources Canada, *Stand-Alone Wind Energy Systems: A Buyer's
    Guide*, 8.

44 "Market for Small Wind Turbines in Canada." See also World Wind
    Energy Association, *2015 Small Wind World Report*.

45 The *Census of Assiniboia 1849*, table 5 (Implements, Mills), reports eigh-
    teen windmills and one watermill in Red River. The *Census of Assiniboia*

1856 (table 5) reports eighteen windmills and nine watermills for Red River. Both censuses are in *Census of Canada 1871*, vol. 4, which reprints all Canadian censuses from 1665 to 1871.

46 Kaye, "Flour Milling at Red River: Wind, Water and Steam."

47 On the first commercial machines, see Righter, *Wind Energy in America*, 66–72.

48 Musgrove, *Wind Power*, 132–3. Musgrove cites the study by the European Wind Energy Association, *Wind Energy in Europe*.

49 Tudor, "A Brief History of Wind Power Development in Canada, 1960s–1990s."

50 Musgrove, *Wind Power*, 96; South and Rangi, *The Performance and Economics of the Vertical-Axis Wind Turbine Developed at the National Research Council, Ottawa, Canada*.

51 Report of the Auditor General of Canada, 1983, chap. 9, s. 9.11.

52 Tudor, "Brief History of Wind Power," n.p.

53 Defenders of VAWTS argue that they are strong, quiet, bird-friendly, and relatively easy to maintain. They can be packed more closely together in wind farms. Researchers at the California Institute of Technology have claimed that a type of VAWT with optimal placement in a wind farm can yield a significant, even tenfold, increase in output: Svitil, "Wind-Turbine Placement Produces Tenfold Power Increase, Researchers Say."

54 Installed wind-power capacity was negligible in the mid-1990s. On the growth of wind power, see also MacDowell, *An Environmental History of Canada*, 184.

55 Rosano, "On with the Wind: Wind Energy in Canada Timeline."

56 Windfacts, "Why Wind Works."

57 The leading countries in 2012 were China (75,564 MW), the United States (60,007), Germany (31,332), Spain (22,796), India (18,421), United Kingdom (8,445), Italy (8,144), and France (7,196). Global Wind Energy Council, "Global Wind Statistics 2012."

58 CanWEA, "Small Wind Energy."

59 CanWEA, "Big Growth in Small Wind: CanWEA."

60 CanWEA, *Wind Vision 2025: Powering Canada's Future*.

61 In 2009 the Danish wind company LM Glasfiber was exporting blades to Brazil and the United States from its plant in Gaspé. Lorinc, "On with the Wind."

62 Battagello, "Wind Turbine Plant Shuts Down Windsor Operations." The ecoEnergy for Renewable Power Program, begun in 2007, helped to finance 4,000 megawatts of wind energy capacity in two years. Lorinc, "On with the Wind," 27. As of 2013 the federal government was committed, under its

Climate Change Plan for Canada, to purchasing 20 per cent of federal government electricity from renewable energy sources such as wind and solar.

63 In 1979 the Danish government offered a subsidy that was worth 30 per cent of the turbine costs. Most purchasers of turbines were individuals or co-operatives. The program was so successful that the subsidy was gradually reduced and then eliminated in 1989. Musgrove, *Wind Power*, 125. The California "wind boom" of the 1980s was stimulated by capital investment tax credits and by federal legislation encouraging renewable energy. The Public Utilities Regulatory Act of 1978 required utilities to buy electricity from "qualifying facilities" – those that produced power from biomass or renewable energy sources. Ibid., 112. The German support was even more dramatic. The 1990 Electricity Feed-In Law required that utilities connect to renewable energy sources and to buy the output of wind energy installations at a price equal to 90 per cent of the average selling price to retail consumers. Ibid., 171.

64 When governments invest and publicize their support, is this not a tacit admission that government support is necessary? Canada, Natural Resources, "Government of Canada Invests in Wind Energy in Southern Ontario."

65 Steward converts energy consumption on a "fuel equivalent basis," and for wind power he follows the precedent of J. Frederic Dewhurst and Associates, *America's Needs and Resources: A New Survey*; Steward, "Energy Consumption in Canada since Confederation." "The work of wind was calculated from the number of sailing ships and their gross tonnage" (239). There is an obvious problem: the number and tonnage of ships is not the same as work. Work is the volume of tonnage times the nautical miles covered times the volume of cargo carried. A meaningful estimate of the work of sailing ships, in equivalencies that allow comparisons with other energy sources, is probably impossible. In the case of windmills, Steward admits that "no data were found for the number of windmills operating in Canada at the time." Ibid., 245. It follows that the estimate of wind power over time in his graphs is speculation.

7

Water Power before Hydroelectricity

Jenny Clayton and Philip Van Huizen

Canada is a water-rich country. More than 7 per cent of its surface area is composed of water, which, in combination with its glaciers, makes for the highest volume of fresh water anywhere in the world, or about 20 per cent of the global supply. A vast system of rivers and lakes, both great and small, stitches the country together, connecting all three of its oceans and providing natural transportation corridors that have been used for thousands of years.[1] This relative abundance has made water a fundamental aspect of the energy history of Canada for as long as humans have called the area home, from pre-contact to Confederation, and during every energy era, from the ages of muscle and wood to the present hydrocarbon era.[2]

In fact, if water power is defined in the most inclusive manner possible, that is, the use of water in any of its chemical states (solid, liquid, or gas) to fuel or perform any kind of work, water is one of the most important and longest-used forms of energy in Canada's history. Ice has been used to control temperature, reduce friction, and break rocks; pressurized steam has been used to pump pistons and turn wheels; liquid water has moved people and their products, turned wheels to run mills, factories, and produce electricity, and washed away unwanted waste and pollution – to say nothing of the fact that water also plays an essential role in sustaining humans and animals alike, not least by acting as the transportation system for nutrients throughout the body.

No single chapter could examine so much in any meaningful fashion. In the interests of both brevity and comprehensiveness, and working within the limits of research on water power in Canada more generally, we focus on the use of water in its liquid state only. Specifically, this chapter is concerned with how the combination of gravity and water has been used to move humans, their machinery, and their products during the period after European contact to the twentieth century. More research needs to be conducted on Indigenous uses of water power before European colonization, and hydroelectricity, the dominant use of water power in the twentieth century, is covered in its own chapter in this volume by Matthew Evenden and Jonathan Peyton. It is important to note from the outset, though, that there are many other uses of water in the energy history of Canada, and that available research into water power really only scratches the surface.

With that being said, to proclaim that water power has been a vital aspect of Canada's energy history, even within our limited definition, is not to say that it has always been equally important everywhere. Like all power sources, the choice and ability to harness water power depended on a number of different factors. Perhaps the most determinative has been the location of waterways. Some historians have argued, for example, that various waters provided a transportation corridor that would literally shape what would later become Canada.[3] Indeed, in 1850 Thomas Keefer, a member of British North America's engineering elite, declared that this configuration was evidence of God's benevolent plan for Britain's colony: "When the Almighty Maker of the Universe 'poured the rivers out of the hollow of His hand,' He gave them that direction which should ultimately ensure the greatest good to the greatest number … we find it impossible to propose any more advantageous position for the St Lawrence than that which was given it 'when the waters were divided from the waters.'"[4]

Canada's wealth of water, though, belies the fact that there have been considerable limitations on its use for power. Not every waterway was navigable or easily harnessed to produce power, and almost all of them froze for good parts of the year, completely halting most uses, especially before technological innovations allowed turbines to turn under a frozen surface. Furthermore, natural limits and configurations would dictate where water mills and later entire factories would be located, which had a considerable impact on power relationships between people, as some were able to harness the potential power of water while others were not. On the Prairies, water typically did not flow with enough power or drop from sufficient heights to even allow for extensive mill set-ups, so direct-drive water power, generated by non-hydroelectric water wheels, was not nearly as prevalent in this part of Canada as it was elsewhere.

The limiting impacts of both location and seasonality were ameliorated by technology. Indeed, each period of transition in Canada's water-power history has been marked by a different suite of technology that both increased the amount of power that people were able to produce from moving water and lessened the constraints of location and seasonality. Rather than seeing water power as transitory or confined to an earlier age – as it is popularly conceived – this chapter argues that the potential energy of liquid water has been constantly harnessed to differing degrees throughout Canada's past, based on changing variables that have been mostly technological. Indeed, as Josh MacFadyen argues for firewood in chapter 5 in this volume, the continuing importance of water power in Canada's history muddies the stark distinctions that are often made between our fossil fuel–based present and the organic sources of energy that were predominantly used in the pre-industrial world.[5] Water power not only runs through the two eras, it has been essential to both, complicating any attempts at neat dividing lines between the two energy regimes.

Lastly, as this chapter shows, changes in technology resulted in new impacts and scales of consequences that affected different groups of people in different ways. Sawmills, factories, and mining operations polluted waters, often adversely affecting water quality, marine life, and the people who relied on both, including those in the fishing industry. The dams and canals that served these industries resulted in complex changes to local environments, altering aquatic ecosystems and flooding out residents, as new forms of water power generated wealth for people far away. Moreover, such impacts would only be repeated into the twentieth century, as water power would continue to be a key aspect of Canadian energy production with the spread of hydroelectricity.

FROM WATER WHEELS TO TURBINES: THE TECHNOLOGY OF DIRECT-DRIVE WATER POWER AFTER COLONIZATION

Although it is not the focus of this chapter, it is worth noting that water power was essential to First Nations communities before European contact. Waterways were of course important transportation routes – in some places more so than overland trails – and First Nations groups built a variety of vessels from different materials to make the best use of these travel routes. The power of moving water was also used to harvest food. On the West Coast, for example, fishers set up nets so that tidal currents would bring in salmon or eulachon, and built fences and stone walls to trap fish at the mouth of a river once the tide had lowered.[6] Coastal groups also practised mariculture, growing clam gardens in estuaries. Hemmed in with rock walls, these gardens were nourished with sediment brought down by the river and nutrients deposited by the tide.[7] Undoubtedly

First Nations groups used water power in many other ways as well, and more research needs to be done on this aspect of Canada's water-power history.

European colonists adopted certain innovations from Indigenous communities, such as the birchbark canoe for the fur trade, but they also brought a whole suite of water-power technologies with them to the New World, most notably the water wheel. Evidence of the first water wheels goes back to 15 BCE when Roman architect Vitruvius published a tract on his use of a water wheel to power a gristmill, and water wheels were commonplace in Europe by the time colonists began crossing the Atlantic. More than 5,500 mills, the majority powered by water, were listed in Britain in the Domesday Book in 1085.[8] In France there were water wheels in use for irrigation from 1100, iron mills from 1116, and sawmills from 1250.[9] The application of direct-drive water power to power machinery was thus introduced to Canada with the earliest European settlement to process grains for consumption and wood for shelter. Establishing mills was also a means to exert physical and symbolic control over Indigenous land and resources.

As old as the technology for harnessing water power was, its basic configuration had changed only slightly by the time it was introduced by French colonists to northeastern North America. In fact, this would remain the case until the nineteenth century when iron and steel components replaced wooden parts, and in many places "traditional" uses of water power would persist alongside industrialized technologies and applications. Mills, whether they ground grain or sawed wood or performed other functions, required a mill dam to raise the water level and ensure a supply of potential energy. Normally, mills would be built at a short distance from the mill dam to control the flow of water and avoid damage from floods, and water flowed from the mill dam through a headrace to the wheel at the mill. Before the introduction of turbines, water mills could use a horizontal tub wheel connected to a vertical shaft that turned a parallel millstone above, or a vertical wheel attached to a horizontal shaft, which would then turn a millstone, or, through a series of gears and levers, other types of simple machinery. Depending on the fall of water at these sites, the vertical wheel would be of the overshot (water hitting the top of the wheel), breast (water entering the wheel around the halfway point), or undershot type (water entering at the bottom of the wheel). After passing by the mill, the water drained to the watercourse, or into another mill, through a tailrace.[10]

The mechanization of a variety of tasks that took place around the millstone became common after the publication of *The Young Millwright and Miller's Guide* by American inventor Oliver Evans in 1795. Evans's inventions, propelled by water power or the weight of flour, moved the flour around the mill and cooled it for storage. These inventions included an elevator to raise grain or flour, a conveyor to move it horizontally, a drill that raked flour from one machine to another, a descender that was powered by the weight of the flour as it fell out of the elevator, and a hopper boy that raked and cooled flour quickly so that it did not need to be stored for twelve hours before being packed in barrels. This was an essential step, as grain ground for export had to be of a high quality and dried enough to be transported and stored for a year. Some of Evans's inventions were installed in Upper Canadian mills within five to ten years of the publication of his book.[11]

The system of mechanization that Evans devised for gristmills peaked in the mid-nineteenth century with the textile mill complexes in New England, most notably in places like Lowell, Massachusetts, a textile factory town founded along canals adjacent to the Merrimack River in the 1820s. Mills in New England added to Evans's innovations by perfecting belt and pulley systems that transferred power from water wheels to machinery throughout the factory. Belts were made of leather, predominantly from the booming buffalo fur trade on the Prairies. Belts and pulleys were more efficient than traditional gearing systems that were more difficult to install, tended to break down, and lost power through friction inefficiency. Belts were also significantly quieter than gear-and-shaft systems.[12]

Such large, water-powered mill complexes used more power than pre-industrial water-powered mills were capable of producing. Increased power came from technological refinements and inventions that made water-power locations like that of Lowell some of the most sought after industrial sites in nineteenth-century North America. An early innovation was the use of iron in water-wheel construction rather than wood, particularly for factory use, during the first half of the nineteenth century. The superior strength of iron and the greater precision with which it could be installed increased both the efficiency and the amount of power that water wheels generated. Even the best vertical wooden water wheels, for example, converted no more than 60 per cent of the potential energy of falling water, with the rest lost to friction. These wheels produced about 20–40 horsepower, which was a far cry from the 50–500 horsepower that large iron water wheels – some of which reached diameters of up to ninety feet – were capable of producing, operating at efficiency ratings of around 80 per cent.[13]

The use of iron in water-wheel technology led directly to an even more significant innovation for direct-drive water power during the nineteenth century – the water turbine. The water turbine was essentially an offshoot of the horizontal wheel, or "tub wheel," an ancient type that lay flat in the water, with a vertical shaft that was generally connected directly to a grist wheel, without the use of gearing. Horizontal water wheels converted energy at a very low rate of efficiency of around 30 per cent, or enough to turn a single grist wheel, but hardly generating more power than what was possible with a draft horse. But horizontal wheels were smaller, cheaper, and easier to install, so they remained popular in smaller mills in certain areas of the world, such as France and the United States.

Experiments with horizontal wheel design in such mills eventually led to the invention of the water turbine in France by Benoît de Fourneyron in the late 1830s. Like the horizontal water wheel, turbines turned on a vertical axis, but at much higher rates. The immense increase in power was made possible by advances in hydraulic engineering and ironworking that manipulated water pressurization and flow within the turbine, thereby reducing the amount of energy lost to friction as the water exited. Fourneyron's first commercial turbine, for example, operated at an efficiency rate of 80 per cent, turning at more than 2,000 rpm and producing sixty horsepower. Its usefulness compared to the much larger vertical water wheel was obvious, for it was smaller (often only thirty to sixty inches in diameter), cheaper, more efficient, and, most importantly for Canada, able to operate completely submerged underwater, including underneath ice.[14]

Some promoters predicted that the invention of the water turbine, in combination with the surplus of water-power locations in Canada, would propel the colony onto the world stage of industrial capability. Alexander Somerville declared in 1860 that water turbines "shall be planted in thousands and tens of thousands at the foot of cataracts and cascades, and in the neighbourhood of the rapids, then shall Canada flourish! Then will she, by the cheapness of motive force, afford higher wages to the operative workers than is afforded by the costly steam engine; then will multitudes of settlers come; then may protective duties be useless."[15]

Somerville's dream of Canadian manufacturing supremacy did not come to pass, but water turbines did become a key component of the country's industrial energy sector, and eventually proved to be an essential aspect in the production of hydroelectricity in the twentieth century.

Although the technology originated in France, North American innovations, such as altering the flow of water both outside and inside the turbine and curving turbine blades to create a vacuum as the water exited to further reduce friction, greatly increased the power capabilities of turbines. Amongst the most popular North American models were the US-made Leffel, Tyler, and Francis turbines, and the Canadian-made "Little Giant."[16] By the end of the nineteenth century, Canadian industries were powered by hundreds of water turbines, some of which reached efficiency ratings of more than 95 per cent and produced as much as 5,000 horsepower, or ten times the power of the best vertical water wheels, although the average amount was closer to 150 horsepower.[17]

THE USE OF WATER-POWER TECHNOLOGY IN CANADA

Social organization, topography, and available technology affected the use of direct-drive water-power technology in different regions at different times. Early European expeditions to North America identified sites with good potential water power in surveys, and settlements grew where they could take advantage of water power, particularly to drive gristmills and sawmills. Such mills became central features of villages. Indeed, their location often determined where towns and villages would be established. Mills also served as trading places where other services sprang up, and their presence increased the value of land. Even road patterns could be influenced by the location of water-powered mills. As Carol Priamo has argued for Canada, "The mill was often the first non-domestic building erected in the new settlement. Its existence led to community development, and attracted settlers and merchants and craftsmen who benefited from the economic progress. It was, therefore, both a cause and an effect of economic growth."[18]

Jacques Cartier asked for material to build windmills, water mills, and hand mills so that members of his third voyage in 1541 would have a way to grind wheat, a Eurasian import, into flour. Cartier's settlement was unsuccessful, and the earliest known water mill in what would become Canada was built at Port Royal in 1607 as a result of Samuel de Champlain's colonization efforts.[19] Seigneurs in New France were required to provide communal mills for grinding wheat into flour, and in return they kept one-fourteenth of the grain ground as payment. Since wheat was the main crop in early Canada, gristmills were essential. They were not necessarily powered by water – more than half of the mills built between 1637 and 1687 were windmills, as Eric Sager points out in this volume, because they did not need to be located next to running water, nor did they stop with ice jams. During the colony's first century, population growth was small, capital was limited, and New France produced grain and timber mainly for local consumption. By 1666 there were eleven water mills in New France, nine gristmills, and two sawmills.[20] New France experienced a period of peace from 1713 to 1744, and the colony began to export lumber and grain to Louisbourg and the West Indies.[21] British seigneurs built mills soon after the Conquest of New France in 1760 and Loyalists established mills at sites in the Eastern Townships starting in the 1790s. Water-powered industries increased in the early nineteenth century, so that Quebec took the lead in industrial production in British North America by the 1840s, most of it driven by water power, only to be surpassed by Ontario in 1851.[22]

The British Crown built some of the earliest mills in Upper Canada following the influx of Loyalists from the United States, including sawmills and gristmills in Niagara in 1783, in Kingston by 1784, on Mohawk lands by the Grand River by 1791, and on the Napanee River by 1787. New settlers also received free seed wheat from the government to establish their farms. Sawmills were usually built first, so that sawn timber was available for building the gristmill. In response to popular demand, settlers in Upper Canada were allowed to own mills privately from around 1788.[23] Mill sites became the nucleus of settlement and transportation routes. Canals, such as the Welland, began leasing water power to industries after the 1820s. Census returns indicate that by the 1850s, Ontario likely had more water-powered mills than other colonies of British North America.[24]

In the Maritimes, the geography of Nova Scotia was particularly well suited for numerous small mills that could be situated close to raw materials, primarily grains, wool, and wood. Indeed, records indicate that in 1861 Nova Scotia had more watermills, although not necessarily more capacity for production, than Ontario, Quebec, or New Brunswick. It is possible that the availability of water-power sites delayed the development of heavy industries in Nova Scotia, which had an abundant supply of coal, but was exporting it to central Canada.[25] Prince Edward Island had a less mountainous topography than Nova Scotia but residents took advantage of numerous mill sites, with the number of mills declining after the construction of the Prince Edward Island Railway in 1871.[26]

In both central and eastern Canada in the nineteenth century, water-powered mills played an essential role in the sawmilling industry.[27] Early sawmills used a single reciprocating blade attached to the wheel by a crank that moved the blade up and down. A simple mill with a single blade could cut about five hundred boards in a day. Single-blade mills were sufficient for local use, and more efficient mills with a gang of saws (several saws working side-by-side to cut the same piece of wood into smaller boards), produced lumber for export. Graeme Wynn estimates that in the early nineteenth century, large mills in New Brunswick produced up to twenty thousand feet of lumber per day.[28]

British North American timber exports skyrocketed after Napoleon blocked British access to Baltic timber in 1808 and grew again after the 1854 Reciprocity Treaty with the United States.[29] The number of sawmills in eastern Canada, particularly in the Ottawa Valley, New Brunswick, and Nova Scotia, increased dramatically as a result. By 1871, for example, there were more than 1,100 water-powered sawmills in Nova Scotia alone (with an additional 18 powered by steam), more than anywhere else in Canada.[30] In New Brunswick, hundreds of mostly water-powered sawmills were producing more than 500 million board feet of lumber each year by the end of the nineteenth century.[31]

Although the majority of sawmills were small-scale operations that still used a combination of human, animal, and traditional water-wheel technology, large-scale, fully mechanized operations also flourished, dwarfing smaller operations in production. According to the 1871 census, for example, 68 per cent of the sawmills in Nova Scotia produced fewer than 100,000 board feet per year, while only 2 per cent produced more than one million board feet per year. The largest, owned by E. D. Davis and Sons on the LaHave River, annually processed nearly eight million board feet, employing 114 workers and using water turbines that produced 200 horsepower.[32] The increased power available to such large-scale mills meant that every stage of lumber manufacturing could be mechanized, from conveyer belts and elevators to move the timber throughout the mill, to the use of high-powered circular saws, which had come into use for crosscutting and edging by 1818 and for cutting lengthwise by about 1850, rather than the single reciprocating saw blades that smaller mills used. Band saws further increased the amount of lumber cut and became common in large Canadian sawmills by 1890. Mills also diversified as they grew bigger, producing a wide range of wood products, including doors, sashes, staves, and pulp and paper. By the 1890s, Canada was home to some of the largest sawmills in the world, most of which still ran on water power. J.R. Booth's mill, one of several large-scale operations that converted the power of the Chaudière Falls on the Ottawa River, for example, had the highest production capacity in the world in

1892, with installed water turbines capable of producing 4,000 horse-power working to produce more than 140 million board feet of lumber per year.[33]

Although grain production and sawmills remained the principal users of water power throughout the nineteenth century, textiles such as wool, linen, and jute were also processed by water power in Canada from the late eighteenth century, and cotton and silk mills were in operation from the nineteenth century. Early mills met local needs, for example carding wool, which would then be spun and woven or knit in the home. By the 1820s for wool, and the 1840s for cotton, Canadian factories were adopting systems developed in Britain and the United States that mechanized the range of steps required to process fibre into cloth.[34]

Water power was also used for working metal, as gears and shafts attached to water wheels drove bellows in the blast furnace, hammers in the forge, and machines that manufactured metal products. An early example was the St Maurice ironworks. Located upriver from Trois-Rivières, the St Maurice Forges began to produce iron in 1738, and it remained the only ironworks in Canada until 1800. Originally, two water wheels were built on the St Maurice Creek by 1738. Under engineer Chaussegros de Léry, construction was undertaken in 1739 to reuse the water and build a series of mill dams with adjacent water wheels. By 1741, the site had three mill dams and seven water wheels. Other mills were added over time, including a stamp mill, sawmill, and two grist-mills. The ironworks remained in operation until 1883.[35] In the nineteenth century, machine shops and foundries on the Welland and Lachine Canals used water power in addition to other forms of fuel to manufacture tools, nails, machines, carriages, boilers, and other metal products.

Water power was not nearly as important on the Prairies as in eastern Canada. British and French fur-trading posts set up mills for local use, but prairie rivers were sluggish and did not fall as dramatically as eastern rivers did, and thus settlers did not use them as often to power mills. Rather, wind and both animal and human muscle power were generally chosen instead, although water-power sites were developed. One example of the temporary role of Prairie water mills can be found at Red River. The first mills at Lord Selkirk's Red River settlement, founded in 1812, were operated by hand and by horse. The first windmill in the area was completed in 1825, and windmills predominated in the 1840s. To process grain in times of year when there was often insufficient wind, settlers built water mills from 1830, but they were not common until the mid-1850s. Freezing in winter, summer droughts, a lack of quick-flowing streams, and the periodic excess of water during spring flooding put water mills at a disadvantage. The first steam mill was built in the area in 1856. As wheat farming became a dominant economic activity in the new province of Manitoba after 1870, wind, water, and steam still ground flour, but soon many of the smaller mills powered by wind and water fell into disuse.[36]

British Columbia, on the other hand, had substantial potential water power, but it was not always easily accessible or in proximity to fertile soil. The Hudson's Bay Company began to establish mills starting in the 1820s, and forts in Victoria, Langley, Hope, Kamloops, and Alexandria, for example, all installed water-powered mills from the 1840s to the 1860s. As in the east, these mills were initially set up to process grains and lumber. As Wallace Liddicoat has pointed out, however, since steam power was readily available by this time, and wood and coal to fuel them was plentiful, water power was not as predominant in British Columbia's mid-century mills as it was in the east, amounting to fewer than twenty known water-powered mills by the time the colony joined Confederation in 1871.[37]

Water power was prevalent in the western mining industry, however. The influx of miners working their way north from California into Indigenous territories along the Fraser River in 1858 brought a range of water-based mining technologies that they applied to waterways and gravel bars to extract gold. The simplest methods of washing gold from surrounding sediment were the human-powered gold pan and the wooden rocker. To process more gravel, miners redirected natural waterways through ditches, flumes, and pipes to achieve a continuous flow of water through sluice boxes. Miners also used water wheels in the British Columbia gold fields to pump water into sluices and out of shafts. By 1859, a thirty-foot water wheel with buckets, called a Chinese pump, lifted water to a sluicing system.[38] Cornish water wheels, powered by water from an overhead flume, were employed in the early 1860s around Barkerville. The water wheel raised and lowered a beam, which pumped groundwater from the shaft, allowing miners like Billy Barker to search for gold deep underground.[39] Other mining industries also employed water power. In the early twentieth century, for example, copper smelters at Brittania Beach and Granby, British Columbia, used direct-drive turbines to force compressed air through Bessemer converters.[40]

During the Yukon gold rush from 1896 to 1899, miners applied their own labour and the energy from wood and water to "disassemble" ecosystems for gold.[41] The climate and environment of the Yukon, including short summers, limited forests located along the edge of rivers, and permafrost, created unique conditions for mining. In the winter, miners sank shafts and set fires so that gravel and mud could be thawed enough to be shovelled out. The gold could not be separated from the dirt until waters flowed in the brief spring and summer. Similar to miners on the Fraser in the 1850s or in the Caribou in the 1860s, Yukon gold miners built dams and reservoirs, dug ditches, and set up flumes to direct the spring runoff to their sluice boxes. As Kathryn Morse writes, gold miners "strung long chains of flumes that crisscrossed the creek valleys and gulches, carrying water from sluice to sluice. The aboveground scaffolds elevated the creeks themselves into high wooden pathways, a mine-to-mine, crazy-quilt circulatory system."[42] Miners depended on running water, sometimes paying for access to dams, pumps, and flumes, and were unable to proceed when waiting for the spring runoff or when creeks dried up.[43]

Lastly, for growing cities in nineteenth-century Canada, public water-supply systems became increasingly necessary to provide clean drinking water and protection against fire, especially where many of the structures were built of wood.[44] Depending on a city's position in relation to its source of water, water power was often used to pump water up to a reservoir. Such was the case in more than thirty-five Canadian cities by 1912.[45] In 1852, for example, the city of Montreal commissioned Canadian engineer Thomas C. Keefer to design a new waterworks system. Keefer had experience with water power from his work as assistant engineer on the Welland Canal from 1840 to 1845 and as principle designer of timber slides on the Ottawa River.[46] Powered by water from the canal, water wheels pumped water to a reservoir on Mount Royal.[47] However, an open canal was not suited to Montreal's cold winter – water in the canal would freeze and ice would slow the flow, reducing the effectiveness of the hydraulic pump. Yet the waterworks made do with water power until 1868, when steam pumps were reintroduced into the system as backup power.[48]

RIVERS, LAKES, AND CANALS

Waterways provided convenient transportation routes to move people and goods. Even when energy was required to propel vessels, moving on a fluid surface made transportation less energy-intensive than going overland. The fur trade travelled inland via rivers and lakes, with the voyageur as "the motive power of this transportation system."[49] The power of water was a force that was often in opposition to paddlers, but could sometimes be harnessed or at least minimized using a variety of strategies. As George Colpitts points out in chapter 2 of this volume, voyageurs expended enormous effort to paddle upriver, driven by the energy of first a corn-based diet, then pemmican. As with other forms of interior water transportation, operations were limited to ice-free months. Especially as the fur trade moved farther inland, furs needed to be moved quickly over extensive distances, leading to long and strenuous days of paddling. Fur traders adopted Indigenous technology, particularly the birchbark canoe for freighting, and used the York Boat after 1821 for lakes and rivers with minimal portages. The more durable York Boat required less crew, could be rowed, and could carry a larger sail than a canoe.[50] Fur traders were aware of and tried to work with, rather than against, currents and high water. For example, David Thompson was frustrated that his guide would lead his crew upriver in a period of high and quickly flowing water in June 1801: "Whoever wishes to attempt to cross the Mountains for the Purposes of Commerce ought to employ a Canoe, & start early in the Spring, say the beginning of May, from the Rocky Mountain House, the Water for that Month being low & the Current not half so violent as in the Summer."[51]

Just as furs moved to markets by river, so did timber. Initially, the transportation of timber relied on pre-existing waterways, as well as human power, animal power, and natural cycles involving the freeze, thaw, and flow of water. Fellers cut logs, and teamsters transported them over frozen sleigh roads to the water's edge during the fall and winter. High water in the spring carried logs down the river. Sometimes splash dams were built to retain additional water to help propel the logs. Log drivers stood on the banks and kept logs moving using pike poles and, after it was invented in Maine in 1858, the peavey hook. Log drivers also navigated around the logs using a pointer boat invented by Canadian John Cochrane. When logs jammed, drivers had the dangerous task of balancing on floating logs to release the key log, or blasting jams with dynamite. Once logs reached a lake, they were boomed and pulled by a steam tug, or winched towards an anchor by animal power.[52]

Rivers in some regions, such as New Brunswick, were better suited than others to transport logs, as rapids occurred more often on rivers in Quebec and Ontario.[53] As a result, logging companies in New Brunswick focused on modifying streams to make them more amenable to the transportation of logs, blasting shallow rapids and large rocks with dynamite, clearing the river of impediments, and building log flumes.[54] Companies in Ontario and Quebec, by comparison, built chutes especially designed to take cribs of timber around rapids. "Single-stick slides," wooden flumes that carried one log at a time down steep inclines, were introduced to Canada from Norway and Sweden. Ruggles Wright, a Canadian lumber merchant, adapted these slides to accommodate a crib of timber, building a twenty-six-foot-wide timber slide on the north side of Chaudière Falls at Bytown (later Ottawa) in 1829. A sheet of water flowed over the slide, enough to carry a crib safely from one water level to the next. For a fee, lumbermen could use timber slides to avoid the risk of breaking cribs that was inherent in shooting them over the rapids, and save the time involved in taking cribs apart and transporting them around the falls by wagon. Other lumbermen soon built similar slides, following the twenty-six-foot width as a standard.[55]

While providing a safe passage for cribs, slides worked only in one direction and were not as useful to lumbermen as canals, which allowed goods to travel upriver as well as downriver. Lumbermen benefited greatly from canals on the Ottawa and Rideau Rivers, transporting finished lumber to markets downstream and upstream, accounting for as much as 80 per cent of canal tonnage up to the First World War.[56] Canals were built in Canada originally as short transportation routes to bypass rapids and avoid the need for portaging, as a way to "improve" upon the natural transportation advantages that Upper and Lower Canada's many rivers and lakes provided. As settlement expanded westward after 1760, for example, the Corps of Royal Engineers of the British Army built four short canals at rapids above Montreal from 1779 to 1783.[57] Throughout the nineteenth century, canal construction expanded exponentially in Canada, the result of improvements in engineering and construction techniques, as well as increased lobbying pressure to both strengthen Canada's defensive position against the United States and to capture the North American grain trade for export to Britain.

Following the War of 1812 and the construction of the Erie Canal in 1825, which opened up an American transportation route from the Great Lakes to the ice-free ocean port of New York, British North Americans constructed a number of canals connecting to the St Lawrence–Great Lakes system. Over the course of the 1820s, the British government constructed canals along the Rideau and the Ottawa Rivers to connect Upper and Lower Canadian military posts. On the St Lawrence, the Lachine Canal was built through Montreal above the Lachine Rapids, while others were built to bypass rapids farther upriver, including the Beauharnois, Cornwall, and Williamsburg Canals. The Welland Canal, Canada's most aggressive example of canal construction until the completion of the

St Lawrence Seaway in the mid-twentieth century, bypassed Niagara Falls and connected Lake Erie with Lake Ontario with four different canals built from the 1820s to the 1920s. The Trent-Severn Canal system connected Lake Huron to Lake Ontario through a series of canals, rivers, and lakes. By the mid-nineteenth century, nearly ten thousand vessels annually carried more than a million tonnes of freight on these canals.[58] This was a far cry from the two million tonnes of freight that the Erie Canal alone transported in the United States by the 1850s, which had captured the North American export trade so coveted by Montreal merchants, but Canada's canals nonetheless created a robust transportation system that was integral to the development of Ontario and Quebec.[59]

Christopher Jones has argued that canals were an essential aspect of North America's energy history, particularly the shift to a fossil fuel–based economy, connecting "landscapes of energy abundance," such as the coalfields of Pennsylvania, with energy-poor urban areas like Philadelphia and New York.[60] As important as canals were for transporting fuel like coal and wood, though, they were also sources of power in their own right, providing sites where water could be leased to industry to drive water wheels and turbines.

The first canal in British North America built to accommodate water power as much as transportation was the Welland Canal; in fact, it was designed to maximize hydraulic power rather than create the shortest route for vessels to travel.[61] At its inauguration ceremony in 1824, a principal promoter and owner of a flour mill, William Hamilton Merritt, declared that the Welland Canal "will afford the best and most numerous situations for machinery, within the same distance in America; wet or dry, warm or cold, we always have the same abundant and steady supply of water ... A general tide of prosperity will be witnessed on the whole line and surrounding country."[62] Completed in 1833, the Welland Canal dropped 327 feet over the course of twenty-seven miles as it crossed the Niagara Peninsula to join Lake Erie to Lake Ontario. The Welland Canal also converted former villages like St Catharines into industrial centres, and, by 1882, more than fifty mills took advantage of this drop in elevation to generate power, most producing flour, but also lumber, steel products, pulp and paper, furniture, and textiles.[63] In 1911, thirty companies still used direct-drive water power from the Welland Canal.[64]

The successful use of canals to provide water power for industry created a way in which canal construction could further be financed and justified. With the merger of Quebec and Ontario into the Province of Canada in

1841, the state took over the construction and improvement of canals, timber slides, and single-stick slides under the Department of Public Works.[65] To help pay for the cost of enlarging the Lachine Canal in the 1840s, the Province of Canada's Public Works engineer, Alfred Barrett, decided to lease the excess water to factories as water power, creating the first factory district in British North America.[66] Three water-power sites were developed on the Lachine Canal in the 1840s and 1850s, including Canal Basin no. 2 near the Montreal Harbour, St Gabriel Locks one mile upstream, and Côte St Paul Locks, two miles upstream from St Gabriel. At Canal Basin, eighteen hydraulic lots were leased to take advantage of the twenty-foot drop to the St Lawrence River, with mills that manufactured nails, flour, and lumber. A more diverse range of industries was established at the St Gabriel Locks, including sash and door mills, textile mills, and rubber mills, in addition to sawmills and flour mills. St Paul Locks was still a rural location in the 1850s, but attracted a few flour mills and tool-making shops.[67]

The Lachine Canal and its water-powered industries played an important role in the industrial development of Montreal. In 1860, the large flour mills on the Lachine Canal produced 57 per cent of Canada East's flour. In 1861, 91 per cent of the nails made in Canada East were produced in water-powered factories on the Lachine Canal. The canal also contributed to the growth of the Montreal workforce and the city's economy. St Gabriel and the Canal Basin employed 10 per cent of Montreal's industrial workers in 1861. A decade later, the value of goods produced at these sites was the equivalent of 28 per cent of Montreal's output.[68]

Canals also provided an ideal location for the construction of textile mills. The Cornwall Canal, for example, became known for its textile industry when woollen and cotton mills were established there to take advantage of the canal's water power. These companies included the Cornwall Manufacturing Company (established in 1868), which processed wool, the Stormont Cotton Manufacturing Company (1870), and, one of Canada's largest factories at the time, the Canada Cotton Company (1872). By 1882, nine mills along the Cornwall Canal employed 2,161 workers, with textile mills employing 90 per cent of these individuals. For mills that relied on water power, work would cease when ice jams in winter restricted the flow of water into the canal and the water wheel failed to turn. Work also stopped when the canal was drained each spring to make repairs to the locks before the navigation season began. To make up for the seasonal loss of water power, the Stormont Cotton Mill also used steam power during periods of low water in the canal.[69]

WATER POWER AND THE STEAM ENGINE

The combination of technological innovations, numerous waterways, and the flurry of canal construction in the nineteenth century in Canada meant that direct-drive water power outpaced steam engines in industrial output until the end of the nineteenth century. Industrialization is generally associated with the ascendancy of the steam engine and fossil fuels as a power source, beginning with coal and followed by petroleum. But water power, especially in the United States and Canada, was a key component of the energy history of every stage of industrialization as well. As Louis C. Hunter states, "The key innovation in the Industrial Revolution was the mechanization of hand operations and the use of mechanical power to drive the machinery. The kind of power adopted was a secondary consideration, the choice depending chiefly upon such basic factors as availability and cost."[70]

Millers and factory-owners consistently chose water power, even after steam engines became common, since water was readily available in Canada and industries tended to be located near waterways for transportation. Indeed, during the industrial era, substantial effort went into improving both the technology of water power and the configuration of Canada's many waterways, which kept direct-drive water power cheap and prominent into the early twentieth century.

Steam engines first came to Canadian industry in the early nineteenth century, and, although still a novelty, had spread throughout Canada by the late 1820s. Indeed, by 1833, Toronto's *Colonial Advocate* predicted that, in short order, "the steam engine should supersede water power for the use of mills."[71] The reasons seemed obvious. Steam engines greatly improved on the kinetic energy of water by heating it to create steam, and then manipulating steam temperature and pressure that in turn moved a piston up and down to drive other machines. The steam vapour did not freeze in winter and could be installed anywhere, countering the two main drawbacks of (liquid) water power.[72] Steam-engine use grew substantially over the course of the nineteenth century as technology improved and coal prices dropped (see Andrew Watson's chapter 8 on coal in this volume), but steam engines did not fulfill the *Advocate*'s prediction until decades later.

The principal reason for the longevity of water power was cost. Setting up water power required far higher start-up costs than steam engines did, particularly if dams and complicated flume and tailrace networks had to be built, and in such instances adopting steam power made more sense (such as in British Columbia). In eastern Canada, however, the infrastructure was often already in place. In addition, because water turbines were so much smaller and could be fabricated and mass-produced relatively easily, they were substantially cheaper to buy and install than traditional water wheels. In 1891, for example, E.D. Davison built a powerhouse with four turbines for his sawmill in Nova Scotia, including all parts and labour, for $1,000. The water turbines, ordered from the Burrell-Johnson Iron Company in New York, cost only $50 each.[73] Vertical water wheels and steam engines alike that were capable of the same amount of power cost about ten times as much.[74] Steam power also required a constant fuel source and fire insurance, and tended to break down more often. All of this made for substantially higher costs, even into the late nineteenth century. When hydraulic engineer Robert C. Douglas assessed the cost of powering the Montreal Cotton Company mill in 1882, for instance, he calculated that it would cost $58 per horsepower for steam (with coal prices at $5 per ton) compared to $7 per horsepower for water power. [75]

In areas where water-power sites were already developed or easily accessible, water power continued to be the motive force rather than steam (or in combination with it) until the turn of the century, because costs were so much lower in these areas. In the early 1890s, Canada's largest manufacturing industry was still flour processing, its second-largest was sawmilling, and, according to the 1891 census, water power continued to be the driving force for both industries, and for most textile mills, especially in Quebec and Ontario. Where water-power sites were all used up (New Brunswick) or not as developed (British Columbia), and where coal was substantially cheaper, steam power surpassed water power by the 1880s. By 1901, steam power had officially surpassed water power in Canada as a motive force in every province but Quebec. The steam engine's ascendancy was short-lived, however, as electricity and internal combustion engines rendered steam engines obsolete by the 1930s. In some respects, though, the use of water versus steam turbines to produce electric power would re-enact the water versus steam contest over the course of the twentieth century.[76]

ENVIRONMENTAL IMPACTS

Water power had a range of environmental impacts. Research has only begun in this area, though, so this section provides only an outline of the many ways in which water power altered landscapes and ecosystems. Harnessing water power often required more than simply putting a boat in the water or erecting a water wheel below a waterfall to cause it to turn. Rather, notwithstanding the enthusiasm of water-power promoters like Thomas Keefer, who, as we saw in the introduction, proclaimed that God had laid out an abundance of rivers so that Canada would flourish, nature generally had to be altered for optimal power or transportation to be realized. Doing so often entailed rearranging entire landscapes with a wide array of intended and unintended consequences.

A mill was a significant investment, and to build and operate a mill one needed access to capital; millers were tightly tied to the economics and politics of their local communities, therefore, and tended to be part of the elite. Often a symbiotic relationship developed between mill-owners and local farmers, but sometimes millers overstepped the bounds of this relationship in search of power and profit. In Ops Township in Upper Canada in the 1830s, for example, William Purdy built a fourteen-foot dam to power a sawmill and gristmill. Residents of Ops turned against Purdy because his dam flooded farmland and stood in the way of the proposed Trent Canal, which would have enabled farmers to transport their produce to markets. The problems were not simply social and economic: by creating additional swampland, the dam exacerbated a pre-existing problem with malaria that returned in 1841 after a wet spring and a hot summer, and fever spread through the township from July to December, causing dozens of casualties. Local residents, frustrated that legal attempts to lower the height of the mill had failed, attacked and destroyed the flume and the dam in two riots that December.[77]

Nineteenth-century gold-mining technologies used massive quantities of water and also rearranged landscapes. By moving earth with water, ground sluicing and hydraulic mining caused significant erosion.[78] The use of mercury in sluice boxes has left deposits in the Fraser River. The removal of topsoil and the disturbance of gravel beds has affected the salmon population, plant ecosystems, and traditional First Nations economies.[79] The water systems established by placer mining had unintended benefits for later settlers as farmers and cattle ranchers were able to take advantage of them for irrigation.[80] Kathryn Morse observes that in the Yukon, when miners disassembled ecosystems for gold by cutting trees, stripping moss from the surface, and setting fires, they inadvertently hampered the steady flow of water. When landscapes were denuded of vegetation by sluicing, water flowed faster and dried up sooner from creeks, resulting in both flooding and drought.[81]

Mills were also a significant source of pollution, as they often relied on the flow of water to flush pollution downstream. This was true of all mills, but became particularly troublesome in areas that were intensely developed. Sawmills, for example, because of their ubiquity and the amount of waste they produced, were particularly problematic sources of pollution. Water-powered sawmills dumped their waste, including sawdust and other small cuts of wood such as edgings and butts, into the waterway under the mill.[82] Regulations in 1793 recognized that mills affected other river uses and stated that they could not block fish or navigation. By 1799, the King's sawmill on Ontario's Humber River had a "wicker stop" at the dam to prevent fish from being caught in the headrace, which directed the water to the wheel.[83] Initial regulations made little difference, though. For most of the nineteenth century, timber production took precedence over fishing, and water pollution was relatively unhindered until the 1890s.

On the Fundy coast in New Brunswick, naturalist and government official Moses H. Perley observed in 1850 that sawmill dams at river mouths prevented salmon from reaching their spawning grounds. Federal legislation in the 1870s and 1880s disallowed the dumping of sawdust, yet exemptions were made for particular mills. While mills in Albert County continued to dump sawdust at river mouths along the coast, the shad fishery started to fail in the 1880s. Even though the government was concerned about the impact of sawmills on the fish population, the timber industry carried more weight than the fishery. As historian Gilbert Allardyce pointed out, "Fish and timber were once equally abundant, but the exploitation of one natural resource drove out the other." Timber was cut at an unsustainable rate in Albert County, leading to the collapse of the logging and sawmilling industries, after they had annihilated the local fishery.[84]

On the Ottawa River, dumping sawdust solved a problem for millowners, but created new problems for fish and shipping. An 1877 report estimated that 12,300,000 cubic feet of sawdust was dumped into the Ottawa River each year. Sawmill refuse obstructed navigation, released gases while decomposing, affected public health, and threatened fish spawning grounds. Methane off-gassing caused explosions in the river, endangering boaters. Public pressure from fishermen and the city of Ottawa's health officer resulted in government co-operation in studying the issue and regulating sawdust pollution. By the turn of the twentieth century, the Department of Marine and Fisheries had achieved compliance from all major mill-owners. Regulations against dumping sawmill waste coincided with diminishing timber resources and increased the willingness of mill-owners to find other uses for mill waste, such as burning it for fuel (transferring the pollution from water to air), or processing it for other uses.[85]

CONCLUSION

The production and use of water power brought about changes to Canadian societies, economies, and environments from the seventeenth century to the turn of the twentieth, and continues to do so as technologies develop. Potential water-power sites formed the nucleus of new settlements. Mills were an important hub in transforming landscapes from forests to farms, processing logs into sawn timber for housing and milling the grain that settlers planted in the place of forests. Reducing the consumption of fuel and other forms of energy, water transportation by lake, ocean, and downstream on rivers was easier and cheaper than moving goods overland. Canada enjoyed low internal transportation costs with its system of rivers, lakes, and later canals that were built wide enough for schooners and steamships to pass through into the Great Lakes. Furs, wood, and wheat were staples shipped downstream and, in the case of the latter two, were often processed by water-powered mills.[86] With the use of iron, the invention of the water turbine, and the construction and enlargement of canals, water power played an important role in industrialization, driving machinery in factories.

But harnessing rivers for water power also altered ecosystems upstream and downstream. Building dams to create millponds barred fish from swimming upriver and flooded land above the dam. The refuse from sawmills clogged rivers, made life difficult for fish, and threatened public health. In the gold rushes of British Columbia and the Yukon, miners re-engineered rivers to direct them through ditches, flumes, and sluice boxes in order to separate out the most profitable element of the ecosystem.

By the twentieth century, water power use had changed again. Steam engines had finally surpassed direct-drive water power in usage in industry in Canada, and railways had eclipsed the use of canals and waterways for commercial transportation. Water power would re-establish its importance in Canada, however, as combustion engines again made water transport profitable on Canada's canal system, culminating in the construction of the fourth Welland Canal in the 1920s and the St Lawrence Seaway in the 1950s. And, even more significantly, hydroelectric development became the principle source of electricity. Such developments would have their own social and environmental impacts that mirrored the ways in which water power had affected Canada before the twentieth century.

The importance of water in Canadian energy history has thus only increased with the use of fossil fuels and the electrification of society, rather than the opposite, as Evenden and Peyton elaborate on in chapter 9 in this volume. As significant as coal and petroleum have been to Canada's history of energy use, it is worth remembering that other energy sources like water remained essential throughout all stages of the fossil fuel era.

NOTES

1 For Canada's freshwater statistics, see Environment and Climate Change Canada, "Frequently Asked Questions."
2 Scholars such as Karen Bakker and John Sprague argue that "water-rich" is a relative term for Canada, since much of the water that Canada has is actually non-renewable and the "myth" of abundance has led to prolific waste. See Bakker, "Introduction"; and Sprague, "Great Wet North? Canada's Myth of Water Abundance."
3 This idea has been most famously explored in Donald Creighton's classic *Empire of the St Lawrence*. For a full exposition on waterways in Canadian historiography, see Armstrong, Evenden, and Nelles, *The River Returns: An Environmental History of the Bow*, 4–23.
4 Keefer, *The Canals of Canada: Prospects and Influence*, 8–9.
5 For examples of such clear-cut distinctions, see Jones, *Routes of Power: Energy and Modern America*; McNeill, *Something New under the Sun: An Environmental History of the Twentieth Century World*, 10–16; Smil, *Energy in World History*.
6 Stewart, *Indian Fishing: Early Methods on the Northwest Coast*, 93–6, 105, 120–3.
7 Williams, *Clam Gardens: Aboriginal Mariculture on Canada's West Coast*.
8 Fox, Brooks, and Tyrwhitt, *The Mill*, 16, 35.
9 Leung, *Direct Drive Waterpower in Canada, 1607–1910*, 32.
10 Ibid., 57. For a thorough description of water wheel technology, see Reynolds, *Stronger Than a Hundred Men: A History of the Vertical Water Wheel*.
11 Leung, *Grist and Flour Mills in Ontario: From Millstones to Rollers, 1780s–1880s*, 56–8.
12 Malone, *Water Power in Lowell: Engineering and Industry in Nineteenth Century America*. See also Steinberg, *Nature Incorporated: Industrialization and the Waters of New England*.

13   Reynolds, *Stronger Than a Hundred Men,* 307–20. Most iron water wheels, however, did not produce more than fifty horsepower.

14   Hunter, *A History of Industrial Power in the United States, 1780–1930.* Vol. 1, *Waterpower,* 292–322; Leung, *Direct Drive Waterpower in Canada,* 57–9; Robertson, *Sawpower: Making Lumber in the Sawmills of Nova Scotia,* 39–42.

15   Alexander Somerville, *Montreal Gazette,* 10 October 1860, reprinted in the *Gazette* (Upper Canada Village, ON), 15 June 1977, quoted in Leung, *Direct Drive Waterpower in Canada,* 59.

16   Robertson, *Sawpower,* 42–5.

17   This was the horsepower rating for "super" turbines installed at Niagara Falls in the 1890s to produce hydroelectricity. The average horsepower for water turbines used in direct drive power installations was far less, around 100 horsepower. See Reynolds, *Stronger Than One Hundred Men,* 345–7; and Hunter, *History of Industrial Power in the United States,* 1:371–96.

18   Priamo, *Mills of Canada,* 27.

19   Leung, *Direct Drive Waterpower,* 10, 33, 43.

20   Macdonald, "Water Power and the Transformation of Canada 1600–1960: A Synthesis of National Museum of Science and Technology Research Reports 1989–1991," 29.

21   Leung, *Direct Drive Waterpower,* 33, 37; Harris, *The Reluctant Land: Society, Space, and Environment in Canada before Confederation,* 73–5.

22   Leung, *Direct Drive Waterpower,* 20–3.

23   Grist refers to grain that is ground in a mill, resulting in a whole wheat flour for local short-term consumption. Flour mills specialized in milling several grades of fine flour that were suitable for storage and export. Leung, *Grist and Flour Mills in Ontario,* 12–21.

24   Leung, *Direct Drive Waterpower,* 25–8.

25   Ibid., 13–15.

26   Ibid., 16–18.

27   Where wood was abundant in nineteenth-century Canada, it was of primary importance to both formal and informal economies. For an assessment of wood as fuel energy in Canada, see Joshua MacFadyen, chapter 6, this volume.

28   Wynn, *Timber Colony: A Historical Geography of Early Nineteenth Century New Brunswick,* 9, 87–9.

29   Ibid., 8–10, 19–27, 33; Gillis, "Rivers of Sawdust: The Battle over Industrial Pollution in Canada, 1865–1903," 265.

30   Robertson, *Sawpower,* 177.

31 Allardyce, "'The Vexed Question of Sawdust': River Pollution in Nineteenth-Century New Brunswick."

32 Robertson, *Sawpower,* 37–38.

33 Leung, *Direct Drive Waterpower,* 73, 126; Benidickson, "John Rudolphus Booth," 329.

34 Leung, *Direct Drive Waterpower,* 84–9.

35 Miquelon, "Les Forges Saint-Maurice," 9–10, 30, 70–3, 112–15, 121–31.

36 Kaye, "Flour Milling at Red River: Wind, Water and Steam." See also Leung, *Direct Drive Waterpower,* 29–30.

37 Liddicoat, *Waterwheels in the Service of British Columbia's Pioneers,* 48–68; Leung, *Direct Drive Waterpower,* 30–1.

38 Kennedy, "Fraser River Placer Mining Landscapes," 41, 44.

39 Liddicoat, *Waterwheels,* 94–7; Forsythe and Dickson, *The Trail of 1858: British Columbia's Gold Rush Past,* 100; Ormsby, *British Columbia: A History,* 186.

40 Leung, *Direct Drive Waterpower,* 79–83.

41 Morse, *The Nature of Gold: An Environmental History of the Klondike Gold Rush,* chap. 4, "The Nature of Gold Mining."

42 Ibid., 105.

43 Ibid., 108–9.

44 Anderson, "Water-Supply," 195–200. Before water supply systems were in place, Canadians accessed water from the nearest natural water source, from private or public wells, or from water vendors.

45 Leung, *Direct Drive Waterpower,* 41.

46 Ross, "Steam or Water Power? Thomas C. Keefer and the Engineers Discuss the Montreal Waterworks in 1852," 51–3.

47 Anderson, "Water-Supply," 202–3.

48 Steam power had pumped water from the St Lawrence through cast-iron pipes to cisterns in the Montreal waterworks system as early as 1816. Ross, "Steam or Water Power?," 51, 60.

49 Morse, *Fur Trade Routes of Canada, Then and Now,* 3, 9.

50 Ibid., 22, 27.

51 Belyea, *Columbia Journals / David Thompson,* 34. Entry for 13 June 1801. Our thanks to George Colpitts for this reference and for his reflections on the role of waterpower in the fur trade, which have informed this section.

52 Rajala, "Forest Industry in Eastern Canada: An Overview," 123–30.

53 Cooper and Clay, *History of Logging and River Driving in Fundy National Park: Implications for Ecological Integrity of Aquatic Ecosystems,* 29, 31. Lee, *Lumber Kings and Shantymen: Logging, Lumber and Timber in the Ottawa Valley,* 57.

54 Cooper and Clay, *History of Logging and River Driving in Fundy National Park*, 31, 32.

55 Lee, *Lumber Kings and Shantymen*, 48–51. River drives predominated in eastern Canada into the early twentieth century, remaining the favoured method as late as 1916, to be gradually replaced by railways, tractors, and trucks. The last log drive took place in New Brunswick in 1964, and in British Columbia there was still some log driving into the 1960s. Rajala, "Forest Industry in Eastern Canada," 135; Cooper and Clay, *History of Logging and River Driving in Fundy National Park*, 31; Rajala, "'This Wasteful Use of a River': Log Driving, Conservation, and British Columbia's Stellako River Controversy, 1965–1972."

56 Lee, *Lumber Kings and Shantymen*, 44–5, 51.

57 Passfield, "Waterways," 113.

58 For statistics on tonnage for each canal, see Legget, *Canals of Canada*, appendix 1.

59 For Erie Canal statistics, see Larkin, "Essay about the Erie Canal." For the hopes and dreams of Montreal merchants, see Creighton, *Empire of the St Lawrence*. Our thanks to Marvin McInnis for pointing out the domestic importance of Canadian canals.

60 Jones, "A Landscape of Energy Abundance: Anthracite Coal Canals and the Roots of American Fossil Fuel Dependence, 1820–1860."

61 In the seventeenth century, the Sulpician Order planned to construct a canal from Lachine to Montreal that would be used for both transportation and milling, but the canal they built was used only for flour mills. Desjardins, "Navigation and Waterpower: Adaptation and Technology on Canadian Canals," 21.

62 As quoted in Jackson and Addis, *The Welland Canals: A Comprehensive Guide*, 33.

63 Leung, *Direct Drive Waterpower*, 143.

64 Desjardins, "Navigation and Waterpower," 26–8; Jackson, *The Welland Canals and Their Communities: Engineering, Industrial, and Urban Transformation*, 246–68. See also Styran and Taylor, *This Great National Object: Building the Nineteenth Century Welland Canals*.

65 Lee, *Lumber Kings and Shantymen*, 51–5; Passfield, "Waterways," 118.

66 McNally, "The Relationship between Transportation and Water Power on the Lachine Canal in the Nineteenth Century," 76. See also Harris, *Reluctant Land*, 272.

67 McNally, "Relationship between Transportation and Water Power," 78.

68 Tulchinsky, *The River Barons: Montreal Businessmen and the Growth of Industry and Transportation, 1837–1853*, 204, 222; McNally,

"Relationship between Transportation and Water Power." For more information, see the Canadian Census of Industrial Establishments, 1871, http://www.canind71.uoguelph.ca/index.shtml.

69 Leung, *Direct Drive Waterpower*, 137; Stein, "Time, Space and Social Discipline: Factory Life in Cornwall, Ontario, 1867–1893," 282–5, 292.

70 Hunter, *A History of Industrial Power*, 1:161.

71 Leung, *Direct Drive Water Power*, 94.

72 For the development and use of the steam engine in North America, see Hunter, *A History of Industrial Power in the United States, 1780–1930.* Vol. 2, *Steam Power.* For Canada specifically, see Robertson, *Sawpower*, 58–81.

73 Robertson, *Sawpower*, 47.

74 Reynolds, *Stronger Than a Hundred Men*, 329–30, 344; Robertson, *Sawpower*, 73.

75 Leung, *Direct Drive Waterpower*, 98.

76 Ibid., 96–112. For more on steam versus waterpower in the Canadian context, see Bloomfield and Bloomfield, *Water Wheels and Steam Engines: Powered Establishments of Ontario.*

77 Forkey, "Damning the Dam: Ecology and Community in Ops Township, Upper Canada."

78 Kennedy, "Fraser River Placer Mining," 37–41.

79 Long, "Emory Creek: The Environmental Legacy of Gold Mining on the Fraser River," 8–10.

80 Kennedy, "Fraser River Placer Mining Landscapes," 66. As Alan Long notes, owners of a campground established in the 1950s at Emory Creek rebuilt flumes to supply a vegetable garden, swimming pool, and an ornamental stream. Long, "Emory Creek," 8–9.

81 Morse, *Nature of Gold*, 108–9.

82 Wynn, *Timber Colony*, 93.

83 Leung, *Grist and Flour Mills in Ontario*, 20–1, 52–3.

84 Allardyce, "'The Vexed Question of Sawdust,'" 129.

85 Gillis, "Rivers of Sawdust," 271, 277–8.

86 Our thanks to Marvin McInnis for sharing his research on wood, wind, and water in Canada.

PART TWO

The Mineral Regime

8

Coal in Canada

Andrew Watson

For over eighty years between Confederation and the end of the Second World War, coal fuelled dramatic economic, environmental, and social changes in Canada. Industrial revolution, unification by rail, and rapid growth in the urban population took place during those eight decades. The concentrated energy available from coal made each possible. Other fuels and forms of energy also provided heat and performed work away from industries, railways, and urban centres, even as they contributed to these developments. Cordwood continued to heat most rural homes, and *charcoal* fuelled several important early and remote industries. Water power drove manufacturing technologies then created hydroelectricity for lighting, heating, and other uses. It was coal, however, that was "powering up" Canada in the early twentieth century, contributing an immensely increasing proportion of total energy consumed, and adding to the remarkable increase in the scale of energy used.

Prior to the Industrial Revolution in England during the late eighteenth century, people around the world, Canada's Indigenous peoples included, had long burned coal to provide energy for heating and cooking. Coal burned hotter and longer, with a greater fuel-to-weight ratio than wood. Industrial applications of coal during the eighteenth and nineteenth centuries, combined with new technologies designed specifically to harness the potential energy contained within the fossilized fuel, made an extraordinary new scale of energy available to humans. From James Watt's first coal-powered steam engine of 1781, coal was used as the main (though not the only) fuel for turning water into steam, which could be built up to tremendous pressures to power even more sophisticated machinery, including turbine engines. During the nineteenth century, this technology was refined for use in trains, ships, and electrical generating power plants.

Coal also produced gas as a by-product of coking. This manufactured or *town* gas was captured and used in city street lights in the United States and Britain in the nineteenth century. The main purpose of coking coal, however, was to create a less volatile fossil fuel with a higher fixed carbon content, called coke. *Coke* was such a concentrated fuel that when burned in a blast furnace (so named because of super-heated air added to ensure a uniform temperature) it was capable of generating enough heat to turn iron ore molten, thereby allowing impurities to be removed to produce iron or steel. Shaped into rails used to convey trains, iron and steel were (and are) the only materials suitable for transporting coal overland.

The significance of these conventional innovations can be seen in the statistical record. In 1871, coal provided just 7.3 per cent of the total energy (measured in petajoules) consumed in Canada. Thirty years later, in 1902, that number had increased to roughly 41 per cent. In 1903, the amount energy consumed from coal surpassed that from wood for the first time in Canadian history. As a proportion of total energy consumed, coal reached an all-time high of 66 per cent in 1918, and continued to account for over 40 per cent of Canada's total energy consumption until 1953. As a proportion of fossil-fuel energy consumed in Canada, coal never dropped below 90 per cent until 1922 and remained well above 50 per cent until 1952.[1] The amount of coal consumed in Canada doubled from just over 2 million tons in 1880 to 4.4. million tons in 1888, then increased by more than 400 per cent to 35 million tons at the end of the First World War. Consumption peaked again just after the Second World War, one year after the Leduc oil strike in Alberta, at more than 47 million tons in 1948.[2] After falling throughout the 1950s and 1960s, coal consumption in Canada rose in the last decades of the twentieth century, peaking at roughly 70 million tons in 2000, with most coal consumed in the country's nineteen coal-powered electrical generation plants.

Enormous coal deposits within the country, and in close proximity to the border in the United States, made it possible for Canada to develop such a reliance on coal (see plate 8.1). Canadians used lots of coal, not because there were no alternatives, but because there was so much coal easily accessible in certain parts of the country. Canada's main coal deposits were in Nova Scotia, Saskatchewan, Alberta, and British Columbia.[3] There were also extremely large deposits of high-quality coal relatively close to southern Ontario and the St Lawrence Valley in Pennsylvania, Ohio, and West Virginia. Generally, Canadian coal tended to supply only regional markets. High transportation costs meant that Nova Scotia coal reached no farther west than Montreal, and Alberta coal no farther east than Manitoba. Central Canada (Quebec and Ontario) was supplied mainly by American coal.

Prior to Confederation, coal production in Cape Breton went from roughly 73,000 tons in 1847, to over 117,000 tons in 1857.[4] By 1867, coal production in the province totalled nearly 600,000 tons (see figure 8.1).[5] By this date, US coal mines were producing tens of millions of tons of coal per year, the greatest share of which came from Pennsylvania, where the largest and most important coal deposits were found. The major portion of American coal found its way to industries and markets along the Eastern Seaboard.[6] A comparatively tiny amount was exported to central Canada, but in terms of Canadian consumption this amount was considerable. In fact, throughout the eighty years between Confederation and the Second World War, roughly half of all coal consumed in Canada was imported (see figure 8.2).[7] During the 1860s, Montreal imported slightly more than half its coal, while Toronto and Hamilton imported all their coal, from the United States.[8] By Confederation, Vancouver Island collieries at Nanaimo were producing 35,000 tons a year,[9] but the enormous coal deposits of the British Columbia interior and the Alberta foothills, as well as the prairie deposits in Saskatchewan, were basically untouched until the construction of the Canadian Pacific Railway during the 1880s. Other than coal from Nova Scotia, the impact of Canada's large deposits on the country's energy budget had to wait until after Confederation.

There is no universal history of coal in Canada. Different parts of the country experienced different coal histories. So while the dominance of coal in Canada's energy budget outlined above applies very well to Nova Scotia, Quebec, Ontario, Saskatchewan, and Alberta, it is somewhat less useful in describing the energy history of New Brunswick, Manitoba, British Columbia, and northern Canada. Nevertheless, coal shaped the history of every region in Canada. The degree to which coal dominated a particular region's energy budget was determined by four factors: (1) the purpose for which coal was used; (2) the price of coal; (3) disruptions to the supply of coal; and (4) the quality of the coal available. All of these factors were closely related to the fact that there was no single type of coal. Different types of coal tended to have limited or preferred applications and in many cases were not interchangeable. The price of different types of coal fluctuated for a variety of reasons, including mining costs, freight rates, and quality, but consumers generally bought the cheapest coal available. Occasionally, access to a particular type of coal was interrupted by labour action or wartime constraints. And consumer preference for a specific type of coal sometimes dictated market demand, even

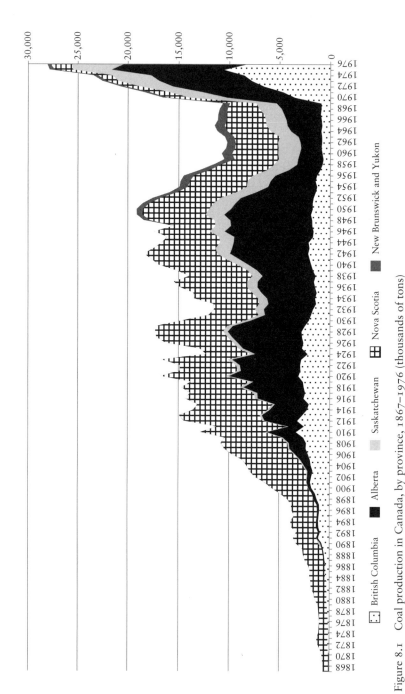

Figure 8.1   Coal production in Canada, by province, 1867–1976 (thousands of tons)

Source: Statistics Canada, Historical Statistics of Canada, series Q1–5, *Canadian Production of Coal, 1867–1976* (thousands of tons), http://www.statcan. gc.ca/access_acces/archive.action?l=eng&loc=Q1_5-eng.csv.

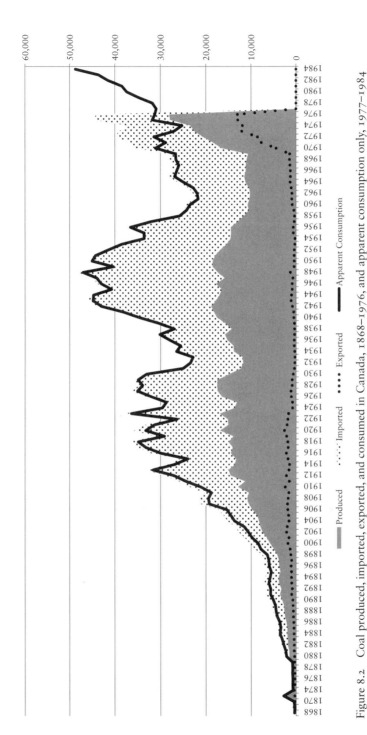

Figure 8.2   Coal produced, imported, exported, and consumed in Canada, 1868–1976, and apparent consumption only, 1977–1984 (thousands of tons)

Sources: Statistics Canada, Historical Statistics of Canada, series Q6–12, *Canadian Utilization of Coal, 1867 to 1976*, http://www.statcan.gc.ca/access_acces/ archive.action?l=eng&loc=Q6_12-eng.csv; and Canada, "Statistical Review of Coal in Canada," 6.

when alternative types of coal could have been substituted. At the same time, these four factors were in turn influenced by geography, protectionist government policies, the exigencies of economic development, the type of mining operation, and competition from other forms of energy. From place to place, different economic, environmental, and social circumstances created the contexts within which coal became an important source of energy.

TYPES OF COAL

Simply put, *coal* is ancient plant matter that was compacted into sedimentary rock. In North America, coal deposits formed during the Carboniferous age (360–300 million years ago) and Permian age (300–250 million years ago) from terrestrial plants located in coastal or swampy regions where the carbon content of the vegetation became trapped in low oxygen settings, preventing decay and entry back into the carbon cycle. Over thousands of years the carbon became locked in sedimentary rock formations where, depending on geological processes from place to place, incredible heat and pressure forced out varying amounts of impurities. The result was different types of coal, arranged in seams anywhere from three to thirty feet thick, located in deposits that corresponded with the strata in which they were laid down millennia ago.[10]

There are three main types of coal: *anthracite*, *bituminous*, and lignite. The differences between them are their carbon content and impurities. Anthracite has the highest fixed carbon content and the fewest impurities, making it the cleanest-burning and therefore most expensive type of coal. Uniquely intense heat and pressure were required to form anthracite, so it is also the rarest type of coal, contributing to its high price. In Canada, bituminous comprises by far the largest percentage of coal consumed and existing as reserves. It contains a medium to high fixed carbon content with varying amounts of impurities, which are often removed through coking, making it useful for the greatest number of different applications. Lignite, less fully formed, has low fixed carbon, so its uses and value are limited.[11]

There is tremendous overlap between different types of coal. Anthracite may contain properties very similar to high-quality bituminous, and high-quality lignite may have the same properties as low-grade sub-bituminous. Generally, coals with high fixed carbon content have a higher heat value and fewer impurities, making them less volatile and smoky. But the location of coal deposits and the quantities they contained were factors just as important as the quality. Where the coal existed greatly influenced how it was extracted and who would consume it.

EXTRACTION AND MARKETING OF COAL IN CANADA: A REGIONAL OVERVIEW

Nova Scotia

Nova Scotia's coalfields were the first in Canada to be commercially exploited. There are four main coalfields in the province: the inland mines at Cumberland and Pictou, and the submarine mines at Inverness and Sydney. All four were accessed via underground mines and contained bituminous with a mid- to high heat value and relatively high volatility, making it useful for a variety of industrial applications after processing into coke, including steam-power generation and metallurgy.[12] Despite its high volatility, Nova Scotia bituminous was widely used for domestic purposes throughout the Maritimes during the nineteenth and first half of the twentieth centuries.

In the eighteenth century, a small number of lease-holders with licences from the Crown mined modest amounts of coal for trade to New England and for use by garrisons in Acadia and Halifax.[13] In 1826, the Crown granted the Duke of York exclusive rights to Nova Scotia coal, which he leased as a monopoly to a London-based firm, the General Mining Association (GMA).[14] Under the GMA, coal production in Cape Breton climbed from 12,000 tons in 1827 to 120,000 tons in 1856. The major portion of Nova Scotia coal was exported to New England during the 1830s, but declined into the 1840s as canals were built to transport Pennsylvania anthracite to these markets. To compensate for the difficulty of competing in US markets, Nova Scotia coal producers marketed their coal in British North America colonies, particularly in Quebec and Montreal. During the mid-1850s, Nova Scotia sent 30,000–40,000 tons to neighbouring colonies, while the US never took more than 20,000 tons.[15] In 1854, the United Province of Canada and the United States signed the Reciprocity Treaty, and by 1865, Canadian coal exports to the United States totalled 640,000 tons. In 1866, the United States abrogated the treaty, and Canadian coal exports to the United States fell to 185,000 tons in 1875.[16] The timing of the US withdrawal from the treaty and the loss of important markets as a consequence provided part of the rationale for the Maritime colonies joining Confederation in 1867.

Supporters of Confederation saw Nova Scotia coal as an important part of interprovincial trade.[17] Quebec markets, especially Montreal, accounted for most sales outside the Maritimes during the 1870s. Demand for rolling stock used in railway construction in central Canada resulted in a rapid increase in the iron and steel industries, and consequently in coal production, in Nova Scotia during the 1880s and 1890s.[18] In 1870, coal production in Nova Scotia was 719,000 tons. Ten years later that number had risen to almost 1.2 million tons, and in 1891 to nearly 2.2 million tons.[19] The discovery of rich iron ore deposits in close proximity to coal deposits and the amalgamation of several smaller mining and iron works during the 1890s enabled Nova Scotia to become Canada's chief centre for both coal and iron/steel production in Canada. During the 1890s, the Nova Scotia Steel Company (later the Nova Scotia Steel and Coal Company – Scotia) and the Dominion Coal Company (later the Dominion Iron and Steel Company – Disco) emerged to dominate the coal industry in Nova Scotia.[20] Logistical challenges getting coal past the Lachine Rapids at Montreal made selling Nova Scotia coal prohibitive in Ontario, but the demand for Nova Scotia–produced iron and steel for building railways as part of the national policy and the growing industrial market along the St Lawrence River provided the stimulus for consistent growth of the coal industry right through to the First World War.[21]

Nova Scotia coal mines are the deepest in Canada, averaging about 0.5 kilometres, with some mines as deep as 1.2 kilometres. Furthermore, while the Pictou and Cumberland deposits were located entirely beneath the mainland of the province, the Cape Breton deposits extended out below the ocean, in some cases for several kilometres.[22] There were two methods of extracting coal from underground mines: room-and-pillar and longwall. The room-and-pillar system involved mining out enormous rooms of coal, leaving wide pillars to hold up the roof. The longwall system involved tunnels in which the coal face was supported by a temporary roof made of timbers, and included both advancing and retreating types (working forwards to the end of a seam, or pushing a corridor to the end of a seam and working back through the seam to the entrance).[23] Until 1926, the room-and-pillar method was employed exclusively in Nova Scotia, after which time mine conditions necessitated adoption of the longwall method.[24]

Production of Nova Scotia coal continued to increase throughout the first decade and a half of the twentieth century, reaching nearly 7.5 million tons in 1915 (a total that would be surpassed only once, in 1940).[25] The combination of mining and ironworks created thousands of jobs and steady growth in industry towns. Between 1891 and 1921, New Glasgow / Trenton's population grew from 4,417 to 11,818, while Sydney's population went from 2,427 to 22,545.[26] The First World War caused significant disruptions to Nova Scotia's coal mining industry. Ships that normally carried coal from Nova Scotia to Quebec were diverted for wartime purposes.[27]

In 1914, more than a third of Nova Scotia coal was shipped to markets in Quebec. Four years later, less than 3 per cent was shipped to Quebec.[28] Quebec sourced its coal from US producers until ships became available after the war, but the temporary loss of this important market, along with falling coal prices and financial troubles experienced by the large coal/steel companies, had serious effects on Nova Scotia's industrial economy.

Economic hardships during the early 1920s hit Disco and Scotia hard. Both companies ended up much larger than the market for iron and steel products could bear after the war. In 1920, Disco and Scotia merged to form the British Empire Steel Corporation (Besco), which experienced significant economic troubles, fell into receivership in 1926, and was reorganized by central Canadian investors with drastically reduced capacity in Sydney and Trenton.[29] In an attempt to forestall the financial hardships of the company, Besco imposed wages cuts on mine employees in 1922.[30] Nova Scotia coal miners were in a unique position to resist these changes.

Initially, miners engaged in "striking on the job" by reducing their productivity when they entered the mines to work. This raised production costs and reduced profits for the company without the confrontation involved in refusing to work entirely. When striking on the job did not have the desired effect, and inspired by the success of a strike by American coal miners earlier that year, Cape Breton miners staged a "100 per cent strike" from August to September 1922.[31] Left unattended, submarine mines quickly filled with water. Seawater did not significantly effect the quality of the coal, but it did render the mines inoperable. The effects of the strike are evident in Nova Scotia's coal production numbers. In 1920, Nova Scotia produced 6.4 million tons of coal. In 1922, that number dropped to 5.5 million tons before rising to 6.6 million tons the next year.[32] Although the miners ultimately agreed to a contract that was less than they demanded, their ability to control the mine environment was used as leverage to gain more favourable terms than other labour unions achieved at the time.[33]

The particular environmental characteristics of Nova Scotia mines, combined with the miners' culture of self-reliance, meant coal mining in Nova Scotia relied heavily on manual labour for much longer than other mines in Canada and the United States.[34] Throughout the interwar years, Cape Breton coal mines continued to use manual tools and nineteenth-century technologies, such as ropes, pulleys, pushcarts, and horses to extract coal. In contrast, most US mines employed digging machines and locomotives to haul out and assemble coal. In one sense, technological advancements were unsuited to Cape Breton mines, which had much steeper grades than locomotives could traverse (16 per cent on average and as much as 38 per cent in some seams) and extended out under the ocean where new digging machines were unreliable.[35] But there was also a more principled opposition to mechanization among miners. Although greater productivity was possible with mechanization, in most cases this translated into job replacement. Thus, while environmental conditions often prohibited these technologies, close-knit mining communities in Nova Scotia tended to reject mechanization even where it was possible.[36] The result was that productivity declined as mines pushed deeper and farther out under the ocean. Although the average miner typically received roughly two hundred shifts per year during the 1940s,[37] the productivity of the Dominion Coal Company dropped from 2.34 tons per man-day in 1939 to 1.58 tons in 1945. During this time, US coal miners, whose product competed directly with Nova Scotia coal in Quebec, produced 5–6 tons per man-day, thanks largely to mechanization.[38] The amount of coal produced in Nova Scotia fluctuated considerably during the interwar years, from a low of 3.8 million tons in 1925 to a high of 7 million tons in 1927.[39] Labour disruptions again in 1924 and 1925, declining coal prices as a result of falling industrial output, as well as competition from US coal and oil created difficult financial conditions for the large coal companies.[40]

The history of postwar coal production in Nova Scotia is one of slow decline until the 1960s, and thereafter of rapid decline. From a postwar high of 6.5 million tons in 1950, production declined to 4.5 million tons in 1960, 2.1 million tons in 1970, and just 1.2 million tons in 1973.[41] The last time production had been so low was 1880. The Dominion Coal Company closed two collieries in 1949, another in 1953, and cancelled ambitious plans to extend new tunnels.[42] In 1957, the Royal Commission on Canada's Economic Prospects attributed the decline of the industry to several factors, including competition from oil, reduced coal consumption by the railway, ongoing challenges of submarine operations, and higher costs compared to those of US producers.[43] In 1960, several mines continued to operate around the province, but as J.R. Donald wrote in *The Cape Breton Coal Problem* (1966), "One is forced to the reluctant but overwhelming conclusion that no constructive solution to unemployment and the social needs of Cape Breton can be based upon coal mining."[44] After eight decades at the heart of Nova Scotia's economy and society, coal ceased to be the province's primary source of energy.

Alberta

Canada's largest coal reserves are located in the prairies and the foothills of the Rocky Mountains in Alberta. The grade of coal in the province declines from a relatively low volatile bituminous with high heat value in the inner foothills between Mountain Park and Crowsnest, to a more volatile bituminous with high heat value in the outer foothills between Prairie Creek and Saunders. Into the prairies surrounding Edmonton and south to Drumheller, coal was sub-bituminous with medium heat value and high volatility, while farther east coal deposits were mainly lignite with low heat value and high volatility.[45] Coal was not mined on a commercial basis in Alberta until the arrival of the railway in the early 1880s. After 1887, the first year records were kept, the amount of coal produced in the province increased steadily from 74,000 tons to over 7 million tons in the years before the Great Depression.[46]

To a certain extent, the consistency of coal production in Alberta is attributable to the geologic structure of the coal deposits. Room-and-pillar mining was used exclusively during the nineteenth century, but all three regions used a combination of underground mining and strip mining during the first half of the twentieth century. Geologic faulting disrupted the continuity of most deposits in the inner and outer foothills, but generally seams ran at steep angles down to about two thousand feet before levelling off to an even grade. Where seams were only a few hundred feet below the surface, companies chose strip mining, which removed the surface rock and soil – somewhat curiously called "overburden" – instead of tunnelling under it.[47] Some mechanization was employed in underground mining, but the important technological advancements in Alberta coal mining came in strip mining.

Alberta coal companies had little trouble getting coal out of the ground, but finding markets was far more complicated. Only a tiny fraction, never more than about 10 per cent, was exported to US markets. As the Prairie provinces were settled by Eurocanadians and production increased during the twentieth century, new markets in western Canada simply could not absorb the surplus.[48] With very little fuelwood available compared to Ontario and Quebec, many households adopted low-grade sub-bituminous and lignite extracted from the outer foothills and prairie coalfields as domestic fuels.[49] In 1910, roughly 30 per cent of Alberta coal was used for domestic purposes. Fourteen years later, in 1924, the proportion was more than 60 per cent. On average, Alberta coal producers sold half their coal, mainly low-grade sub-bituminous and lignite, as household fuel during the interwar years.[50] Less volatile, higher heat value bituminous from the inner and outer foothills was used mainly for generating steam to perform work, a small proportion of which was used within Alberta. As production increased significantly, Alberta coal producers began promoting their coal in Winnipeg. With the help of favourable eastbound freight rates, Alberta coal captured the Winnipeg market, essentially displacing US coal that arrived via the Great Lakes.[51] In fact, while Nova Scotia coal failed to produce any profits without government aid between 1930 and 1944, Alberta coal producers fared quite well.[52]

From the earliest days, however, the single most important and stable market for Alberta coal was the railway. With no other source of fuel between Port Arthur (where US coal was available via the Great Lakes) and the Rocky Mountains, Alberta coal producers had no competition for Prairie railway fuel until locomotives began switching to oil near the end of the Second World War. Full statistical series are unavailable for railways in Canada, but data available between 1911 and 1930 suggest that coal consumption by railways in Canada rose steadily until the First World War, after which point it remained stable until the Great Depression.[53] In 1930, prairie railways used approximately 30 per cent of the coal consumed by railways in Canada.[54] If this proportion is accepted as representative of the amount of coal consumed by the railways on the Prairies in a typical year, railways consumed roughly 40–50 per cent of coal produced in Alberta every year, with higher proportions consumed during the First World War.[55] Apart from the railways, however, Alberta coal producers faced many challenges.

According to experts at the time, the main problem was that Alberta had too many mines. Few of them produced coal consistently, but combined they produced more coal than there were markets.[56] Unlike Nova Scotia, which had a few dozen large mines, Albert had hundreds of mines of all sizes. In 1925, economist M.J. Patton claimed that Alberta's 400 mines produced three to four times as much coal as the province could consume. Patton blamed "a policy of unrestricted leasing of coal lands" for the over-production.[57] Compounding the problem was that demand for domestic coal declined to almost nothing during the warm months of the year.[58] The result was that most mines simply could not afford to stay open year round. Most operated for just six months at a time when demand for coal justified the costs. Therefore coal companies could not provide consistent employment for prairie miners. On the one hand, this complemented the farmer's seasonal schedule. Many seasonal mines hired farmers who supplemented their agricultural income with wages working in the coalfields during the winter. On the other hand, the inconsistency produced labour tensions among full-time miners who were unable to make ends meet when mines closed for several months. Like their peers in Nova Scotia, underground miners had a great deal of autonomy over their work environments, but the seasonality of the market in Alberta removed the workers' leverage when demand disappeared.[59]

Although the transition was not as sudden or complete as it was in Nova Scotia, Alberta experienced a fairly consistent decline in coal production during the 1950s and 1960s. In 1950, Alberta produced 8.1 million tons of coal. Eleven years later, production had declined more than 75 per cent to 2 million tons. After the Leduc oil strike in 1947, petroleum replaced almost every market that coal had once dominated in the Prairies. Yet, unlike in Nova Scotia, coal production in Alberta made a comeback during the late twentieth century. In the wake of concerns over oil shortages and demand for coal from overseas markets, such as Japan, Alberta coal production reached record highs in the 1960s and 1970s.[60] As we saw above, the increase in coal-powered electrical generation boosted demand across Canada as well.

British Columbia

Commercial exploitation of coal in British Columbia began when the Hudson's Bay Company hired Aboriginal and Eurocanadian labourers to mine coal at Nanaimo on Vancouver Island during the 1850s. The coalfields around Nanaimo continued to be the only ones mined in British Columbia until the end of the nineteenth century. Vancouver Island coal was a high-grade bituminous characterized by high volatility and heat value. It existed in a variety of dispersed deposits starting as submarine seams several hundred feet under Nanaimo harbour and tiered at increasing elevations by geologic upthrust moving inland.[61] In 1862, the Vancouver Coal Mining and Land Company was established in Nanaimo, followed eleven years later by Robert Dunsmuir's company slightly northwest of Nanaimo in Wellington. Both companies were financed with some combination of capital from the United States, Britain, and eastern Canada.[62] In the years after Confederation, the production of the coal around Nanaimo rose from just 35,000 tons in 1867 to over 1 million tons in 1897.[63] Vancouver was basically the only Canadian market for Nanaimo coal. During the nineteenth century, most British Columbia coal was exported to markets in the United States, particularly California.[64] In the 1890s, several thousands, and occasionally tens of thousands, of tons of Nanaimo coal found markets in the Hawaiian Islands, while a few thousand tons were exported to East Asian countries, including Japan.[65] Many of these patterns continued after 1898 when the Canadian Pacific Railway built a new line through the Rocky Mountains and several new collieries were opened in the Crowsnest Pass area on both sides of the Alberta–British Columbia provincial border.[66] Crowsnest deposits consisted of high heat value bituminous with lower volatility than Vancouver Island coal and were located in deposits at depths similar to most inner foothills Alberta coal but separated from adjacent deposits by faulting.[67] Although major labour strikes in 1890–91, 1912–14, and 1924 caused noticeable production declines, with the addition of Crowsnest coalfields, British Columbia coal production climbed to over three million tons per year in 1910, 1912, and 1920.[68] The US Pacific Coast continued to be an important market for British Columbia coal, regularly receiving a third of annual production before and during the First World War.[69] During the first thirty years of the twentieth century, emerging industries, domestic consumers, and railways within the province formed the main markets for Crowsnest coal.

Coal production on Vancouver Island peaked in the 1920s and declined in the 1930s. In the decade after the Depression, annual production in the province fluctuated, but remained well below 2 million tons. By this date, Nanaimo seams were commercially exhausted, and the vast majority of British Columbia coal came from Crowsnest and a variety of other small interior collieries.[70] Some mechanization was introduced to Vancouver Island mining, but carts were still being hand-loaded as late as 1946, when production per man-day was just 2.4 tons. Earlier mining may have utilized room-and-pillar methods, but by the 1940s desperate companies employed 300-foot longwall methods to access the remaining coal.[71] By the 1940s, Crowsnest deposits were mined using mainly room-and-pillar methods, while tunnels with more conveniently arranged seams and modest inclines enabled more mechanization than on Vancouver Island.[72] At the same time as the Vancouver Island coalfields were nearing exhaustion, added competition from oil and hydroelectricity eliminated several markets, including domestic and industrial consumers, bunker fuel on ocean-going ships, and even the railway.[73] After a brief period when coal production was buoyed by the war, annual production declined to under one million tons during the 1960s. Yet, as was the case with Alberta coal, British Columbia coal experienced a production renaissance during the 1970s in response to oil shortages and overseas demand from Japan.

Saskatchewan

Coal production in Saskatchewan shared very few of the features that defined the industry in Nova Scotia, Alberta, and British Columbia. Annual production was negligible during the nineteenth and early twentieth centuries, reaching 500,000 tons for the first time in 1929. After 1936, production fluctuated between one and two million tons per year until the 1970s.[74] Early mining in Saskatchewan employed underground room-and-pillar methods to extract high volatile, low heat value sub-bituminous and lignite coal found in the southern region of the province around Estevan. But Saskatchewan's coal seams occurred in shallow deposits, only a few hundred feet below the surface. As a result, coal companies in Saskatchewan employed strip mining methods much earlier and more extensively than mining operations elsewhere in the country.[75] After about 1930 strip mining operations expanded in scale until they accounted for over 70 per cent of provincial production in 1945.[76] Strip mining necessitated heavy mechanization, such as diesel-fuelled power shovels, bulldozers, and trucks that removed the overburden, dug out the coal, and hauled it to cleaning and screening stations. During the early twentieth century, most coal produced in Saskatchewan was used for domestic purposes within the province. After 1930, however, the use of digging machines caused slacking (crumbling/deterioration of coal), and most was used for industrial purposes in Saskatchewan, Manitoba, and Northern Ontario.[77] Unlike other coal regions in the country, Saskatchewan did not experience significant production declines after the Second World War. Instead, production experienced a relatively steady climb between the Second World War and the 1970s.

Competition from US Coal

At Confederation, Canada produced several hundred thousand tons of coal, almost all of which was extracted from Nova Scotia mines. By 1900, that number had climbed to over 5.7 million tons, split roughly 70/30 between Nova Scotia and British Columbia. During the next thirty years, as Alberta coalfields were opened, annual production in Canada rose dramatically, reaching an interwar high of 17.5 million tons in 1928.[78] To truly understand the importance of these numbers, however, Canadian coal production must be placed within the context of US coal production. In 1903, for example, the United States produced 359 million tons of coal. Pennsylvania alone produced 103 million tons – almost thirteen times as much coal as all of Canada.[79] Twenty-five years later, in 1928, total US coal production was more than thirty-four times that of Canada.[80] In other words, competition with American coal was a constant reality for Canadian producers. Tariffs on US bituminous coal helped protect regional markets in Canada. East of Montreal, Canadian markets received most of their coal from Nova Scotia, while markets in western Canada and the West Coast received their coal almost exclusively from Alberta and British Columbia. In central Canada, however, even tariffs, government assistance, and freight rate subventions could not create the conditions within which Canadian coal could compete with cheap, abundant, and high-quality coal from the United States. The proximity of southern Ontario and Montreal to the enormous coalfields of Pennsylvania, Ohio, and West Virginia resulted in two sets of energy anxieties in Canada: one on the part of coal producers and another on the part of coal consumers.

Competition with US coal producers over New England markets was a key consideration for the Maritime provinces joining Confederation. To entice these provinces, the terms of Confederation guaranteed favourable trade conditions for Nova Scotia coal in central Canada as a way of compensating for the loss of New England markets. Yet, despite the tariffs imposed on US coal imports to Canada, Nova Scotia coal was unable to compete with American coal farther west than Montreal.[81] Quebec became an important market for Nova Scotia coal in the early twentieth century, but the First World War temporarily disrupted this, causing renewed anxieties about markets amongst Nova Scotia coal producers.[82] Similar concerns emerged in Alberta where overproduction became a challenge for the industry. The Great Depression heightened these concerns in both Nova Scotia and Alberta as regional economies stagnated during the 1930s. In an effort to alleviate these anxieties and make Canadian coal more competitive with US coal, the federal government established a series of transportation subventions in 1931 to reduce the freight costs of shipping coal to central Canada.[83] These efforts to connect Nova Scotia and Alberta coal with central Canadian markets were also informed by anxieties amongst central Canadian consumers about access to coal.

CONSUMPTION OF COAL IN CANADA

Ontario and Quebec contain no coal deposits.[84] Historically, the costs and logistical challenges have proven too great to justify transporting significant amounts of Nova Scotia coal any farther west than Montreal, and Alberta coal any farther east than the top of Lake Superior. Thus, while Montreal received roughly half its coal from Nova Scotia throughout the first half of the twentieth century (excluding wartime), southern Ontario has only ever received very small quantities of coal from anywhere in Canada. Nevertheless, coal was critical to central Canada's energy budget. According to the authors of the 1946 *Report of the Royal Commission on Coal*, as late as the 1930s and 1940s, "about one half of the energy from all sources in central Canada was obtained from coal ... much less than in the Maritimes, about the same as on the Prairies and much greater than in British Columbia."[85] Nearly all the coal consumed in southern Ontario came from the United States. The reason, quite simply, was geography. As economist M.J. Patton pointed out in 1925, "The Ontario peninsula, which is the most highly industrialized section of Canada, is thrust southward almost into the large coal-producing area of the United States, with which it is connected by numerous, well-organized and highly efficient lines of transportation."[86]

In the decade prior to and after Confederation, Ontario imported tens of thousands of tons of coal from the United States via Lake Ontario and Lake Erie by way of the Welland Canal.[87] In 1870, Ontario imported almost 115,000 tons. In 1880, that number had increased to over 752,000 tons. And by 1890 it reached almost 2.3 million tons. In 1900, the last year for which Ontario is recorded separately from the rest of the country, coal imports to the province exceeded 4.2 million tons.[88] The coal imported was both anthracite from eastern Pennsylvania and various grades of bituminous from several states along the western slopes of the Appalachian Mountains. In 1880, roughly half of the coal imported to Ontario was bituminous, used mainly to generate steam power for manufacturing and railway locomotives, while the other half was anthracite, used mainly as domestic heating fuel.[89] Over the next decade, as manufacturing expanded in the province, bituminous increased to 61 per cent of total coal imports by 1890. The establishment of manufacturing and heavy industries in Toronto and Hamilton during the 1890s caused the proportion of bituminous to climb to roughly 69 per cent in 1900.[90] It is within this context of dependency on US coal that Ontario and Quebec became among the first, and remain among the most substantial, producers of hydroelectricity. Hydroelectricity became an important component of Ontario's energy budget during the interwar years, replaced steam power for many manufacturing and industry purposes, and from an early date meant the province could avoid using coal to generate electricity.[91] Consequently, the proportion of bituminous as a share of total coal imports from the United States to central Canada shrank to less than 20 per cent in the years immediately preceding the Great Depression, before rising again to 30–40 per cent during wartime.[92]

As an energy source, coal was primarily an urban and industrial phenomenon. As a share of total coal consumption, rural communities used very little coal. Large quantities of coal were shipped up the Great Lakes to fuel resource industries after the turn of the century,[93] but most coal

Table 8.1   Prices paid by Toronto dealers for various coals, 1931 ($/ton)

	American anthracite		Alberta coals	US bituminous		Sydney bituminous	
	Stove	Buckwheat		Lump	Slack	Run of Mine	Slack
Price at mine	8.00	3.25	3.00–4.25	1.10–1.50	0.25–1.00		
Transportation (tariff)	4.08	3.62	12.50	3.24	3.24		
Transportation (special rate)			-5.75				
Duty	0.40	0.40		0.75	0.75		
Excise	0.08	0.03		0.02	0.02		
Average price to dealers	12.56	7.30	9.75–11.00	5.11–5.51	4.26–5.01	6.70	6.20

Source: Thomson, "Some Economic Aspects of the Canadian Coal Problem," 378.

imported to central Canada was consumed in cities. Prior to industrialization, Toronto imported an average of 40,000 tons per year during the 1860s – approximately a quarter of Canada's total coal imports.[94] In 1903, Toronto imported 581,000 tons of anthracite (42 per cent of total anthracite imports from the United States), and 444,000 tons of bituminous (12.2 per cent of bituminous imports from the United States).[95] By 1941, at least 64 per cent of households in Toronto used anthracite to heat their homes,[96] and bituminous imports from the United States to Toronto had risen to over 1.3 million tons per year.[97] Prices were determined by many factors, including the pit head price of coal, transportation, import duty, and local distribution charges (see table 8.1).[98] Industrial, manufacturing, and commercial consumers who bought quantities in the hundreds or thousands of tons obtained coal from wholesalers who distributed orders directly from centralized docks or depots, while households and small businesses acquired their smaller orders from retailers who made coal available at different points around the city.[99] By the 1940s, a large retailer handled 50,000–100,000 tons of coal per year and generally sold coal used to heat apartments, commercial buildings, stores, and offices.[100] Small retailers, meanwhile, dealt mainly with individual households and sold only a few hundred tons per year.[101]

At both the industry and household levels, consumer preference and technological lock-in created market preferences for certain types of coal, depending on use. Its high carbon content made anthracite the preferred fuel for use in homes and small businesses, which accounted for a very sizable but difficult to quantify portion of total coal consumption.[102]

Although it was hard to ignite, its high heat value and extremely low volatility meant anthracite produced almost no smoke. This attribute was important because most stoves and furnaces in homes and businesses were hand-fired equipment with natural draft from which smoke could easily escape.[103] The difficulty in getting it to ignite was overcome by the mid-nineteenth century when anthracite coal-producers in Pennsylvania developed equipment specially designed to burn anthracite.[104] As this equipment was continually improved, consumers became locked in to the technology, and consequently dependent on anthracite for heating. It was so popular that consumers in central Canada basically refused to switch to an alternative heat source until natural gas became available in the late 1950s when the infrastructure was built to deliver it to market (see appendix 3).[105] As energy historian Christopher Jones argues, "For those with access to anthracite, life was better with coal than without it."[106] Since anthracite came entirely from coalfields in eastern Pennsylvania, disruptions to Canadian deliveries, while limited to certain areas of the country, created significant anxiety.

Throughout the first fifteen years of the twentieth century, coal imports from the United States to Canada increased steadily. In 1914, annual imports from the United States reached a pre–Second World War high of over 19.3 million tons before dropping by a quarter over the first two years of the war.[107] This created particular challenges for large markets in Quebec, which simultaneously experienced a near-total reduction of shipments from Nova Scotia due to wartime constraints and raised the issue of energy security in Canada for the first time.[108] This first supply crisis in central Canada related to bituminous coal, not anthracite, when annual imports to Canada dropped from 13.7 million in 1914 to 9.6 million in 1916, before recovering to 16.4 million in 1918.[109] The resumption of US coal imports at the end of the war permitted the director of the Mines Branch of the Department of Mines in Ottawa, John McLeish, to conclude, "There is no actual shortage of coal resources in either the United States or in Canada, at least in so far as bituminous coal is concerned, so that the problem relates primarily to questions of production and distribution."[110] McLeish's insinuation that access to anthracite had greater potential for crisis than bituminous, and that the production and distribution of coal, not the existence of sufficient reserves, were the main cause for concern, were proven correct less than two years later when Appalachian coal miners staged a massive strike in 1922.

The 1920s were a raucous decade for labour relations in mines north and south of the border. A series of strikes followed economic decline in the steel and coal-mining industry in Cape Breton, and in 1922 American coal miners staged one of the largest strikes in American history. The effects of the 163-day strike had an almost immediate impact on imports of anthracite into central Canada and lasting effects on energy politics. The same day the strike came to an end (21 September 1922), Congress created the Federal Fuel Administration to deal with the acute anthracite shortage. Almost immediately, distribution of anthracite coal came under federal oversight.[111] Although anthracite production in Pennsylvania resumed over the next several months, imports to Canada fell off sharply. In 1923, total imports of anthracite to Canada dropped from 4.4 million tons to 2.9 million tons, a 34 per cent decline.[112] This created a serious shortage of domestic heating fuel during the 1922–23 winter and instigated serious debate about a national fuel policy to make Canada independent of foreign sources of energy.[113] Adding to these anxieties was an unofficial warning issued by the US Bureau of Mines to their Canadian counterpart that Canada would have to find potential alternatives to American anthracite. At one point, three bills were even introduced to Congress that would have placed an embargo on anthracite exports to Canada. None of the bills passed, but this nevertheless demonstrated to Canadian politicians that energy issues demanded close attention.[114] Inspired by the creation of a similar body in the United States, the Canadian government created the Dominion Fuel Board in November 1922.[115] The fuel board immediately secured agreements for increased imports from Britain. In 1922, anthracite imports from Britain totalled 110 tons, while in the next year it was almost 248,000 tons.[116] After a tariff was placed on US anthracite for the first time in 1932, imports from Britain climbed to over 1 million tons each year until the Second World War.[117] Eventually anthracite imports from the United States recovered. By the 1940s, federal bodies on both sides of the border felt confident enough to claim there was sufficient anthracite to meet the level of demand for another century. Anthracite continued to be widely available throughout central Canada, but its price rose as mining challenges increased.[118] These kinds of costs encouraged consumers to switch to natural gas when widely available in the 1960s.

Railways played an important and consistent role as consumers between the 1870s and 1950s. Few other statistics in the history of coal in Canada show as much stability as the quantity of coal used by railways during the first half of the twentieth century. Between 1911 and 1930, railways consumed an average of 29 per cent of all coal consumed in Canada, and at no point did the proportion drop below 25 per cent or exceed 34 per cent.[119] During the nineteenth century, the coal used by railways in the

Maritimes was almost entirely Nova Scotia coal, while in the Prairies and the Rockies it originated primarily from the inner foothills of Alberta and British Columbia. In central Canada, it came mostly from the United States. In 1939, railways in Ontario and Quebec accounted for more than 98 per cent of bituminous coal imports used by railways in Canada.[120] Railways consumed 35–40 per cent of the coal used in the Prairies and roughly 70–75 per cent of the coal produced in Alberta.[121] Oil began displacing coal as railway fuel during the late 1920s in British Columbia and throughout much of the rest of the country by the 1950s.[122] In Nova Scotia, the decline was precipitous, from 1.2 million tons in 1950 to just 35,000 tons in 1959.[123] In addition to their important role as consumers, railways were essential to the distribution of coal.

DISTRIBUTING COAL IN CANADA

Apart from mining operations and the Nova Scotia steel industry, coalfields tended to be located far from markets in Canada. Connecting producers and consumers of coal was the earliest and greatest challenge in the fuel's history.[124] Coal was shipped by both railways and boats everywhere in Canada, except the Prairies and the Rocky Mountains, where railways were used exclusively. Distance and geophysical obstacles dictated freight rates and how far coal could be shipped and still generate a profit. During the interwar years, concerns about energy security informed the federal government's 1931 decision to introduce transportation subsidies on Canada coal. These subventions allowed Canadian coal to better compete with American coal until oil and natural gas became competitive during the 1950s.

Nova Scotia coalfields were close to shore, especially in Cape Breton. From a very early date Nova Scotia coal was brought to market by ship. After New England markets were eliminated with the resumption of tariffs in 1866, Nova Scotia producers found new water-accessible markets within the Maritimes and Quebec. Small shipments by rail occurred, but were generally uneconomical until subventions were introduced in the 1930s. Rail did have certain advantages over water transport. Not only were shipments by water disrupted by war, but the annual winter freeze-up of the waters around Quebec ports created seasonal constraints to shipping.[125] These cycles are also evident in the monthly exports of US bituminous to Canada. Most coal exports to Canada were shipped by rail, because the amount shipped by water fluctuated 25–33 per cent between winter and summer (see figure 8.3).[126]

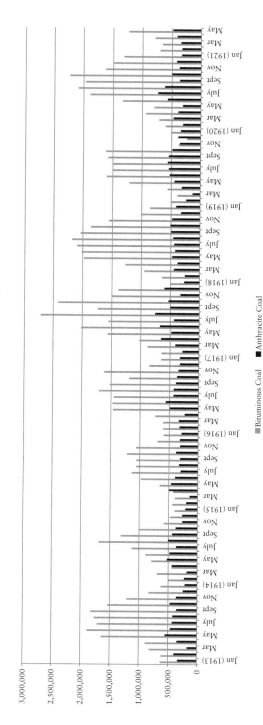

Figure 8.3   Bituminous and anthracite coal exported from the United States to Canada, January 1913–May 1921 (tons)

Sources: For 1913: McLeish, *The Production of Coal and Coke in Canada during the Calendar Year 1916*; for 1914–18: McLeish, *The Production of Coal and Coke in Canada during the Calendar Year 1918*, 12–13; for 1919: McLeish, *The Production of Coal and Coke in Canada during the Calendar Year 1919*, 9; for 1920–21: McLeish, *The Production of Coal and Coke in Canada during the Calendar Year 1920*, 7–8.

The Lachine Rapids presented an obstacle to shipping large quantities of Nova Scotia coal to Ontario markets. Transferring coal into smaller boats degraded the coal at the same time as it created a bottleneck to its flow into Ontario. Throughout the first half of the twentieth century, government officials and industry experts came up with several ideas to overcome the transportation challenges of marketing Nova Scotia coal in Ontario, including one that proposed building a canal between the mouth of the Ottawa River and Georgian Bay.[127]

In part to address the energy anxieties that emerged in central Canada during the 1920s, and in part to provide aid to coal producers and consumers during the Depression, the federal government passed a bill in May 1931 to provide subventions on coal shipped from Nova Scotia to Quebec and Ontario. In 1928, the government provided just $65,000 to help move 114,000 tons (1.7 per cent of total coal production). In 1932, $546,000 in subventions were applied to 710,000 tons (17.4 per cent of total production), and in 1939 $3 million was applied to 2.4 million tons (34.3 per cent of total production).[128] By the time the St Lawrence Seaway was completed past Montreal in 1959, Nova Scotia's coal industry was in decline, and oil and natural gas were quickly replacing coal in Ontario's energy budget.

Getting coal from Appalachian coalfields to markets in Ontario involved a variety of routes by both rail and water. Between the 1820s and 1860s, US coal companies established a network of canals and railways to get Appalachian coal to East Coast markets.[129] Several American industrial centres on Lake Erie inaugurated a flow of coal towards the Great Lakes, which then put large quantities within reach of Canadian markets on Lake Ontario via the Welland Canal. Geographically, Ontario became a logical market for Appalachian bituminous. River valleys ran out of the mountains towards the west before curving north to meet the plains along Lake Erie. These valleys prevented railway companies from building direct lines between mines and markets. Instead, coal was transferred from railcars to ship holds at places such as Toledo, and shipped in large quantities by water to market.[130] As Ontario's railway network intensified during the late nineteenth century, several American railway companies built north to meet at several border locations, including Sarnia, Windsor, Niagara Falls, Fort Erie, and Prescott.[131] Bituminous and anthracite coal producers in northern and eastern Pennsylvania also sent coal by rail to ports on Lake Ontario, such as Oswego.[132]

For most of the nineteenth century, coal was shipped down the Welland Canal or across Lake Ontario in relatively small vessels. There were

direct rail connections at the Niagara Falls Suspension Bridge and the
Fort Erie–Buffalo International Bridge. All other US-Ontario rail connections were made by car ferries (barges that carried train cars across the
lake).[133] The total amount of coal entering Ontario from the United
States by rail and water are unavailable, but the annual reports and year-
books from the Toronto Board of Trade provide some idea of the proportional split between the two.[134] Until 1900, roughly equal amounts of
anthracite entered Toronto by both rail and water. After that time, anthracite imports by rail climbed from 175,000 tons in 1900 to 579,000 tons
in 1910, and 780,000 in 1914. Over the same years, bituminous imports
to Toronto by rail went from 422,000 tons in 1900 to 1.1 million tons in
1914, while the amount brought in by water never exceeded 56,000 tons
per year. Over the thirty years prior to the First World War, rail shipments
never accounted for less than 64 per cent of total coal imports to Toronto,
and reached a high of 93 per cent in 1912–14. If these proportions reflect
the general picture of coal imports to southern Ontario, shipments by
water diminished in importance relative to rail. A significant factor shaping this trend was that coal deteriorated when transferred into and out of
ship holds. This was particularly important for anthracite, which was
used primarily as household fuel and therefore needed to remain in lump
sizes suitable for handling and slow burning. But rail shipments were also
vitally important because they could get coal delivered much closer to
their ultimate destinations than boats could by water.[135] Bituminous coal
was shipped in large quantities from Lake Erie ports during the interwar
years, but much of this went to markets on the Upper Great Lakes, including Sault Ste Marie and Port Arthur.[136]

Transportation freight rates accounted for a significant proportion,
in some cases more than half, of the total price of coal to the consumer.
On the East and West Coasts, the cost savings of water transport meant
freight rates affected the price of coal only a little, while on the prairies
the adjustment for freight charges was linear, based strictly on distance
from the mine. In central Canada, however, the presence of the Great
Lakes created unique conditions. Not only were direct railway routes
complicated by the geography of the lakes, but the costs involved in shipping coal by water were also much less than by rail.[137] Competition from
water transport meant Ontario railways rarely, if ever, profited directly
from transporting coal. During the late nineteenth and early twentieth
centuries, railway companies in Ontario kept freight rates on coal artificially low to encourage industry to develop in the province. By absorbing
some of the costs of transporting coal, Ontario railways benefitted in the

long run with even more traffic generated by industry and manufacturing. Artificially low rail freight rates were, therefore, essential for large-scale consumption of coal in Ontario. One chief way these costs were absorbed by railway companies was by offering lower rates for longer hauls. Without regulated rates, coal would have been prohibitively expensive and Ontario's economy would have had difficulty industrializing.[138] Coal continued to be shipped to Ontario markets by water throughout the first half of the twentieth century, but ironically, the real importance of water transport resided mainly as a check on rising freight rates by rail.

CONCLUSION

Coal was not Canada's only source of energy, but for roughly eighty years it was the most important. Prior to the 1950s when even more energy-dense and abundant fossil fuels became widely available as conventional fuels, coal dominated the country's energy budget. Its availability, adoption, and applications were not universal, but coal was used by more people, for more reasons than any other fuel in Canada between 1870 and 1950. In other words, coal helped create modern Canada in this period. In 1867, small amounts of coal were mined in Nova Scotia and Vancouver Island, and even smaller amounts of American coal were being consumed in Quebec and Ontario. Industrialization, urbanization, and the building of the railways across Canada greatly expanded the need for coal within the Canadian economy. And hundreds of thousands of Canadians came to know coal intimately as a material substance of their everyday lives through labour and the home. Throughout this period, Canada's history of coal was mediated by four main factors: the use to which coal was put, its price, disruptions to the supply, and the quality of the coal.

Geology and natural forces determined what kind of coal was located where, and to a certain extent this shaped the ways people utilized coal as a fuel. Nova Scotia's abundant bituminous coal was used for many purposes, from steel-making to heating homes, but was easily replaced by other fossil fuels after the Second World War. On the West Coast, bituminous coal from Vancouver Island, the British Columbia interior, and Alberta's inner foothills supplied Canada's only significant export markets, while also meeting the relatively limited needs of consumers within the province. Alberta's deposits of bituminous, sub-bituminous, and lignite offered the greatest diversity of coal in Canada. With few markets, however, its applications were limited to the railways, mining operations, and domestic heating on the Prairies. Although Ontario and Quebec contained no coal deposits, central Canada consumed more coal than any other region in the country.

Roughly half (and often much more than half) of the coal consumed in Canada every year between the 1880s and 1970s was imported. For several years during the 1930s a significant portion came from Britain. But this paled when compared to the millions of tons of coal Canada imported from the United States. The vast majority of US coal was imported to central Canada, particularly southern Ontario. Bituminous always accounted for most coal used in central Canada, but the preference amongst household consumers for high heat value, clean-burning, and relatively expensive anthracite distinguished the pattern of coal use in this region of the country. Despite the unevenness and variety of coal consumption in Canada during these eighty years, the use of coal by Canada's railways provides the only real commonality among the regions. The importance of coal as a source of domestic, industrial, manufacturing, or electrical energy varied considerably from region to region. But railways consumed coal everywhere in Canada.[139]

This chapter would not be complete without a discussion of the unexpected and unwanted consequences of such a heavy reliance on coal in Canada during the nineteenth and twentieth centuries. They were, and continue to be, extraordinary. The rapid pace and enormous scale of coal extraction and consumption wrought serious human and environmental tolls. Coal turned out to be particularly deadly to the men who spent their lives extracting it from the ground. Despite safety precautions, coal dust occasionally "caught a spark," exploded, and killed miners. More than 1,600 miners were killed between 1879 and 1939 in Nova Scotia alone – more than twenty-three deaths every year.[140] Those who were not killed in mining accidents often endured many of their final years in poor health and suffering from diseases caused by the inhalation of toxic coal dust. Even after mines were closed, the environmental legacy of mining operations continued for generations in the form of toxic tailings and other mineral solid wastes.[141] Burning coal also produced myriad by-product waste and pollution problems, including serious smog and air quality issues, acid rain, and the contamination of soils and groundwater by chemical leaching from slag pits and tar ponds. The most profound ecological damages resulted from the leaching of contaminants such as selenium, arsenic, cadmium, and lead as a consequence of improper disposal of coal ash in rivers and streams.[142] In fact, if one industry expert from the 1920s is correct, Canadian coal consumption produced millions of tons of coal ash every year.[143] Canada's fossil fuel legacy – which began with coal – promises to endure in the form of greenhouse gases and an altered climate for many generations after the last of these finite coal resources have been burnt. Coal accounts for approximately one-third of the fossil fuel carbon emissions that are set to warm the planet's global average temperature by two degrees Celsius, resulting in higher incidents of extreme weather events, prolonged heat waves, and severe drought, not to mention the melting of the polar ice caps and a rise in sea levels of 0.6–1.8 metres by 2100.[144]

Canadian coal enjoyed a renaissance in the wake of global energy crises in the 1960s and 1970s. After experiencing rapid declines during the 1950s and 1960s, Alberta and British Columbia coal production sky-rocketed to all-time highs during the 1970s.[145] Very little of this coal was consumed in Canada. Most was exported to Asian markets, particularly Japan.[146] Within Canada, coal lost its dominance after the Second World War as it was replaced by oil and natural gas in industrial, manufacturing, and transportation applications. Yet even after it began its decline in real terms and as a proportion of Canada's total energy budget, coal continued to fit important niches in the steel industry, manufacturing, and electrical generation. For example, coal declined from 53 per cent of total energy consumed in Canada in 1948 to 19 per cent in 1960, and roughly 12 per cent in 1975. At the same time, it also became important for electrical generation for the first time in Canadian history. Of the total electricity generated in Canada, the proportion from thermal sources fuelled by coal rose from less than 5 per cent in 1950 to 22 per cent in 1975.[147] Indeed, the number of kilowatt hours generated by coal rose from just 1.7 million in 1950 to over 53 million in 1975. While the consumption of coal declined in real terms over the fifteen years following the end of the Second World War, starting in the early 1960s, the amount of coal consumed in Canada began to increase again. By the 1980s Canada, and particularly Ontario, was consuming more than ever before. Thus, even as coal's place in the country's overall energy budget declined significantly after the Second World War, its use for certain applications increased. And given the bountiful reserves in Canada, there is every reason to expect that Canadians will continue to utilize coal for various purposes as both producers and consumers for several generations to come.

NOTES

1 Unger and Thistle, *Energy Consumption in Canada in the 19th and 20th Centuries: A Statistical Outline*, 127–32. Also see Steward, "Energy Consumption in Canada since Confederation," 240–1; McDougall, *Fuels*

and the National Policy, 15; Canada, *Report of the Canada Royal Commission on Coal* (hereafter RCC), 379.

2 This statistic is the annual "apparent consumption of coal," which is the quantity produced plus the quantity imported minus the quantity exported. See Statistics Canada, Historical Statistics of Canada, series Q6–12, *Canadian Utilization of Coal, 1867 to 1976,* http://www.statcan.gc.ca/access_acces/ archive.action?l=eng&loc=Q6_12-eng.csv. After falling in the 1970s and 1980s, by 2009 coal consumption in Canada had risen to almost 53 million tons (48 million metric tonnes), mainly because of the nineteen coal-fired electricity generating plants, which consumed 42 metric tonnes of coal in that year. IndexMundi, "Coal Consumption in Canada." Consumption rose from about 41.3 million tons in 1980 to a high of 69.6 million tons in 2000.

3 Minor coal deposits were also present in New Brunswick, Yukon, and Northwest Territories, and a very tiny amount in northern Ontario, but for the purposes of this chapter those quantities had little or no impact on the history of coal in Canada and so will be left out of the discussion that follows.

4 Brown, *The Coal Fields and Coal Trade of the Island of Cape Breton,* 99.

5 Statistics Canada, Historical Statistics of Canada, series Q1–5, *Canadian Production of Coal, 1867–1976 (Thousands of Tons),* http://www.statcan. gc.ca/access_acces/archive.action?l=eng&loc=Q1_5-eng.csv.

6 American coal production was several decades ahead of Canada's. During the 1830s and 1840s, owners of anthracite coal mines in Pennsylvania built canals to get coal to New England markets and introduced new technologies for the use of coal in manufacturing, which resulted in the industrialization of the northeastern United States half a century before similar developments took place in Canada. Chandler, "Anthracite Coal and the Beginnings of the Industrial Revolution in the United States"; Jones, "A Landscape of Energy Abundance: Anthracite Coal Canals and the Roots of American Fossil Fuel Dependence, 1820–1860."

7 Statistics Canada, Canadian Historical Statistics, series Q6–12, *Canadian Utilization of Coal, 1867 to 1976.*

8 Tables of the Trade and Navigation of the Province of Canada, *Sessional Papers of the Province of Canada,* for the years 1860–1866; Tables of the Trade and Navigation of the Dominion of Canada, *Sessional Papers of the Dominion of Canada* for the years ending 30 June 1867 and 30 June 1868.

9 Statistics Canada, Historical Statistics of Canada, series Q1–5, *Canadian Production of Coal, 1867–1976.*

10 Mackenzie, *Our Changing Planet: An Introduction to Earth System Science and Global Environmental Change,* 248; Christopherson, *Geosystems: An Introduction to Physical Geography,* 335.

11 *RCC*, 1–3.

12 Patton, "The Coal Resources of Canada."

13 Brown, *The Coal Fields and Coal Trade of the Island of Cape Breton*, 44–59, 73.

14 Gerriets, "The Impact of the General Mining Association on the Nova Scotia Coal Industry, 1826–1850"; MacEwan, *Miners and Steelworkers: Labour in Cape Breton*, 4–5.

15 Brown, *Coal Fields and Coal Trade of the Island of Cape Breton*, 99.

16 McDougall, *Fuels and the National Policy*, 18; Tables of the Trade and Navigation, *Sessional Papers of the Dominion of Canada*, for the year ending 30 June 1875. The 1865 number taken from McDougall may not be reliable, because McDougall cites 90,000 tons in 1875, half of what the Tables of Trade and Navigation record for exports from Canada to the United States that year. In later years, new Smoke Laws in New England also contributed to the reduction in exports to the United States. Nova Scotia bituminous was much more volatile than Pennsylvania anthracite and did not meet smoke regulations. Dominion Coal Company, *Memorandum Respecting Reciprocity in Coal with the United States of America*.

17 McDougall, *Fuels and the National Policy*, 20–1.

18 *RCC*, 65–6.

19 Statistics Canada, Historical Statistics of Canada, series Q1–5, *Canadian Production of Coal, 1867–1976*.

20 Heron, *Working in Steel: The Early Years in Canada, 1883–1935*, 16–24; MacEwan, *Miners and Steelworkers*, 6–7.

21 McDougall, *Fuels and the National Policy*, 32.

22 *RCC*, 12–22.

23 Ibid., 76–7.

24 Ibid., 85.

25 Statistics Canada, Historical Statistics of Canada, series Q1–5, *Canadian Production of Coal, 1867–1976*.

26 Heron, *Working in Steel*, 31.

27 McDougall, *Fuels and the National Policy*, 32; *RCC*, 65–6.

28 McLeish, *The Production of Coal and Coke in Canada during the Calendar Year 1915*, 19; McLeish, *The Production of Coal and Coke in Canada during the Calendar Year 1920*, 15.

29 Heron, *Working in Steel*, 25–6; MacEwan, *Miners and Steelworkers*, 6–61.

30 MacEwan, *Miners and Steelworkers*, 83.

31 Frank, "Class Conflict in the Coal Industry Cape Breton 1922," 173–7.

32 Statistics Canada, Historical Statistics of Canada, series Q1–5, *Canadian Production of Coal, 1867–1976*.

33 Timothy Mitchell argues that the unique underground environmental conditions of coal mining provided miners with the position to demand more
   equitable treatment within labour relations. Mitchell, "Carbon Democracy";
   Frank, "Class Conflict in the Coal Industry Cape Breton 1922," 173, 177.
   Although the Pictou and Cumberland miners enjoyed the same autonomy
   over their workplaces as Cape Breton miners did, comparatively dry conditions in the land mines resulted in gas buildups. These conditions prevented
   Pictou and Cumberland miners from engaging in the same 100 per cent
   strike as the Cape Breton miners in 1922.

34 The only use of machines were steam engines used to pump seawater out
   of the mines.

35 RCC, 89.

36 Ibid., 90, 295–6.

37 Ibid., 306.

38 A man-day is the amount of work a single miner performed in a single
   day's work. MacEwan, *Miners and Steelworkers*, 267. Again the comparison is not simply one of resistance to mechanization, since US mines
   tended to be much better located to drive mine shafts into coal seams at
   a grade more suited to locomotives. RCC, 89.

39 Statistics Canada, Historical Statistics of Canada, series Q1–5, *Canadian
   Production of Coal, 1867–1976*.

40 In fact, without operating subsidies from the federal government between
   1930 and 1944, Nova Scotia coal producers would have suffered a net loss
   of $0.145 per ton. Even with assistance, net profit was only $0.003 per
   ton. RCC, 281–2.

41 Statistics Canada, Historical Statistics of Canada, Series Q1–5, *Canadian
   Production of Coal, 1867–1976*; Statistics Canada, Historical Statistics of
   Canada: Main Index, http://www.statcan.gc.ca/pub/11-516-x/index-eng.htm.

42 MacEwan, *Miners and Steelworkers*, 285.

43 Ibid., 309.

44 Ibid., 333.

45 RCC, 32–42.

46 In fact, coal production in Alberta reached all-time highs twenty-five out of
   the forty-three years between 1887 and 1929. And only once, in 1911, did
   production drop by more than 20 per cent of the previous year's total. In
   1911, production declined by 48 per cent as a consequence of a prolonged
   strike by miners in the province. A smaller strike in 1919 resulted in a production decline of 17 per cent. RCC, 68; Statistics Canada, Historical
   Statistics of Canada, series Q1–5, *Canadian Production of Coal, 1867–1976*.

47 RCC, 103–10.

48 McLeish, *The Production of Coal and Coke in Canada during the Calendar Year 1909*, 18; McLeish, *Production of Coal … 1919*, 24; McLeish, *Production of Coal … 1920*, 21.

49 *RCC*, 103–10.

50 Ibid., 69. A small portion of this was sent to domestic markets in British Columbia, mainly Vancouver. Actual amounts are difficult to find. In 1916, roughly 90,000 tons of coal from Alberta was sold in British Columbia. McLeish, *The Production of Coal and Coke in Canada during the Calendar Year 1916*, 30; *RCC*, 322, 375.

51 *RCC*, 68. With assistance, western coal producers made a profit of $0.195 per ton. Without operating subsidies they would still have made a net profit of $0.18 per ton.

52 Ibid., 281–2.

53 It is not clear whether the Great Depression had a significant impact on the amount of coal consumed by railways. McLeish, *The Production of Coal and Coke in Canada during the Calendar Year 1919*, 11; Thomson, "Some Economic Aspects of the Canadian Coal Problem," 353, also see appendix 15 on page 464.

54 *RCC*, 451.

55 To calculate this percentage, the amount of coal consumed by railways on the prairies was extrapolated from the total amount of coal used by railways in Canada (see notes 50 and 51), and then this amount was converted into a percentage of total coal production in Alberta. Stats Canada Historical Statistics Index, http://www.statcan.gc.ca/pub/11–516-x/index-eng.htm.

56 This overproduction informed the efforts of Alberta coal producers to market their coal in southern Ontario during the 1930s and 1940s. Even with freight rate subventions and government assistance, however, Alberta coal never became an economical option for consumers in central Canada. McDougall, *Fuels and the National Policy*, 37.

57 Patton, "Coal Resources of Canada," 83.

58 Bercuson, *Alberta's Coal Industry, 1919*, x.

59 One of the only logistical issues underground miners could use to their advantage was that underground mines still needed maintenance. Although flooding was a much lesser problem than it was in Cape Breton, mines still had to be pumped free of volatile gases that built up when left unattended. Bercuson, *Alberta's Coal Industry, 1919*, x.

60 The last year for which records are available from Stats Canada lists 12.2 million tons of coal produced in 1976. Statistics Canada, Historical Statistics of Canada, series Q1–5, *Canadian Production of Coal, 1867–1976*.

61 *RCC*, 49.

62 Gidney, "From Coal to Forest Products: The Changing Resource Base of Nanaimo, B.C." 19–20.

63 Statistics Canada, Historical Statistics of Canada, series Q1–5, *Canadian Production of Coal, 1867–1976.*

64 Gidney, "From Coal to Forest Products," 21; McDougall, *Fuels and the National Policy*, 36; *RCC*, 459; Williams, *Energy and the Making of Modern California*, 46.

65 Tables of the Trade and Navigation, *Sessional Papers of the Dominion of Canada* for the years ending 30 June 1880, 1890.

66 McDougall, *Fuels and the National Policy*, 36; *RCC*, 71.

67 *RCC*, 43–4.

68 Statistics Canada, Historical Statistics of Canada, series Q1–5, *Canadian Production of Coal, 1867–1976.* For more on these strikes, see Norris, "The Vancouver Island Coal Miners, 1912–1914: A Study of an Organizational Strike"; Mouat, "The Politics of Coal: A Study of the Wellington Miners' Strike of 1890–91."

69 McLeish, *The Production of Coal and Coke in Canada during the Year 1910*, 21; McLeish, *The Production of Coal and Coke in Canada during the Calendar Year 1913*, 29; McLeish, *Production of Coal … 1916*, 32.

70 None of the Vancouver Coal Company's mines remained open for more than fifteen to twenty years. Once the seams were exhausted, the company closed the mine and moved labour and equipment to a new location. By the 1930s, viable mines had nearly all run their course in this way. Gidney, "From Coal to Forest Products," 38.

71 *RCC*, 111–12.

72 Ibid., 111.

73 Ibid., 71–3.

74 Statistics Canada, Historical Statistics of Canada, series Q1–5, *Canadian Production of Coal, 1867–1976.*

75 *RCC*, 28–30, 100–1.

76 Ibid., 76.

77 Ibid., 67; Saskatchewan (Wynne-Roberts), *Report on Coal and Power Investigation*, 22.

78 Statistics Canada, Historical Statistics of Canada, series Q1–5, *Canadian Production of Coal, 1867–1976.*

79 Milner, *Coal: Analysis of the Trade between Canada and United States*, 5, 7.

80 Thomson, "Some Economic Aspects of the Canadian Coal Problem" 341, 432. In 1928, the United States produced 600 million tons of coal. The most important reason the United States produced so much more coal than

Canada was simply that their deposits were larger and more conveniently located to bigger markets. Moreover, US coal deposits, especially those in the Appalachian Mountains, tended to be more easily accessible, and therefore better suited to mechanization. This had the effect of significantly enhancing productivity. Between 1934 and 1944, the average number of tons per man-day produced by American miners in West Virginia, Kentucky, Pennsylvania, and Ohio rose from 4.32 to 5.69. By comparison, the productivity in Nova Scotia, Alberta, and British Columbia went from an average of 3.08 to 3.21 tons per man-day over the same years. Only Saskatchewan operations, which used heavy machinery for strip mining, had higher productivity than Appalachian coal miners, 8.66 tons per man-day in 1944. RCC, 83.

81 In fact, without the tariff, US coal would likely have displaced Nova Scotia coal entirely in the Quebec market. Dominion Coal Company, *Memorandum Respecting Reciprocity in Coal*, 41. A primary reason Nova Scotia coal never penetrated the Ontario market in a meaningful way was the natural obstruction of the Lachine Rapids, which prevented large coal ships from reaching Lake Ontario. The rapids both restricted the flow and increased the price of Nova Scotia coal in Ontario. Under these circumstances, Nova Scotia coal simply could not compete with the cheaper, abundant, more easily accessible, and higher-quality Pennsylvania anthracite and bituminous coals. McDougall, *Fuels and the National Policy*, 20–9.

82 See note 23. Since coal companies in Nova Scotia were also steel manufacturers, these anxieties were closely tied up with concerns over depressed steel markets during the 1920s.

83 RCC, 66, 320–2; McDougal, *Fuels and the National Policy*, 11, 38–9.

84 This is not entirely true, as there is a very small amount of lignite in northern Ontario. But the quantity and quality of this coal make these deposits irrelevant to the discussion that follows.

85 RCC, 425.

86 Patton, *Coal Resources of Canada*, 73.

87 Tables of the Trade and Navigation, *Sessional Papers of the Province of Canada*, for the years 1858–1866, 1867. Tables of the Trade and Navigation, *Sessional Papers of the Dominion of Canada*, for the years ending 30 June 1868, 1870, 1871, 1872, 1873, 1874, 1875; Zercher, "The Economic Development of the Port of Oswego," 125.

88 Tables of the Trade and Navigation, *Sessional Papers of the Dominion of Canada*, for the years ending 30 June 1870, 1880, 1890.

89 Ian Drummond argues that demand for anthracite in Ontario was fairly inelastic (apart from population growth) because of its importance as

domestic fuel, while bituminous serves as a better barometer of industrial growth, since bituminous imports fluctuated in line with economic recessions. Drummond, *Progress without Planning: The Economic History of Ontario from Confederation to the Second World War*, 108–9. As late as 1943, only 200,000 tons of anthracite were used for industrial purposes in Ontario or Quebec. RCC, 433. According to the authors of the RCC, more than 90 per cent of anthracite imported to Canada by the 1940s was used for domestic purposes in Ontario and Quebec. RCC, 428.

90 For more on the rise of metallurgical and steam-powered industries in Ontario, see Drummond, *Progress without Planning*, chapters 7 and 10; Heron, *Working in Steel*. Tables of the Trade and Navigation, *Sessional Papers of the Dominion of Canada*, for the year ending 30 June 1900. Statistics by province are discontinued after 1900.

91 By the 1930s, advocates made it clear that, given the option, hydro power was much more economical than coal power for generating electricity. Combe and Farmer, "The Relation of Electricity to Coal." According to Drummond (*Progress without Planning*, 147), "Almost everywhere [in Ontario], by 1941, hydroelectricity had driven out old thermal installations [to generate electricity]." In 1946, the authors of the *Report of the Royal Commission on Coal* reported, "At least four-fifths of the power machinery installed in Central Canadian industry is fed by hydraulically generated electricity." RCC, 433. For an overview of the development of hydroelectricity in Ontario, see Armstrong and Nelles, *Monopoly's Moment: The Organization and Regulation of Canadian Utilities, 1830–1930*.

92 Statistics on coal imports by region (Maritimes, central Canada, Prairies, British Columbia) between 1928 and 1945 are available in RCC, appendix D, 659–63.

93 In 1901, the Algoma Steel Company was formed in Sault Ste Marie. Heron, *Working in Steel*, 21–2. Over the next several decades, large commercial sawmills and pulp and paper companies developed on the Great Lakes. By the 1940s, wood and paper products companies were the largest consumers of coal in Ontario, more than twice as much as the next two categories of consumers combined (non-ferrous metal products and iron and steel products). RCC, 433.

94 Tables of the Trade and Navigation, *Sessional Papers of the Province of Canada*, 1860–1866, 1867; Tables of the Trade and Navigation, *Sessional Papers of the Dominion of Canada*, for the years ending 30 June 1868, 1869, 1870.

95 Milner, *Coal: Analysis of the Trade*, 30–1.

96 Canada, *Census of Canada 1941*, vol. 9, *Housing*. In all likelihood this number was higher, since this number was actually the national average for urban areas with populations over 30,000. Coal used for domestic purposes in Toronto was almost exclusively anthracite. So while the census did not specify the type of coal used, this can be inferred for the Toronto case.

97 RCC, 434.

98 Thomson, "Some Economic Aspects of the Canadian Coal Problem," 354, 378; RCC, 357.

99 RCC, 360.

100 Ibid., 473.

101 Ibid., 365.

102 Indeed, few energy historians have spent much time exploring this important aspect of coal consumption. As Christopher Jones points out, scholarship on the adoption of coal has been "biased towards the industrial sector." Jones, "The Carbon-Consuming Home: Residential Markets and Energy Transitions," 791.

103 RCC, 6, 428.

104 Chandler, "Anthracite Coal and the Beginnings of the Industrial Revolution," 158–63; Jones, *Routes of Power*, 46–49. Equipment used to burn bituminous coal also created aspects of technological lock-in, but to a much lesser extent than domestic reliance on anthracite. The variety of different types of bituminous, and the overlap between them, allowed more flexibility for large-scale manufacturing, industrial, and railway consumers. What tended to be more important for more volatile types of bituminous were aspects such as ash and sulphur content, which created by-products that could damage equipment. High ash content from spent bituminous, for example, had a tendency to fuse as "clinkers" after combustion. If temperature and air conditions were not right, these clinkers could form into hard slag that fixed itself to equipment, essentially ruining the combustion chamber of high-capacity furnaces. RCC, 461–3.

105 Almost as soon as hydroelectricity was introduced in the province, it was clear that electric heat was not a viable alternative to coal as a source of domestic heating. Barnes, *The Heating of Houses, Coal and Electricity Compared*; Combe and Farmer, "The Relation of Electricity to Coal," 328.

106 Jones, *Routes of Power*, 74.

107 Canada, Department of the Department of Trade and Commerce, "Report of the Department of Trade and Commerce Part 1," *Sessional Papers of the Dominion of Canada*, for the year 1914.

108 McDougall, *Fuels and the National Policy*, 33–4; McLeish, *Production of Coal ... 1919*, 1.

109 Canada, Department of the Department of Trade and Commerce, "Report of the Department of Trade and Commerce," *Sessional Papers of the Dominion of Canada*, for the years 1914, 1915, 1916. During the same years, anthracite imports remained constant or increased slightly.

110 McLeish, *Production of Coal ... 1919*, 1.

111 Kanarek, "The Pennsylvania Anthracite Strike of 1922," 221.

112 Canada, Dominion Bureau of Statistics, Department of Trade and Commerce, "Trade of Canada," *Sessional Papers of the Dominion of Canada*, for the year 1923.

113 McDougall, *Fuels and the National Policy*, 5–6, 37–8.

114 *RCC*, 429.

115 Thomson, "Some Economic Aspects of the Canadian Coal Problem," 380; McDougall, *Fuels and the National Policy*, 37–8.

116 Canada, Dominion Bureau of Statistics, *Sessional Papers of the Dominion of Canada*, Trade of Canada, for the year 1923. There is considerable discrepancy between the numbers presented in the Tables of Trade and Navigation and those presented in by the 1946 *Report of the Royal Commission on Coal*. The numbers provided here are from the Tables of Trade and Navigation.

117 For several years during the 1930s, Canada also imported a considerable amount of anthracite from Nazi Germany. *RCC*, 428. Exact numbers are not provided, because their accuracy have not been cross-referenced with official government records. See previous note.

118 In 1946, the Royal Commission on Coal reported that approximately twenty tons of water had to be pumped out of the ground for every ton of anthracite extracted. *RCC*, 58.

119 McLeish, *Production of Coal ... 1919*, 11; Thomson, "Some Economic Aspects of the Canadian Coal Problem," 353 – see appendix 15, 464.

120 *RCC*, 436.

121 Ibid., 120, 320–2, 451.

122 Ibid., 73.

123 MacEwan, *Miners and Steelworkers*, 286–7.

124 As Unger and Thistle argue in a recent survey of energy consumption in Canada, "The geography of energy sources and location of consumers, personal and industrial, was to be a constant problem for [Canada]." Unger and Thistle, *Energy Consumption in Canada*, 58.

125 Frank, "Class Conflict in the Coal Industry Cape Breton," 163.

126 McLeish, *Production of Coal ... 1916*, 11; McLeish, *The Production of Coal and Coke in Canada during the Calendar Year 1918*, 12–13;

McLeish, *Production of Coal … 1919*, 9; McLeish, *Production of Coal … 1920*, 7–8.

127 Canadian Federation of Boards of Trade and Municipalities, *The Georgian Bay Canal and Nova Scotia Coal.*

128 *RCC*, 566–7. Subventions on Nova Scotia coal dropped during the war, but resumed for another decade after 1945.

129 Jones, *Routes of Power*, 23–87.

130 Lezius, "Geographic Aspects of Coal Cargoes from Toledo."

131 Walker, "Transportation of Coal into Southern Ontario, 1871–1921," 18.

132 Zercher, "Economic Development of the Port of Oswego."

133 Walker, "Transportation of Coal," 18–20.

134 The statistics are reproduced in Walker, "Transportation of Coal," 29–30.

135 Until the 1920s, railways were used to distribute large quantities to markets at the periphery of major trade hubs like Toronto and Hamilton. By the 1930s, however, trucks were used to distribute coal to smaller retail markets. *RCC*, 320–4.

136 Between April 1944 and March 1945, for example, ports on Lake Superior and Lake Huron received 41 per cent (6.7 million tons) of all the bituminous coal imported to Ontario by water. A significant portion of this was used by railways, with a smaller amount sent on to Winnipeg. *RCC*, 651–3; Canadian Federation, *The Georgian Bay Canal*, 9; Lezius, "Geographic Aspects of Coal Cargoes," 376.

137 Walker, "Transportation of Coal," 24.

138 These reduced rates applied only to the main lines. Branch or minor lines received comparatively high rates, which had the effect of focusing industrial and manufacturing along the main lines of the province. These practices were established and monitored by the railway companies until 1904 when the Board of Railway Commissioners was created and took over regulating freight rates in the country. Walker, "Transportation of Coal," 22–3.

139 Trains in British Columbia were the first to transition to oil in the 1930s. *RCC*, 73, 459.

140 Frank, "Class Conflict in the Coal Industry Cape Breton," 167.

141 Sandlos and Keeling, "Zombie Mines and the (Over)burden of History," 80–3.

142 Parks, "The Ecological Costs of Coal Ash Waste," 400.

143 In 1925, British coal expert Rudolf Lessing estimated "the mean percentage of ash in coal put on the market to be 10%." Lessing, "Coal Ash and Clean Coal. Lecture I, Delivered November 23, 1925," 183.

144 Marland, Boden, and Andres. "Global, Regional, and National CO_2 Emissions," 244 (1994). For contemporary data, see Marland, Boden, and Andres, *Global, Regional, and National Fossil-Fuel CO_2 Emissions.*

145 Statistics Canada, Historical Statistics of Canada, series Q1-5, *Canadian Production of Coal, 1867–1976*.

146 Exports of Canadian coal to Japan climbed from less than ten million tons in 1974 to over sixteen million tons in 1984. Canada, Energy, Mines and Resources, *Statistical Review of Coal in Canada*, 2; Boon, *A Preliminary Analysis of Factors Affecting Modal Selection for the Movement of Western Canadian Coal to Ontario*, 1–2.

147 Unger and Thistle, *Energy Consumption in Canada*, 127–32. As late as 2007, 21 per cent of Canada's electricity was still generated by coal. Canada, Statistics Canada, *Electric Power Generation, Transmission and Distribution*, 11. In fact, the amount of coal used for generating electricity in Canada increased from 63 per cent of total coal consumed in 1974 to 83 per cent in 1984. Statistics Canada, Historical Statistics of Canada, series Q81–84, *Electric Generating Capacity by Type of Prime Mover, 1917 to 1976*. See figure 9.1 in this volume. Canada, Energy, Mines and Resources, *Statistical Review of Coal*, 11.

9

Hydroelectricity

Matthew Evenden and Jonathan Peyton

Dreams sometimes shed light on reality. When Harry Warren, a UBC geology professor, promoted the construction of the Moran Dam on the Fraser River in the late 1940s, he was dreaming – of massive power, of new industries, of a new regional powerhouse. The dam he imagined would remake British Columbia and carry it along a familiar and enviable path of hydroelectric development. He had witnessed the process to the south in the United States on the Columbia River; in Canada, Ontario, and Quebec had shown the way. Of course, Warren was dreaming in a second sense; he was imagining an unobtainable goal. What he proposed would annihilate the Fraser salmon fishery. A politicized commercial fishery, aligned with scientists, federal civil servants, American fisheries interests, and indigenous organizations formed a powerful opposition. In time, other developments on the Columbia and Peace Rivers would fulfill the developmental ambitions of the province without risking Fraser salmon. The Moran dam would be abandoned, though the dream it represented lingered.

Although Warren's Moran proposal would remain unfulfilled, it highlights several dimensions of the history of hydroelectricity in Canada. Conceived at a time of surging postwar growth, with the experience of wartime dam building in the background, the Moran Dam signalled a new era of large projects, conceived and designed by experts, pursuing what they understood to be the public good, transforming rivers into power generators. The negative response to Warren's proposal and the complex regional, national, and international politics that shaped alternative developments on the Columbia and Peace Rivers also suggest the extent to which the big dam era encountered complications from the start. High modern designs found early critics. Large projects also encountered new challenges occasioned by the politics of international river development and expanding transmission grids. While the project faltered and failed, the fact that it gained temporary traction should remind us of the value assigned to hydroelectricity in the mid-twentieth century, building on a foundation of early growth and pointing to a future of great promise in which electricity held the key.[1]

Hydroelectricity has been the primary source of power in Canadian electrical systems since the late nineteenth century. Although rival thermal and nuclear sources have made it relatively less important in the last thirty years, in the mid-twentieth century, hydro accounted for 98 per cent of all electricity generated in Canada.[2] Today, hydro still supplies over 60 per cent of the country's electricity (see figure 9.1). In Quebec, Manitoba, Newfoundland, and British Columbia, the figure remains above 90 per cent. This heavy reliance on hydro makes Canada an energy anomaly internationally. In most Western states, hydroelectricity typically represents a minor component of power supply. In the United States, for example, only about 9 per cent of electrical generation comes from hydro sources.[3] Despite its small population, Canada is the third-largest hydro power producer in the world, following China and Brazil.[4] Viewed in comparative terms alone, hydroelectricity represents a singularly important dimension of Canada's energy history.[5]

This overview of hydroelectricity in Canada seeks to map the shift from small, local systems to the era of large dams and continental grids. Generalizations quickly run into problems, however, because hydroelectricity is regulated and planned primarily at the provincial level and because the physical geography of Canada gives rise to very different hydro resource endowments, east to west.[6] We can best cut across an uneven provincial experience and a diverse country by focusing on three themes that are central to the story: system development, transmission, and consumption. Looked at over the long term, these themes suggest some of the ways in which the growth of hydroelectricity in Canada unfolded within expanding geographical scales shaped by environmental circumstances, technology, markets, and politics. Although it is difficult to relate the history of hydroelectricity in Canada without reverting to a narrative of progressive expansion, along the way we will pause to consider the environmental transformations and contests that this expansion precipitated and ask what were the social benefits and costs.

Hydroelectricity harnessed the power of rivers in a new way. Whereas waterwheel technologies captured the force of flowing water to turn

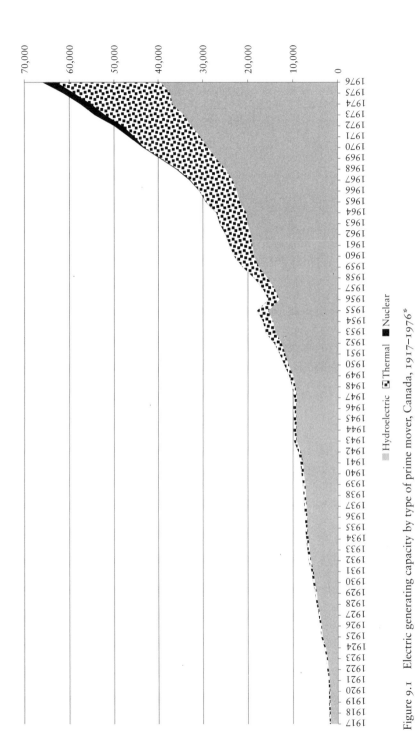

Figure 9.1   Electric generating capacity by type of prime mover, Canada, 1917–1976*

■ Hydroelectric ▣ Thermal ■ Nuclear

*Generating capacity in thousands of kilowatts, 1956–1976, in thousands of horsepower, 1917–1955.

Source: Statistics Canada, Historical Statistics of Canada, series Q81–84, *Electric Generating Capacity by Type of Prime Mover, Canada, 1917–1976*, http://www.statcan.gc.ca/access_acces/archive.action?l=eng&loc=Q81_84-eng.csv.

wheels attached to gears, hydroelectricity harnessed new technologies to convert kinetic energy into electricity. Instead of deploying a wheel to capture force, hydro systems used a turbine, which was attached to a shaft within a generator. As the water passed the plant, the turbine spun. A rotor on the spinning shaft carried electromagnets; when these revolved around a conductor, they produced electricity. Electricity was generated as the turbine spun. If the water flow was stopped, the turbine slowed and electricity was no longer produced. From the powerhouse, wires stretched outwards, distributing the electricity along a network.

Hydroelectricity emerged in Canada in the last decade of the nineteenth century in a great burst of technological enthusiasm. Spectacular displays of arc lighting in urban centres bespoke a new age of illumination. Hydroelectric plants arrived in Canada after the heavy lifting of invention and system design had occurred in the 1880s in the northeastern United States and Europe.[7] In Canada, early adopters backed by domestic and foreign investors formed companies to obtain water rights from provincial governments, construct dams and power houses, and generate electricity for local markets.[8] The earliest sites of innovation and expansion were in southern Ontario, along the Niagara frontier, and on the St Lawrence and its tributaries in Quebec, where swift flowing rivers tumbled off the Canadian Shield.[9] Smaller projects emerged slightly later in the Canadian west on the Winnipeg River in Manitoba and the Bow River in Alberta, in the mining districts of south central British Columbia, and on small streams at the edges of Vancouver and Victoria.[10] Maritime centres also adopted hydro systems in their turn but more slowly, owing partly to the availability of coal.[11]

In the earliest phase of development, hydro systems using direct current were highly circumscribed. Power could not travel efficiently over distance without major losses off the lines. As alternating-current systems were developed in the 1890s, power could be transmitted at higher voltages and over greater distances. When an eleven kilovolt line was constructed in 1897 between St-Narcisse and Trois-Rivières, Quebec, a distance of 27 kilometres, it proved to be the longest in the British Empire. Six years later, a line stretching 136 kilometres was built between Shawinigan and Montreal.[12] Other innovations in transmission would follow to meet the classic Canadian problem of distance.

The earliest sites of hydro development shared some characteristics. First, as a result of the transmission problem they were all located close to urban centres or resource settlements. Unlike fuels such as coal or oil, electricity could not be loaded onto trains or stored in warehouses. As hydro-powered systems adopted alternating current and expanded in scale, however, they needed a complex delivery infrastructure. Transmission lines suspended from wooden poles and later lattice metal towers carried electricity from powerhouses to transformer stations where the voltage was "stepped down" or lowered. From there, distribution lines strung along roadsides and hanging on porcelain insulators atop wooden poles carried electricity to factories, businesses, and homes. From the site of generation to the electrical outlet, emerging systems formed a whole united by wires and the electrical current they carried.

Second, hydro developments all arose at sites where there were several important material conditions: a flowing river, a firm foundation (preferably granite bedrock), and a physical space that included a topographical differential (a falls), and possibly a canyon for flooding. These material conditions allowed engineers to design water control systems that included a dam to impound the river and create a reservoir, canals and penstocks (pipes that carried a controlled flow through a powerhouse), and a tailrace where the excess water flowed past.

These common characteristics occurred unevenly across Canada. Not only did provinces approach hydro development differently from a legal and political perspective, they also faced different resource endowments and opportunities. Sections of Canada well-endowed with rushing rivers (the Canadian Shield and mountainous regions of the West) turned to hydroelectricity early and gained a lead that would endure; in areas without obvious water power sources, such as Saskatchewan and PEI, options were limited and developments few; in areas with productive coal deposits and mining industries, such as Nova Scotia and Alberta, thermal generation competed with hydro.[13] The energy map of Canada looked one way through the prism of a coal-based industrial revolution. It looked quite differently through the prism of hydroelectricity and the second industrial revolution.

The earliest hydro systems faced design problems that would not go away. Unlike most energy forms, electricity cannot be stored, except in small amounts. Once electricity has been generated, it must be utilized almost immediately, or it will be wasted. Hydro developers had to consider the best means to generate power predictably and steadily through the day and seasons, notwithstanding variations in river flow. The best means to avoid flow variation was by tapping a major river (like the Niagara or the St Lawrence) and diverting some of its flow through a power canal, or by damming upper basin tributaries and lakes as storage reservoirs, which could be released downstream in periods of low flow. From a hydro developer's perspective, freshets and droughts were problems; the best river was an even river, steady and predictable. Of course, rivers rarely fit this ideal type and had to be trained.

Hydro developers not only had to consider the best means to redesign rivers and build wired systems, they also had to think about their markets and variations in demand. Electrical engineers conceived of use as utilization and markets as load. Hydro systems worked most efficiently when the utilization matched system output and when the output could adapt to load or vice versa. Utilization and load could be affected by numerous factors. In all early Canadian systems, demand and load dropped on Sundays, at night, and during the summer. The ups and downs of demand created swings in hydro systems that had to be faced in real time by people managing instrumentation. When demand declined, they turned off turbines and saved water. When demand spiked, they set all of their turbines spinning and let the river roar. Because the variations of markets were as troubling to electrical engineers as the variations in rivers, they worked closely with marketing departments to attract predictable customers with large appetites for power. Streetcar systems offered one excellent and relatively predictable form of demand. As a result, many early power companies developed their own urban transportation systems, opening up new acres for residential and industrial expansion. Hydroelectricity was, therefore, one of several conditional factors making suburban expansion possible.[14] Power companies also sought out a range of industries that relied on large blocks of energy, such as smelters and pulp and paper mills. They needed contracts that could justify the initial large capital investments in the system and then moved on to market excess power to other consumers. While power for commercial businesses and households added to the mix, before 1950 it accounted for a relatively low proportion of total demand.

Domestic demand grew slowly for a variety of reasons. For one, power companies had to make the effort to build the web of distribution lines in urban markets to create the possibility of a domestic market. The speed with which this happened depended on the city and the calculated capital costs versus profits. Beyond city limits, extension occurred slowly. In addition to the infrastructural, supply-side cause of delays, there were a range of barriers at the scale of the household. For households to become consumers, houses had to be reconceived, wires had to be integrated into the design and light fixtures and plugs installed. This all took time and money and raised questions about the risk of electrical fires. In addition, domestic consumers had to be convinced that the additional expense of electricity consumption bought something worthwhile. For decades, most domestic households relied on electricity primarily for lighting. The adoption of appliances like electric ranges, irons, kettles, and washing machines was neither immediate nor ensured.[15] Appliances had to be demonstrated and sold. In many instances, utilities provided customers with appliances at a fraction of their real cost to spur consumption in off-peak hours.[16] Consumers who had been raised on other technologies and domestic labour practices did not immediately see the point, or at least could not justify the cost. Urban utilities ran a range of sales campaigns including mobile electric kitchen demonstrations to convince consumers, but the barriers of cost and custom only began to dissipate across society after the Second World War, assisted by rising income levels and new ideals and patterns of domesticity.

From the start, questions emerged about how to organize hydro systems as elements of democratic society. Since most hydro systems encompassed a regional market and since most such markets did not contain competitive systems, hydroelectricity was understood to be one of a number of utilities that existed as "natural monopolies." As rates of usage rose, the idea that electricity was also a necessity of modern life complicated the issue further. How to control a monopoly delivering a public good? In Ontario in the late nineteenth century, a highly organized group of municipal politicians collaborated to insist that hydroelectricity ought to come under the purview of the provincial government working in collaboration with a range of municipally controlled utilities. To avoid corporate exploitation and to liberate the magical powers of electrification, these civic populists argued for a strong state role. The Hydro-Electric Power Commission of Ontario (HEPCO) that emerged from their efforts proved to be a vanguard institution not only in Canada but also in North America. Hitched to wider goals of public policy, it pursued rural electrification programs and sought to stimulate consumer demand.[17] In other parts of Canada, where private corporations dominated hydro development and utilities, the cultivation of the consumer market occurred much more slowly. In Quebec, private power firms sold most of their power to large industries. In smaller markets, like British Columbia, urban-based utilities like the British Columbia Electric Railway Company disposed of most of their power in the streetcar system. In Vancouver, the BCER also made large sales to the grain elevators, light manufacturing, and resource processing firms.[18] City and provincial governments sought to cajole private power firms to pursue wider public goals but without much success. When utilities commissions were invented in different jurisdictions starting in the 1910s, they worked to monitor and regulate rates and ensure against price gouging.[19] But this was a far cry from the public policy goals pursued by HEPCO. With time, other provinces would create their own power commissions, or provincial utilities, but the process varied by region and extended well into the 1960s.

Hydroelectric development also changed rivers and conflicted with a host of other river uses and interests. When hydro developers looked at rivers, they counted cubic feet per second. Rivers mattered in terms of their volume, flow, and control. In the earliest phase of expansion, hydro engineers made no account of other human uses or understandings of rivers. Since hydro systems invariably redesigned rivers and their flows, however, hydroelectric development affected fluvial geomorphology and ecology, riverine aquatic ecosystems, and the range of animals and plants that found sustenance at the river's edge and in the watersheds beyond. Because other people and interests depended on rivers for other purposes, this development process could be threatening, even ruinous. As a result, the early promise of hydro expansion ran headlong into social groups who demanded respect for prior uses and claims. Sometimes this involved contending industrial interests, such as timber harvesters who valued rivers as log transportation systems. Sometimes it involved farmers who discovered that their lands would be flooded by reservoirs. And sometimes it involved indigenous peoples who found traditional harvesting and economic uses of rivers displaced or undermined. Chief Johnnie of the Coquitlam Band put this point to federal fisheries officials in 1899 when he heard rumours that the BCER would dam the Coquitlam River: "If the creek is taken away from us," he wrote, "it is like a man taken the food out of my cupboard."[20] Although there were some legal constraints in the early twentieth century to block or temper dam development, particularly the federal fisheries act, which explicitly forbade riverine obstructions, they were rarely exercised.[21] Hydro developers proved able to exploit jurisdictional conflicts and provincial developmental ambitions to brush protests aside.

How did things look by the 1920s? Although systems were slowly developing through the first two decades of the twentieth century, expansion surged in the 1920s in lockstep with a wider phase of growth in the continental economy. New dams were constructed east and west. Some sectors, like the burgeoning export trade in pulp and paper, were premised on the availability of cheap hydro power. Following coal shortages during the First World War, a number of utilities began to enhance their interconnections with neighbouring systems to reduce dependence on local generation sources and to spread the risk of demand peaks. This practice, known as load balancing, increased the reliability of systems while reducing capital expenditures. Across the country, utilities began to investigate new generation projects, assuming that the growth would never end. In the most developed provinces, regional systems now encompassed wide territories. These systems integrated a range of generation sources (dams and power plants) into regional systems of transmission. In southern Ontario, for example, the HEPCO operated the Niagara system, extending around the Niagara peninsula, west to Sarnia, and north past Toronto where it abutted a neighbouring system on Georgian Bay. In Quebec, one of the largest systems run by the Shawinigan Water and Power Company encompassed the St Maurice Valley, but diverted wholesale power to neighbouring systems in Montreal and south across the St Lawrence River.[22] By the end of the decade, after a rapid phase of growth, many of the country's largest systems had drawn their developmental outlines, wired together key markets, and prepared blueprints for another phase of expansion. The stock market crash in the fall of 1929 brought these plans to a halt. As Canada's resource and manufacturing economy contracted, so did power demand. It would not revive significantly until the Second World War forced a new hydro mobilization.

The Second World War drove a wide-ranging expansion of hydroelectricity in Canada and laid the foundations for postwar growth. Over six war years, hydroelectric output increased by 40 per cent. Under a federal program of power control, interconnections were executed, reservoirs built, dams raised, and power conservation programs introduced. Canada and the United States cooperated to expand power production on the transboundary Kootenay River, to release limits on water-taking at Niagara Falls, and to agree on a St Lawrence Seaway program – which would ultimately be deferred as a postwar project. Canada's munitions and supply role in the Allied war effort led to major expansions of mineral smelting and other energy-intensive chemical production. In Calgary, a single ammonium nitrate plant was established in 1940 that consumed the equivalent of half the existing urban demand. In Quebec, the aluminum industry increased its output twenty-fold and required a suite of new dams in the Saguenay Basin as well as power diversions from neighbouring systems delivered over new or expanded transmission networks.[23] After a decade of no or slow growth during the Depression, the war triggered a new wave of expansion. Looking towards the postwar period, utilities and governments began to envision a wider diffusion of the benefits of electrification across society through a range of rural electrification programs, mandated rate reductions, and public utilities. As the postwar economy expanded, and as lines extended and prices dropped, industrial, commercial, and household demand increased, fuelling further rounds of development. A new era of megaprojects lay on the horizon.[24]

The postwar phase of hydro expansion witnessed a scale shift in market demand. Consider figure 9.2 and the steady climb in demand after 1950 that dwarfed pre-war levels. Increases occurred in all classes of sales. Energy-intensive industries remained the major factor spurring growth, but with the emergence of almost universal electrification, population growth, and increased consumption levels emerged as significant drivers. After the war, developmental pricing policies lowered unit costs based on volume purchases. Households paid less per kilowatt as they used more and more. Concerted efforts were also made to electrify small-town and rural Canada beyond the bounds of the urban markets that had served as the core of pre-war electricity markets.[25] As a component of total demand, consumers came into their own after 1960. Residences and farms consumed about a quarter of total power generated in 1960, and about a third in 1970 (see figure 9.2). After mid-century a host of new electrically powered domestic goods became fixtures of everyday consumption, from kitchen ranges to vacuum cleaners to washing machines.[26] In markets with low electricity costs, electrical space heaters were installed in new homes and condominiums, which built load to new heights.[27] As air conditioning technologies emerged in the 1980s, a further change in consumption patterns occurred, shifting the seasonal rhythms of peak loads. In the mid-1990s, the search for cool air in Ontario drove up electricity demand past wintertime peaks. In the first decade of the twenty-first century, the top twenty demand peaks in Ontario all occurred between June and August.[28]

Market demand also became a more complex, long-distance affair. Although transboundary exports were executed along the Canada-US border in the first phase of hydro development, the federal government moved in quickly to regulate their operation.[29] Power export agreements expanded during the Second World War in the context of continental resource collaboration, but as a temporary measure. A shift began in the 1950s when Canada and the United States collaborated on the St Lawrence Seaway development; it came to completion in 1964 when British Columbia forced the federal government to allow power exports in the context of the Columbia River Treaty.[30] New long-distance transmission systems made such possibilities imaginable, as did rising demand in all North American markets.[31] In the late twentieth century, different

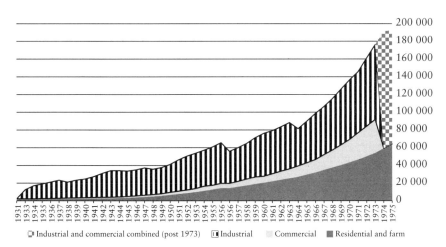

Figure 9.2   Electric utility sales by class of customer, Canada, 1930–1975 (millions of kWh)

Source: Statistics Canada, Historical Statistics of Canada, series Q102–106, *Electric Utility Sales by Class of Customer*, selected years, 1930–1975 (millions of kWh), http://www5.statcan.gc.ca/access_acces/archive.action?l=eng&loc=Q102_106-eng.csv.

provincial utilities deepened grid connections to neighbouring US utilities, profiting from spot sales in peaking US markets and from long-term supply contracts. These export sales never crept above 10 per cent of total demand, but they could be highly profitable. In Quebec, exports to neighbouring jurisdictions in Canada and New England accounted for 20 per cent of power generated.[32] The north-south axis of development may well have pursued the provincial logic of hydro development, but as Karl Froschauer argues, it ruled out anything like a national electrical policy.[33]

To meet rising postwar demand, new investments had to be made in dams, transmission lines, and infrastructure. Some of this effort was undertaken by private developers, for specific resource-processing facilities, as in Alcan's Nechako project in northern British Columbia, completed in 1952 to power an aluminum smelter at Kitimat.[34] For the most part, however, the emerging big-dam era turned on public investments. In the immediate postwar period a series of provincial power commissions in British Columbia, Manitoba, and New Brunswick extended electrical systems into previously under-served rural areas and began to build a range of generation projects.[35] In Quebec a more sweeping expropriation of the Montreal Light, Heat and Power Company in 1944 set the stage for a new provincial power commission, focused initially on the Montreal region but setting out the parameters for future intervention.[36] In Ontario, HEPCO became more centralized as a Crown corporation and pursued new generation projects on the Ottawa River, the St Lawrence, and at Niagara.[37] Ontario's expansion required international agreements with the United States over water-taking from the Niagara and navigation and power on the St Lawrence. None of this was easy to work out. International development on the St Lawrence proceeded only once Canada threatened to build the project on its own.[38]

In the 1960s, provincial hydros drove development at a new scale. Between the mid-1960s and the mid-1970s Canada's generating output doubled (see figure 9.1). British Columbia's Social Credit government created BC Hydro in 1961 when the privately owned BC Electric would not pursue ambitious new dam developments on the Peace River that the provincial government favoured. BC Hydro quickly moved to build new installations on the Peace River, which, coupled with developments on the Columbia, significantly increased the province's generation and export of hydroelectricity.[39] On the Peace, the WAC Bennett Dam created the largest reservoir in North America, Williston Lake.[40] In Quebec the Liberal government of Jean Lesage directed Hydro-Québec in 1963 to cannibalize every significant private electricity utility in the province, thereby increasing its generating capacity by several orders of magnitude. Hydro-Québec also moved forward with a new keystone development in eastern Quebec, the Manicouagan-Outardes Project (often called Manic-Outardes).[41] Hydro-Québec not only nationalized power in Quebec, it also bound together hydroelectricity with the emerging themes of Quebec nationalism. Technological modernism was infused with a spirit of emancipation and the claims of "maîtres chez nous."[42]

Although British Columbia and Quebec were significant postwar builders, power development also played an important role in the Prairies and Atlantic Canada. The Gardiner Dam, Saskatchewan's first foray into hydroelectric development, was completed in 1967 on the South Saskatchewan River.[43] In Manitoba, studies were undertaken to extend the generating potential of the Nelson River by diverting the flow of the Churchill River to supplement the Nelson as it made its way to Hudson Bay. When completed in the 1970s, these installations effectively turned Lake Winnipeg into a giant reservoir.[44] In New Brunswick, the Mactaquac project on the St John River was conceived in the 1960s as another step towards providing power for industry but was turned into a regional modernization scheme, modelled partly on the Tennessee Valley Authority.[45] In Newfoundland, an outlier in provincially controlled hydro development, the privately held British Newfoundland Company (or Brinco) built a hydroelectric generating station at Churchill Falls in Western Labrador rated at over seven million horsepower. The electricity was then directed to southern markets via Hydro-Québec transmission lines at prices negotiated through a long-term contract. The Churchill Falls project remains controversial because the Newfoundland and Labrador government contends that the province has not received fair market value on its electricity exports; Hydro-Québec, by contrast, notes the service of its transmission system and stands by the original terms of the 1969 contract.[46]

The greatest hydro expansion of the late twentieth century was realized in northern Quebec, on James Bay. Initiated by Premier Robert Bourassa in 1971, Hydro-Québec's James Bay Project was conceived as a massive undertaking, unparalleled at the time in endeavour, geographical scope, or landscape alteration. Built primarily on the La Grande River, construction of the various installations proceeded over two decades. Currently, the generating capacity of the entire project reaches over 16,000 megawatts, making it one of the largest systems in the world. The scope of this project and its diverse social and environmental effects caused Cree leaders, activists, and community members in the region to seek measures to protect their homelands of Eeyou Istchee. During the initial phase of development, the Cree and Inuit negotiated the first modern comprehensive land claims in Canada, the 1975 James Bay and Northern Quebec Agreement, which secured limited self-government, regional land management authority, and cash compensation in exchange for allowing James Bay I to proceed. This was an important development but did not ameliorate the negative environmental effects of dam construction. In the mid-1990s, under the charismatic stewardship of Grand Chief Matthew Coon Come, the Cree successfully sued to block the planned extension of the Hydro-Québec dam network, and James Bay II, or the Great Whale, was temporarily cancelled.[47] This was made possible in part by the Cree's strategy of working closely with environmentalists in Canada and the United States to force US states that purchased Quebec power to reconsider.[48] Quebec nationalists who associated Hydro-Québec's mission with nationalist emancipatory goals viewed these protests as acts of sabotage. Despite the cancellation, parts of this expansion program have since been revived and implemented.

The Cree protest may have been one of the most successful anti-dam protests of the postwar period, but it was not without precedent. Dams had been debated and fought on the Fraser River during the 1950s, in response to the potential consequences of large dam development on the river's salmon fishery.[49] In the 1960s protesters in British Columbia called into question the damming of the Arrow Lakes as storage for American dams downstream and sought to defend "wilderness" against flooding from Seattle Lights' Ross Lake high dam.[50] In all of these cases, protest and resistance to dam construction was diverse, often mixing direct action with protracted policy advocacy. It was also diffuse, with local protest often supplemented by campaigners working in metropolitan centres. Nor was protest uniform. In northern British Columbia in the 1970s BC Hydro's Stikine-Iskut plan to dam a tributary of the Stikine River produced mixed local reactions. Members of the Tahltan Nation who feared the impacts of dams on indigenous territories and economies fought BC Hydro's attempt under the banner, "No Dam Way," while others in the community worked for BC Hydro contractors to make the dams a reality. The Friends of the Stikine, an advocacy group based in Victoria, was an occasional ally of the Stikine-based dam dissenters but was often antagonistic as well, reflecting different values about nature and outsider knowledge.[51] Notions of a pristine nature under threat were troubled by the lived presence of indigenous peoples.

While dam developments were abandoned in a few rare and important cases, protests often led only to project deferrals or did little to affect the course of development. This led rural and indigenous communities to bear the environmental brunt of hydro development: from the loss of farmlands in the upper Saguenay in the 1920s to the relocation of whole villages and communities along the St Lawrence, Nechako, and Columbia Rivers in the 1950s and 1960s.[52] As developments turned northward in the 1960s, an increasing number of displacements affected indigenous communities in particular. In Manitoba, the construction of the Grand Rapids Dam on the Saskatchewan River in the early 1960s forced the relocation of the Cree community of Chemawawin to a new town at Easterville. While Easterville was planned to include a range of modern conveniences, it lacked a good resource base and problems emerged with sanitation infrastructure and water quality. As materials decomposed in the reservoir, mercury was released and stored in the fatty tissues of fish. This led government officials to close the fishery, which merely exacerbated the social burden of relocation.[53]

These examples point to the blunt geographical consequences of postwar construction. Dams not only displaced settlements, they also changed the land and rivers: altering fish habitats and riparian ecosystems, triggering bank erosion and new ecosystem dynamics in reservoirs.[54] People's material circumstances changed, as well as their sensuous experience of place.[55] Hydro megaprojects were built in remote locations, far from the metropolitan contexts of electricity consumption and the centres of a bourgeoning environmental awareness and organization. There was a certain environmental determinism (inevitability?) to this phenomenon, as the most suitable sites for dam construction were located in the north, but the distance also eased the politics of development from the board-rooms and backyards of urban Canada. As a result of this uneven geography, the negative environmental consequences of dam construction and electricity generation were experienced primarily by rural and indigenous peoples living in peripheral locations.

The transmission lines that linked dams and markets produced their own environmental footprints. Transmission lines required pathways, many of which were cut across Crown lands, but others required the dispossession of agricultural and other rural and urban lands. To give transmission towers and lines sufficient unobstructed space, vegetation and trees were removed in a corridor forty to sixty metres wide. Once cut, these pathways required regular maintenance and/or herbicide applications. In Ontario in the mid-twentieth century, Agent Orange was applied as a highly effective defoliant.[56] Since transmission lines cut across vast distances, linking northern dams and southern markets, the scope of corridor construction and maintenance became significant. By 2013, BC Hydro maintained about 18,000 kilometres of transmission corridors across British Columbia.[57] Where possible, utilities have tried to integrate secondary uses alongside transmission line corridors, such as parking lots and soccer fields, but this accounts for only a limited section of the infrastructure, which passes primarily outside of urban areas.

By the 1980s, several factors combined to slow the pace of hydro expansion and call into question some basic tenets.[58] In part this was in response to the protests waged by anti-dam organizers and indigenous peoples. Provincial governments and provincial hydros responded to local concerns by integrating new protocols into the approval process in environmental and socioeconomic impact assessments. In most jurisdictions assessments became common practice only in the late 1970s.[59] As a form of social licence, they offered a measure of transparency and accountability and provided venues for public consultation. Assessments also produced a range of studies to provide guidance about assumed impacts. Reports on such issues as geological features, animal habitats and populations, river characteristics, regional economies, and employment figures, itemized, catalogued, and enumerated the environment that was to be used or altered by dam construction. The economies that would result from dam construction were then compared to other potential land revenues, like archaeological sites or mineral values. The results left room for cynicism. As one BC Hydro contractor suggested, environmental assessment was designed to prove the "technological feasibility and economic attractiveness" of a proposed generation scheme.[60] Nevertheless, once instituted, new regulatory processes could lead to diverse results. In British Columbia, the newly constituted BC Utilities Commission was given a mandate to provide an independent review of hydroelectric developments. In its first decision in 1981, it did not grant an energy project certificate to BC Hydro to build a third dam on the Peace River. The utility, it claimed, had overestimated future power demands. Another dam would be unnecessary.[61] Adding to the complications arising from this new web of regulation, provincial hydros faced increased borrowing and debt-servicing costs as interest rates spiked in the early 1980s, putting a brake on a range of projects and causing sober second thoughts in provincial capitals. Increasingly, large dam projects did not seem like the uncomplicated blessings that they had appeared to be in earlier decades.

As dams became harder to organize and build, provincial hydros sought to manage their assets more carefully and took new directions to expand generation capacity. By the early 1980s, several provincial utilities were well along the path to developing thermal and nuclear power facilities for lack of hydro opportunities within their own provincial boundaries. This shifted the dependence away from hydro but also produced its own environmental consequences in fossil fuel emissions and the range of social and environmental concerns associated with nuclear waste and risk, as Laurel Sefton McDowell describes in chapter 12 on nuclear power. Provincial utilities also sought to better position themselves in cross-border power trading. BC Hydro developed its export business into an art, selling power into the US market when prices rose, and purchasing power back during market troughs. PowerEx, the utility's export arm, proved to be a key component of a provincial strategy to hedge against the market and exploit the benefits of an integrated grid stretching as far south as California. In several provinces, utilities also began to explore seriously a range of demand-oriented policies to dampen rising consumer and commercial electricity needs. These included public education campaigns, appliance exchange programs, household power audits, and a staggered pricing system to encourage consumers to monitor and reduce their electricity demands.[62] The savings generated by such programs were significant and helped to curb the pace of expansion.

It also put an end to the unquestioned and long-term goal of demand stimulation, which had reigned since the late nineteenth century.

Canada is a hydroelectric giant on the international scene, a country that has profited handsomely from the benefits furnished by hydroelectric development, supplied by large rivers and a large land base. Over the course of a century, entrepreneurs, multinational businesses, and government corporations built large technological systems that converted the kinetic energy in falling water to electricity at one end of the energy system and produced flashing lights and humming appliances at the other end, often at great distance. Dams and powerhouses drove the system and transmission and distribution lines carried power to markets. All Canadian regions felt the impact of these processes. By the turn of the twentieth century, over 930 large dams studded Canadian rivers along with thousands of small dams and plants.[63] To the night sky, Canada revealed its electrified face, glittering and bejewelled, a riot of lights emphasizing the settled ecumene and urban network. Invisible against the dark night remained the costs of development and their uneven burden.

Although the scope and scale of hydro-powered electrical grids presented enormous possibilities, they also introduced profound vulnerabilities. Predictably unpredictable environmental changes in rivers and weather conditions could upset the carefully calibrated functions of hydro systems with long-distance effects. Floods, ice storms, low precipitation levels, wind events, lightning strikes, and riverine ice jams could all wreak havoc with the fundamental processes of hydro systems, triggering temporary shutdowns in their mildest forms to regional blackouts in the worst-case scenario. In recent times, the 1998 ice storm that caused transmission towers to bend like tinfoil under the weight of ice, stands out as the most remarkable example, one in which millions of customers lost power over wide sections of eastern Ontario and Quebec for many days. Outages have always marked hydro systems, however, and so the story of hydro's progressive expansion that we have related here needs to be understood also in the context of a constant background noise of breakdown, repair, and risk.

Although our analysis ends with the rise of demand-side management and the decrease of large projects in the 1980s, the story continues to unfold. In the last decade, for example, British Columbia, Manitoba, Newfoundland, and Quebec have all launched new large dam projects, some deferred from the era of cancellation in the 1980s. Newfoundland's lower Churchill River project, for example, promises to resurrect a dam project conceived in the 1970s but delayed because of the longstanding dispute with Hydro-Québec over the contracted price of Churchill power. The project will proceed in the twenty-first century by circumventing Hydro-Québec's transmission system entirely using an underwater maritime transmission link between Labrador and Nova Scotia. With the increasing flexibility of grids to alternative generation hookups, a range of small projects have also been pursued, from wind power to geothermal projects to micro-hydro, a neologism for hydroelectricity on smaller rivers that seeks to emphasize minimal impacts. Coal and gas power plants also continue to be built to meet the rising peak demands of North American markets. Notwithstanding the potentially detrimental environmental and social consequences of dam development, compared with fossil fuels and nuclear power many have now recast hydro as a "green" renewable energy form. Governments wary of the unpredictable response of environmental communities to new hydro developments have begun to embrace this language, as in British Columbia's Site C development.[64] Hydro-rich provinces have also begun to exploit the presumed virtues of hydro power to extract political gains from jurisdictions that rely more significantly on fossil fuels. With great self-satisfaction, for example, commentators in Quebec compare the province's hydro-powered economy to the coal-based thermal plants and oil sands of Alberta, forgetting the extent to which this position of high virtue is a product of geography as much as choice.[65]

NOTES

1 The Moran dam idea is discussed in Evenden, *Fish versus Power: An Environmental History of the Fraser*, 179–230.
2 Canada, Dominion Bureau of Statistics, Canada, "Central Electric Stations," 3.
3 United States Environmental Protection Agency, "Fuel Mix for U.S. Electricity Generation."
4 World Energy Council, *Energy Resources, Hydropower*.
5 A fascinating comparative perspective is provided in Hausman, Hertner, and Wilkins, *Global Electrification: Multinational Enterprise and International Finance in the History of Light and Power, 1878–2007*, which includes a Canadian dimension because of H.V. Nelles's research contributions to the volume.
6 For readers unfamiliar with the constitutional history of Canada, it is well to point out that many of the important legal and political dimensions of hydroelectric development were framed in the British North America Act

(1867), Canada's founding constitutional legislation. Those who framed the act did not anticipate hydroelectric development, of course, but the division of powers they outlined fundamentally shaped the process. The division of powers, principally outlined in sections 91 and 92 of the act, allocated control over natural resources to the provinces, yet granted the federal government authority in a range of spheres that would bear on river management and hydroelectric development. The federal government held authority over navigable rivers, for example, as well as fisheries. Raising dams in rivers would thus introduce the possibility of federal intervention. The federal government also held authority over rivers that crossed provincial borders or Canada's border with the United States. This introduced another area where federal and provincial jurisdictions intersected. In a related way, the federal government held authority over issues of trade and export, which became relevant when Canadian utilities began to sell power into the United States market. While, in general terms, hydro was a provincial matter, in a host of cases the scope and extent of hydro development involved a multi-jurisdictional exercise in contested governance.

7 The classic work on the development of electrical systems remains Hughes, *Networks of Power: Electrification in Western Society, 1880–1930*; see also Nye, *Electrifying America: Social Meanings of a New Technology.*

8 Armstrong and Nelles, *Monopoly's Moment: The Organization and Regulation of Canadian Utilities, 1830–1930.*

9 Massell, *Amassing Power: J.B. Duke and the Saguenay River, 1897–1927*; Massell, "A Question of Power: A Brief History of Hydroelectricity in Quebec"; Nelles, *The Politics of Development: Forests, Mines, and Hydroelectric Power in Ontario, 1849–1941.*

10 Armstrong, Evenden, and Nelles, *The River Returns: An Environmental History of the Bow*; Mouat, *The Business of Power: Hydro-Electricity in Southeastern British Columbia, 1897–1997*; Roy, "Direct Management from Abroad: The Formative Years of the British Columbia Electric Railway."

11 King, "The Electrification of Nova Scotia, 1884–1973: Technological Modernization as a Response to Regional Disparity."

12 Fleming, "Hydroelectricity."

13 King, "Electrification of Nova Scotia, 1884–1973."

14 Harris, *Creeping Conformity: How Canada Became Suburban, 1900–1960*; Nye, *Electrifying America.*

15 Sandwell, "Mapping Fuel Use in Canada: Exploring the Social History of Canadians' Great Fuel Transformation," and "Pedagogies of the Unimpressed."

16 Stadfeld, "Electric Space: Social and Natural Transformations in British Columbia's Hydro-electricity Industry to World War II," 190.

17 Nelles, *The Politics of Development: Forests, Mines and Hydro-electric Power in Ontario, 1849–1941*.

18 Roy, "The British Columbia Electric Railway Company, 1897–1928: A British Company in British Columbia."

19 Armstrong and Nelles, *Monopoly's Moment*, 187–210.

20 Quoted in Evenden, *Fish versus Power*, 73.

21 Ibid., 53–83.

22 Bellavance, *Shawinigan Water and Power, 1898–1963: Formation et décline d'un groupe industriel au Québec*.

23 Massell, "'As Though There Was No Boundary': The Shipshaw Project and Continental Integration"; Massell, *Québec Hydropolitics: The Peribonka Concessions of the Second World War*; Evenden, "Aluminum, Commodity Chains and the Environmental History of the Second World War"; Evenden, *Allied Power: Mobilizing Hydroelectricity during Canada's Second World War*.

24 Evenden, "Mobilizing Rivers: Hydro-electricity, the State and the Second World War in Canada."

25 Dorion, "L'électrification du monde rural québecois"; King, "The Electrification of Nova Scotia, 1884–1973," 287–8; Dolphin and Dolphin, *Country Power: The Electrical Revolution in Rural Alberta*; Fleming, *Power at Cost: Ontario Hydro and Rural Electrification, 1911–1958*, 221–47; Kenny and Secord, "Public Power for Industry: A Re-examination of the New Brunswick Case, 1940–1960"; Netherton, "From Rentiership to Continental Modernization: Shifting Policy Paradigms of State Intervention in Hydro in Manitoba, 1922–1977," 111–223.

26 Parr, *Domestic Goods: The Material, the Moral and the Economic in the Postwar Years*, 243–5.

27 A new demand peak reached in Toronto in 1988 was linked by Toronto Hydro officials specifically to new residential construction and electrical heating demands. Moloney, "Electricity Demand Tops Record."

28 Independent Electricity System Operator (IESO), "Ontario Demand Peaks."

29 Martin-Neilsen, "South over the Wires: Hydro-electricity Exports from Canada."

30 Macfarlane, "'Caught between Two Fires': St Lawrence Seaway and Power Project, Canadian-American Relations, and Linkage"; Parr, "Movement and Sound: A Walking Village Remade: Iroquois and the St Lawrence Seaway," in Parr, *Sensing Changes*. The Columbia case was complicated by the fact that BC was not initially selling power generated in BC but power that would have been returned to Canada under the terms of the downstream benefits

principle of the Columbia treaty. As BC could not then absorb this power, the BC government wished to sell the power and use the funds earned thereby to help finance the Peace River power project. Swainson, *Conflict over the Columbia: The Canadian Background to an Historic Treaty.*

31 Hirt, *The Wired Northwest: The History of Electric Power, 1870s–1970s.*

32 Massell, "Question of Power," 141.

33 Froschauer, *White Gold: Hydroelectric Power in Canada.*

34 Windsor and McVey. "Annihilation of Both Place and Sense of Place: The Experience of the Cheslatta T'En Canadian First Nation within the Context of Large–scale Environmental Projects"; Coates, "The Power to Transform: The Kemano Power Project and the Debate about the Future of Northern British Columbia"; Evenden, *Fish versus Power,* 149–78.

35 Evenden, *Fish versus Power,* 119–48; Netherton, "From Rentiership to Continental Modernization"; Manitoba Power Commission, *22nd Annual Report of the Manitoba Power Commission, Year Ended 30 November 1941,* 6; Manitoba Electrification Enquiry Commission, *A Farm Electrification Programme: Report of Manitoba Electrification Enquiry Commission, 1942.*

36 Bellavance, Levasseur, and Rousseau, "De la lute antimonopoliste a la promotion de la grande entreprise l'essor de deux institutions: Hydro-Québec et Desjardins, 1920–1965," 551–78.

37 Freeman, *The Politics of Power: Ontario Hydro and Its Government, 1906–1995,* 87–118.

38 Macfarlane, *Negotiating a River: Canada, the United States and the Creation of the St Lawrence Seaway.*

39 Mitchell, *WAC Bennett and the Rise of British Columbia.*

40 Loo, "Disturbing the Peace: Environmental Change and the Scales of Justice on a Northern River."

41 Caron, "De Manic-Outardes à la Baie James: la gestion des choix techniques à Hydro-Québec."

42 Desbiens, *Power from the North: Territory, Identity, and the Culture of Hydroelectricity in Quebec.*

43 See Rasid, "The Effects of Regime Regulation by the Gardiner Dam on Downstream Geomorphic Processes in the South Saskatchewan River."

44 Waldram, "Hydro-electric Development and the Process of Negotiation in Northern Manitoba, 1960–1977."

45 Kenny and Secord, "Engineering Modernity: Hydro-electric Development in New Brunswick, 1945–1970."

46 Bédard, "Le Frontière du Labrador et le Contrat de Churchill Falls Corporation"; Churchill, "Pragmatic Federalism: The Politics behind the

1969 Churchill Falls Contract"; Feehan, "Smallwood, Churchill Falls, and the Power Corridor through Quebec."

47 Desbiens, *Power from the North*; Atkinson and Mulrennan, "Local Protest and Resistance to the Rupert Diversion Project, Northern Quebec"; Carlson, *Home Is the Hunter: The James Bay Cree and Their Land*; McCutcheon, *Electric Rivers: The James Bay Project*; Dufour, "Le Projet Grande-Baleine et L'Avenir des Peuples Autochtones au Québec."

48 See "Power," dir. Isaacson.

49 See Evenden, *Fish versus Power*.

50 Loo, "People in the Way: Modernity, Environment, and Society on the Arrow Lakes"; Van Huizen, "Flooding the Border: Development, Politics and the Environmental Controversy in the Canada-US Skagit Valley."

51 Peyton, "Corporate Ecology: B.C. Hydro's Stikine-Iskut Project in the Unbuilt Environment."

52 Parr, *Sensing Changes: Technologies, Environments and the Everyday*; Loo, "People in the Way."

53 A similar story of dislocation occurred at South Indian Lake as a result of the diversion of the Churchill into the Nelson River. Loney, "The Construction of Dependency: The Case of the Grand Rapids Hydro Project"; Waldram, *As Long as the Rivers Run: Hydroelectric Development and Native Communities in Western Canada*; Martin and Hoffman, *Power Struggles: Hydroelectric Development and First Nations in Manitoba and Quebec*.

54 Rosenberg et al., "Large-scale Impacts of Hydroelectric Development."

55 Parr, *Sensing Changes*.

56 "Agent Orange 'widely used' in Ontario."

57 BC Hydro, "Right of Way Management."

58 Netherton, "The Political Economy of Canadian Hydro-electricity: Between Old 'Provincial Hydros' and Neoliberal Regional Energy Regimes."

59 Gibson, "From Wreck Cove to Voisey Bay: The Evolution of Federal Environmental Assessment in Canada"; Usher, "Northern Development, Impact Assessment and Social Change"; Hoberg, "Sleeping with the Elephant: The American Influence on Canadian Environmental Assessment."

60 Quoted in Peyton, "Corporate Ecology," 362.

61 Smith, "Taming B.C. Hydro: Site C and the Implementation of the B.C. Utilities Commission Act."

62 On the BC case, see Dusyk, "The Transformative Potential of Participatory Politics: Energy Planning and Emergent Sustainability in British Columbia, Canada," 48–9.

63 Environment Canada, "Dams and Diversions."

64 BC Hydro, "Site C Clean Energy Project."

65 The rhetoric has reached a point where specific attempts have been made
to promote interprovincial understanding. See in particular Kelly-Gagnon,
Belzile, and Chassin, "A Plea for a Quebec-Alberta Dialogue." The telling
cover of this document shows a hydro project and a pair of hands holding
oil sands. Recent discussions about building a pipeline to Quebec to process oil sands may well change the nature of the discussion.

10

Petroleum Liquids

Steve Penfold

Running between Windsor and the Ontario-Quebec border, Highway 401 is an almost pure expression of modern engineering vision. Built to the highest standards, it spans 500 miles across southern Ontario, unencumbered by anything approaching scenery. Earlier phases of agricultural settlement had mostly dispensed with forests and wetlands, but where farmers hadn't cleared the way, engineers did the rest. They took account of natural obstacles, but typically to avoid them, smooth them over, blow them up, or (where convenient) use them to the highway's advantage. Learning from the nearby Queen Elizabeth Way, engineers eliminated inconveniences like median landscaping (which disrupted snow removal) and intersecting roads (which would have plagued the efficient running of traffic), while planning officials smoothed the way more metaphorically by deciding against tolls, managing the politics of municipal complaint, and providing freeway-side service centres for hungry cars and drivers.[1] The result was the multi-lane landscape of speed that forms the spine of southern Ontario's automobile culture.

The 401 is also a landscape of petroleum. Most obviously, it is a place to consume. Even before the last section was opened in 1967, tens of thousands of vehicles streamed along its route every day, driven by commuters, tourists, cottagers, shoppers, and at least a few academics. They were joined by trucks, buses, motorcycles, RVs, and snowplows. But vehicles were only the most obvious connection between petroleum and the new highway, whose very construction expressed and depended upon hydrocarbon civilization. During a decade of construction, machines burned *diesel* and *gasoline* to clear land, pour concrete, smooth road surfaces, haul material, and drive workers to construction sites. Parts of the path itself were (quite literally) petroleum: much of the highway was

Table 10.1   World car ownership:
ratio of cars to population, 1927

United States	1:5.3
New Zealand	1:10.5
Canada	1:10.7
Australia	1:16.0
Britain	1:44.0
France	1:44.0
Germany	1:196.0

Source: Flink, *The Automobile Age*, 129.

made of concrete, but sections (and the many roads that linked to it) were *asphalt*, one of the heavier results of crude oil refining. It was a Lincolnesque story – a road built of petroleum, for petroleum, and by petroleum – and a first clue to the ever-presence of oil in modern Canadian life. This chapter charts the rise of petroleum to hegemonic energy status in the twentieth century. This story is at once simple and complex. On the one hand, oil's quantitative and symbolic rise is stunningly clear, a point examined in the first section, "More!" At the same time, the upward curve of consumption was the result of several technological, economic, social, and political – in a word, human – choices. The second section, "The Social Life of Oil," examines these choices at many levels.

MORE!

It is an exaggeration, but not much of one, to say that the car saved oil. Illumination (and, secondarily, lubrication) drove Victorian-era demand for petroleum,[2] but in the early decades of the twentieth century, the industry faced stiff competition from electricity. The decline was slow, incomplete, and uneven, but an oil industry built on its Victorian markets seemed doomed to secondary status. By happy coincidence, Canada's automobile revolution began in earnest just as *kerosene* began losing ground. By the First World War, semi-artisanal auto pioneers gave way to more stable companies and more systematic production. Gordon McGregor brought the Ford Model T across the Detroit River; Sam McLaughlin of Oshawa linked up with General Motors; eventually, Chrysler arrived in Canada as well. By 1930, branch plants of the American Big Three constituted almost 90 per cent of Canadian production, and while none of the Canadian firms had yet fully embraced the efficient mass-production techniques of their American parents, or the standardized one-colour-fits-all approach that

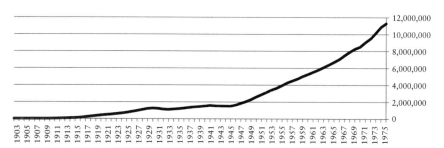

Figure 10.1   Number of registered cars in Canada, 1903–1975

Source: Canada, Statistics Canada, *Historical Statistics of Canada*, series T147–194a, *Motor Vehicle Registrations, 1903–75* (Canada, Newfoundland, Prince Edward Island, Nova Scotia), http://www. statcan.gc.ca/access_acces/archive.action?l=eng&loc=T147_194a-eng.csv.

lay behind Henry Ford's economic philosophy, prices fell quickly, transforming the car from a plaything of the rich and the techno-geek into something bordering on a common product. Even in Moose Jaw – hardly the pioneering centre of Canada's burgeoning consumer culture – a Model T retailed for over $1,000 in 1911 but less than $600 by 1917. There were 2,700 cars in that city by 1929 (up from a dozen or so two decades before), and almost half the cars in Saskatchewan at the time were Fords.[3] It was the same story across Canada, with many local and regional variations, as the car gathered considerable national momentum. Even before the postwar boom, Canada was one of the most car-devoted nations in the world (table 10.1), and while registration rates dipped at some moments, they always recovered in time and resumed their upward trend. By the 1970s, every province averaged one car per household; Ontario had almost one automobile per driver. If anything, the oft-quoted Arthur Lower complaint about the Great God Car seemed understated.[4]

Almost all of these cars burned gasoline. Even when McGregor brought the Model T to Canada, the motive force of cars remained undecided. The gasoline-powered, *internal combustion engine* (ICE) was only one technology competing for supremacy under the hood, with the electric motor and (to a lesser extent) steam holding their own until at least 1910. If anything, as Donald Davis reports, "Canada's main technological advantage at the turn of the century lay in electric motors," but the industry north of the border nonetheless followed the American path toward to the ultimate triumph of the ICE.[5] Scholars still struggle to explain this outcome. In technological terms, each motor had advantages and limits, but many social and cultural factors also intervened, and the proper analytic

Table 10.2   Canadian refinery shipments of selected petroleum products, 1950–1975

	1950	1955	1960	1965	1970	1975
Gasoline and naphtha	46,682	78,258	104,144	131,106	163,756	223,876
Aviation turbo fuel		2,631	4,879	6,734	13,835	23,763
Kerosene and stove oil	6,459	11,064	14,611	18,861	22,295	22,231
Diesel fuel	6,882	15,398	28,192	39,470	52,841	75,685
Light fuel oils (nos 2, 3)	11,546	28,759	48,621	58,657	70,499	90,988
Heavy fuel oils (nos. 4, 5, 6)	21,460	34,035	38,043	48,093	70,245	110,709
Lubricants	1,665	1,540	1,754	3,273	4,017	3,717
Asphalt	4,165	7,201	9,557	11,161	14,772	18,727
Petroleum coke	88	154	178	1,156	603	1,020
Liquefied refinery gas	169	1,851	4,075	5,018	8,020	8,734
Petrochemical feed stocks			3,619	8,298	12,461	

Note: 1960 and earlier figures report "output" rather than "shipments."
Source: Statistics Canada, series Q149–159, *Canadian Refinery Shipments of Petroleum Products, 1949–1976 (Thousands of Barrels)*, http://www.statcan.gc.ca/access_acces/archive. action?l=eng&loc=Q149_159-eng.csv.

balance remains unclear.[6] For the energy historian, the debate about causes (while important) shouldn't distract from the result: the ICE triumphed, and the effect was an almost constant increase in demand for gasoline, a fortuitous development for oil capitalists facing extinction in the face of electric lights. By 1910, for the first time, gasoline and *naphtha* produced more revenue for the Canadian petroleum industry than kerosene; seven years later, gasoline alone represented fully half of the industry's revenue, and by 1921 almost 60 per cent. The ratio of gasoline to other petroleum sales showed considerable variation but remained high throughout the century. Gasoline, then, remained the most significant contributor to dollar volumes of the many petroleum products (table 10.2).[7]

Indeed, at one level of approximation,[8] you could reduce the entire history of Canadian gasoline to a single keyword: *more*. Essay finished. Over time, Canadians bought more cars, generally larger and heavier ones, and drove them farther and more often. Certainly, gasoline consumption did not increase in every single year – there were micro-dips and recoveries at many points, often following the fate of the economy – but the overall increase is undeniable. Only war, regulation, and crisis delayed that dynamic. Gasoline consumption slowed in the early years of the Depression, but nonetheless recovered its upward trend by the late 1930s. Comprehensive wartime rations reduced consumption in the early 1940s, but once controls were lifted during reconstruction, the relentless climb resumed. Even with the so-called energy crisis of the 1970s – set off, at least symbolically, by the OPEC oil boycott in 1973 and called a key "hinge" moment by one American historian[9] – shockingly little long-term change materialized. Certainly, demand for smaller cars increased and fuel economy became one of many factors shaping buying decisions. "Even Cadillac buyers want to know what kind of mileage they can expect," one General Motors dealer commented in 1977.[10] But the federal government's "two-price" policy (controlling domestic prices while allowing export prices to rise, an approach that understandably enraged Alberta) and the short-term nature of the crisis tended to distract Canadians from any sort of energy reckoning. Both polling data and actual behavior suggested that, in general, Canadian consumers were surprisingly unconcerned, at least compared to Americans, who lined up, moaned, phoned congressmen, wrote country songs, and even rioted at service stations. Canadian consumption didn't drop until 1979 and the subsequent recession of the early 1980s.[11]

Sadly, then, long-term "progress" after the "energy crisis" flowed less from fundamental changes in driving practice than from designed-in economies and efficiencies. Cars got shorter and lighter, and governments stepped in with new regulations and guidelines. In the United States, corporate average fuel economy (CAFE) regulations mandated miles-per-gallon targets for a company's total fleet, while Canada instituted a similar but voluntary regime in 1977. Nonetheless, Canadian plants faced many external pressures for improvement. Where corporate officials were not worried about the potential for regulation (in 1982, the federal government passed but did not proclaim mandatory targets), the economic logic of following American standards in a thoroughly continental industry was hard to resist. Still, while fleet efficiency showed some improvement, and companies invested heavily in technological innovation, progress slowed in the 1980s and 1990s, when prices dropped and larger cars returned in new forms.[12] Gas-guzzling minivans and SUVs (the latter being, bizarrely, classified as a light truck and therefore exempt from the most stringent fuel economy regimes) flooded Canadian roads, and gasoline consumption returned to its seemingly inevitable rise by 1986. What could have been an energy reckoning – even Pierre Trudeau warned Canadians of "the revolution of rising expectations" – ultimately confirmed the long history of rising demand.

Of course, the triumph of the ICE went well beyond cars. Any driver on the 401 will know that trucking has become a key economic sector. The trend has a deep history. Intra-city trucking was common enough by the First World War, but few truckers were brave enough to venture onto Canada's shamefully poor intercity and rural roads. With road improvements in some areas after 1920, the estimated radius of economically viable truck delivery began to increase.[13] Postwar growth was even more notable. By 1950, the industry was developed enough that, when Canada faced a national rail strike, the expected calamity did not emerge.[14] Over time, trucking culture intensified, with the advent of "just-in-time" production by the 1970s (which shifted inventories from warehouses to the backs of eighteen-wheelers) and, more recently, the acceleration of a consumer economy built on hyper-efficient distribution (ranging from centralized logistics of Wal-Mart to the decentralized auctions of eBay, both ultimately reliant in their own way on delivery by truck).[15] It is only a slight exaggeration to say that on any normal day you could walk across the top of heavy trucks for the length of the 401.

Moreover, if we pull off the road, as many Canadians did in the age of oil, we would find many other gasoline-powered vehicles gaining momentum, particularly after the Second World War. On land, railways abandoned coal, which alongside trucking produced a steep demand curve for diesel fuel (see table 10.2). On farms, gasoline-powered tractors had been available since the time of the Model T but had been crucially limited by technological problems, cost, and the good sense of farmers (who recognized in horses a much more flexible machine). After 1945, tractors burst out of those confines. By 1951, one study reports, "almost all serious commercial farms were operating with tractors," and five years later there were more tractors than farms in rural Canada. Motivations and incentives were many and complex – technological improvement, falling machine costs, cheap oil, labour shortages, marketing board policies that favoured extensive operations, and the increasing size of the average farm – but the upward trend seemed unstoppable.[16] Other vehicles added to the ICE's off-road triumph. In winter, snowmobiles became suddenly popular in the mid-1960s as both work and leisure machines, with well over 100,000 registered in Ontario alone by 1970, an ascent so sudden that provincial officials struggled to adapt safety and regulatory regimes. On water, gasoline-powered pleasure boats added to the ICE revolution, much to the chagrin of many traditional cottagers, who made sound politics a crucial fulcrum of debate about nature, modernity, and machines. Finally, in the air, the jet age added to our now tedious fuel count, with commercial aviation eating up four billion litres of fuel by 1984 (figure 10.2).[17]

Once we leave the road, we would have to include small machines that used petroleum to serve domestic and industrial purposes, particularly after the Second World War. Take two common examples: chainsaws and

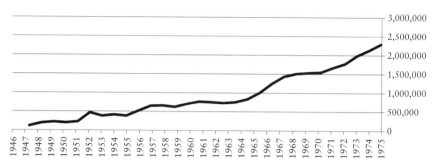

Figure 10.2   Hours flown by Canadian airline industry, 1946–1975

Source: Statistics Canada, Historical Statistics of Canada, series T195–198, *Canadian Commercial Aviation Activity, 1946 to 1975*, http://www.statcan.gc.ca/access_acces/archive.action?l=eng&loc= T195_198-eng.csv.

lawn mowers. The timber industry had been experimenting with power cutting since the First World War, but only a few two-worker models had trickled into forests in the 1930s. By the following decade, improvements in durability and portability produced the chainsaw, a portable and multi-use machine operated by a single worker and driven by a small gasoline motor. Nor was the chainsaw the only gasoline-powered motor entering Canadian forests in this period. In British Columbia, Rick Rajala reports, the ICE transformed almost every stage from forest to consumer, driving a stunning improvement in capacity and transforming the industry's energy regime dramatically. Between 1949 and 1961 alone, industry productivity in British Columbia doubled.[18] Meanwhile, cutting the lawn joined the ICE revolution along a parallel technological path. Motorized mowers actually predated the First World War, but as well-trimmed grass became the common-sense yard in burgeoning suburban neighborhoods in the 1950s, and technological innovation and plastic parts produced lighter and more reliable machines, power lawn mowers became virtually ubiquitous household items. In 1963, tractor-titan John Deere recognized this trend by entering the consumer market. By this time, most lawn mowers were gasoline-powered, as the early century diversity in energy systems (as with cars, gasoline, electric and steam engines had initially competed for lawn-mowing supremacy) gave way to the triumph of ICE, despite the stubborn persistence of lighter electric models. By 2006, about two-thirds of lawn-owning households used a gasoline-powered mower.[19]

Alas, the grim petro-enumeration must continue. Gasoline and motor fuels are the most prominent but hardly the only products that flow from liquid petroleum. Indeed, products are so varied that oil companies came to practise a sort of virtual economics, openly admitting (often at rather convenient moments, like in front of commissions of inquiry) their inability to accurately assess specific costs.[20] As early as 1917, Victor Ross found an astonishing number of uses for oil: "*benzene*, gasoline, kerosene, *paraffine*, paraffine wax, vaseline, white mineral oil, axle grease, *lubricating* oil, fuel oil, *gas oil*, road oils, asphalt, and *petroleum coke*," but noted that "a complete and accurate record would embrace a list not less than four thousand individual employments."[21] Some of these would become more important over time, while others would decline in relative or absolute terms. Kerosene waned in relative importance, though the trend was uneven, and it remained a key product on many farms well into the twentieth century.[22] Fuel oil, for its part, had many industrial and commercial purposes, while asphalt concrete – a mix of gravel and the heaviest fraction of crude oil – became the most common road-paving material in Canada by the 1960s.[23]

Canadians brought oil into their homes as well. Canada gets cold, of course, and oil eventually made serious inroads into domestic heating, though it remained a minority fuel for both cooking and hot water. As this volume makes clear, Canadians traditionally burned copious amounts of wood and coal, with both showing remarkable tenacity well into the twentieth century. But liquid petroleum had several obvious advantages for household use, which marketers were eager to exploit. It flowed from truck to tank to furnace, unseen and untouched by the consumer. Its flame was not pollution-free, to be sure, but oil burned more cleanly than either wood or coal. And with a thermostat installed and the level set, an oil furnace regulated temperature automatically. "No Ashes, No Dust, No Shovelling, No Shivering, No Smoke," promised one ad for Oil-Right Burners in 1926, a sentiment echoed by a *Globe and Mail* story almost three decades later: "So you now have an oil furnace? You've said good-by to ashes, cinders, and dust. Henceforth all you're going to do is adjust the thermostat when you arise in the morning and a flood of warmth is going to take away that morning chill."[24] By 1954, oil was the most common home-heating fuel in Canada (although the transition came later in the Atlantic provinces), a lead it maintained even through the 1960s and 1970s, when natural gas mounted a serious challenge. Gas had always dominated households in Alberta, and with the completion of major pipelines in the late 1950s, it entered markets farther east and west. In raw number of households heated, gas didn't surpass oil until 1980, but it had already taken over most furnaces east of Quebec by the 1960s.

Still, at a broader level, even this partial shift to natural gas in heating didn't blunt the upward charge of total petroleum consumption.[25]

Then, of course, there are the petrochemical uses of petroleum. Fossil fuel, sometimes in liquid form and sometimes as natural gas, became a key input for plastics, synthetic rubber, solvents, and many other products. Here, a publicly owned sector grew in parallel with private ventures. Polymer, a wartime Crown corporation, extracted hydrocarbons from oil to make synthetic rubber, struggling to replace the natural version, which had been lost to Japan's military expansion in the Pacific. The company did well, surviving the postwar purge of public companies, though its historian is at pains to point out the degree to which it operated like a capitalist venture.[26] Outside of Polymer, the industry was dominated mostly by private enterprise. In 1942, Dow Chemical opened a polystyrene plant in Sarnia, which became one key hub of the emerging plastics industry. Montreal became another centre of oil-based petrochemical development, beginning with Union Carbide in 1957. With the emergence of a petrochemical cluster around Edmonton after 1953 (based largely on natural gas), the basic geography of the industry became clear, centred in Ontario, Quebec, and Alberta and focused largely on the American market. In 1995, petrochemicals used about 5 per cent of Canada's oil supply (20 per cent of its gas), and shipped almost all its production south of the border.[27] A driver on the 401 speeds along in a car riddled with these petroleum-derived substances: synthetic rubber in the tires and in multiple engine parts, plastic throughout the cab, and Styrofoam in the cup that holds a coffee or Coke.

Sigh, in charting petroleum's Canadian history, one runs out of synonyms for growth. By around 1950, F.R. Steward reports, petroleum surpassed coal as Canada's dominant energy form.[28] By that time, almost every conceivable symbol of the North American good life became stained by oil, sometimes in obvious ways (gasoline pumped into cars and lawn mowers) but often in unseen and hidden forms (helping to make a tree into lumber to build suburban homes, forming the chemical foundation of the plastic that lined kitchen tables, or driving the motors that transported almost every key commodity, from food through toys to appliances). Vaclav Smil points out that, today, even alternative-energy technologies remain rooted in the fossil-fuel economy. In wind, for example, coal and oil are crucial in making steel and plastic parts, transporting turbines to market, and at other points in the commodity chain.[29] It is no surprise, then, that petroleum consumption charged forward and upward with such tenacity (see figure 10.3).

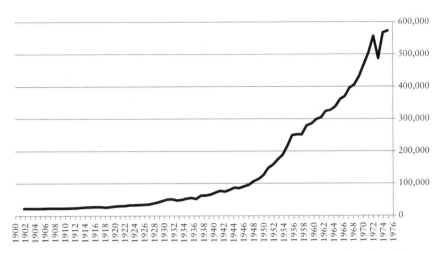

Figure 10.3   Petroleum consumption in Canada, 1900–1975 (thousands of barrels)

Source: Statistics Canada, Historical Statistics of Canada, series Q19–25, *Production and Trade of Crude Petroleum, 1868–1976* (thousands of barrels), http://www.statcan.gc.ca/access_acces/archive.action?l=eng&loc=Q19_25-eng.csv.

But at what point does momentum become inevitability? At one level, a case might be made that petroleum was likely to succeed as an energy source in some way. Victorian oil capitalists never found a good use for gasoline, which they considered an inconvenient by-product of the search for illumination (and, in fact, often dumped it into nearby ponds and streams). Yet they lived in an energy-intensive society for the time. Coal and kerosene had already generated a modern energy system in many ways, and since petroleum produces several energy-dense forms of fuel, it seems likely that, eventually, some clever tinkerer would have found an energetic use for it outside illumination. If cars had gone electric, for example, demands on that energy source would have been tremendous, while petroleum (in the midst of losing the illumination market) might have been stunningly cheap. It seems doubtful that any energy-rich fuel would have been dumped into rivers when it could be burned in power plants and thereby pumped into electric car batteries. Even if such techno-counterfactuals are best left for Star Trek conventions and drunken debates at the bar, the broader point remains: *at one level of approximation,* the history of petroleum culture (and high energy society more broadly) was more about the momentum of machine life than what particular modern fuel was fed into what particular technology at what particular point. Still, as historians, we do need to understand the particular energy regime we have inherited, the importance of human choices in making it, the level of intellectual effort and material work that the process required, and the place of the physical properties of oil itself in these developments. In between the ICE revolution and those trucks on the 401 lay considerable industrial development, technological innovation, political choices, social adaptations, machine work, and human labour, but also many natural facts and dynamics. Charting the relentless climb of petroleum consumption can cloak the significant complexities that produced it, just as keywords like *more* capture the "what" but say little of the how, why, when, and where.

THE SOCIAL LIFE OF CANADIAN OIL

Oil began with nature. Its origins are in buried marine organic matter that was fossilized over hundreds of thousands of millennia of oxygen deprivation and decay.[30] Once oil formed, it migrated underground, either reaching the surface as seepage or getting blocked by a layer of non-porous rock, where it would collect in a "reservoir." (The term should not imply an actual pool, since the oil is in fact trapped as droplets in porous rock). These processes produced a substance that, once released from the ground, had two key material advantages: it could be processed into several energy-dense fuels, and it flowed easily. Both features helped make oil the fuel of the twentieth century. Every fuel faces limits, either economic or energetic. High costs of development or transportation can make it too expensive for general use and, at a certain point, more energy is required to develop and transport a fuel than the fuel itself produces. Just as coal helped humans break out of a biomass energy regime, so oil helped them surpass coal, though both those developments (as should be clear from this volume) were highly uneven and complex.

In the twentieth century, almost all oil reached the surface as a commodity. No one was soaking oil up with rags off the top of rivers anymore, few users produced their own, and you didn't drill wells like you chopped wood. Even before the age of the car, Big Oil had asserted its hegemony over key choke points in the industry. By the 1890s, Imperial had emerged as the dominant player in the oil lands of southwestern Ontario, and its Google-like leadership of the Canadian industry continued well into the twentieth century. In 1920, the company refined 91 per cent of the Canadian crude. Other companies began to challenge this virtual monopoly in the interwar years. British-American, Shell, and McColl-Frontenac joined Imperial in

Table 10.3. Percentage share of
Canadian refining capacity, 1920–1960

Year	Imperial	Big Four[a]
1920	91	91
1930	74	92
1940	56	88
1950	51	88
1960	35	72

[a] Imperial/Standard, British American/Gulf,
McColl-Frontenac/Texaco, Shell
Source: Grant, "Canada's Petroleum Industry."

the so-called Big Four by mid-century, forming what Hugh Grant calls an "overcrowded oligopoly." The term is apt beyond simple market dominance: the companies cooperated in several joint ventures (pipelines in particular), and the smaller three continued to explicitly admit (submit?) to Imperial's price leadership until at least the 1960s. The Big Four exerted particularly strong control over refining (see table 10.3), but remained highly integrated both "upstream" (exploration, extraction, transportation to refineries) and "downstream" (wholesale distribution, retail, and marketing). Independent companies (so called because they weren't tied to the Big Four, not because they were necessarily hand-to-mouth operations) found many open spaces in exploration, but the Big Four absolutely dominated downstream operations, despite the efforts of "unbranded" discount sellers in the 1960s. Each of these Big Four, moreover, became part of larger global structures: Imperial was taken over by American behemoth Standard in 1898, British-American became part of Gulf after the Second World War, McColl-Frontenac was swallowed by Texaco, while Shell had always been a branch plant of its parent company.[31]

These companies engaged in geographically extensive operations, riding and reinforcing a series of spatial shifts in the broader oil industry. In the United States, the rise of automobility coincided with a geographic shift from Pennsylvania to the West (Texas, Oklahoma, and later California). A similar but less dramatic shift occurred in Canada, with the Turner Valley region of Alberta forming the core oil lands by the interwar years. Still, Canadian production remained small, never amounting to more than 10 per cent of demand, with the rest flowing from lucrative fields in the United States, Mexico, and South America. Then came the iconic strike at Leduc in 1947, setting off a rush of exploration and development in the West, centred mostly on Alberta. Between 1946 and

1952, the number of exploratory wells increased over ten times and total output almost sevenfold.[32] Over the 1950s, Alberta became a petroleum province. Revenues from oil shipped out of the province surpassed those from agriculture in 1960, and by that time the province supplied over one half of Canada's needs.[33]

In the three decades after about 1970, western conventional *crude* had to compete with interest in three new Canadian sources: offshore wells in the Atlantic, the oil sands, and the Arctic. These developments, each in its own way, were a product of higher prices and a new atmosphere of crisis after 1973, though each had roots in an earlier period as well. Offshore exploration on the Atlantic, which had begun in earnest in the 1960s, intensified in the 1970s and eventually led to the discovery of several large fields. Hibernia was first, discovered in 1979, though oil was not commercially pumped until almost two decades later. Facing perennial underdevelopment and a collapsing cod fishery, Newfoundland made oil central to provincial development, a commitment that was sustained through prominent disasters, battles with the federal government, and wrangling with multinational oil companies.[34] The situation could not have been more different in the West, where early moratoriums on Alaska oil tanker traffic were broadened to exploration by the BC government in 1981 and subsequently extended – thanks in part to the massive *Exxon Valdez* oil spill off Alaska in 1989 – into the new millennium. There were no such environmentalist barriers in Alberta, where years of academic and industry research produced the first commercial oil sands development in 1967. Serious and sustained activity, however, awaited changes in royalty and tax regimes in the 1990s and sustained high prices in the following decade. For the ten years after 2002, oil sands production increased almost 10 per cent per year, passing 1 million barrels a day by 2004 and 1.5 million by 2010, about triple the level of conventional oil. By contrast, Arctic sources remained more in the realm of promise and possibility. Interest peaked after a massive oil find in Alaska in 1968, subsequent gas discoveries in the Canadian Arctic, and high prices in the 1970s, but local resistance delayed development, while complex geology and difficult climate made exploration expensive. Once prices dropped and government subsidies melted away in the 1980s, little real development occurred. The big international companies have now returned to the area, however, with massive investments in land and exploration.[35]

Over time, Canada aspired to the status of energy superpower. But until declining conventional reserves and startlingly higher prices in the first decade of the twenty-first century made Alberta's oil sands appear viable,

Canada remained an energy minor-power and continued to depend on the complex continental and international politics of oil. Even after Leduc, Canada acted as both importer and exporter. Alberta sent oil to the United States and to Canadian markets west of the Ottawa River, while Quebec and Atlantic Canada were served by imported "overseas" oil (mainly from Venezuela, though Middle Eastern supplies became somewhat more prominent over the 1960s and 1970s). This regional oil geography made considerable economic sense. Alberta oil was expensive to move east but could be sold in the American Midwest at high prices, while so-called overseas oil was cheap to import up to Montreal through the St Lawrence system. But it also implicated Canadian consumers in a whole series of international developments. Petroleum politics occupied considerable diplomatic energy, particularly in relations with the United States. Meanwhile, the dollars and cents of Canadian consumers flowed to support the modernizing plans of Venezuelan governments, who hoped to "sew the oil" to produce a more diverse industrial economy and more progressive political and social system. And, of course, Peter Lougheed's Progressive Conservative government in Alberta had similar dreams, directing oil revenues into a Heritage Fund that was intended to provide for future projects and to support increased provincial activity in several sectors.[36] Drivers along the 401, heading west from Montreal on a tank refined from Venezuela crude or stopping at Kingston to fill up on Alberta sources, supported modern state projects in a variety of widely dispersed locations.

In the aggregate, Canadian oil was regionally diverse and jurisdictionally complex, but all production areas shared one feature: they lay far from their key markets. Standing before the Empire Club of Canada in April 1951, Eugene Holman, the president of Standard Oil of New Jersey, made this point clearly and succinctly: "There is no relation between the places where oil is found and where it is used," he noted. "Oil is found where nature put it. It is consumed where populations have concentrated and industry flourishes."[37] As other contributors to this volume have demonstrated, distribution became a key – indeed, the key – problem for the energy industries of the mineral regime. It is no surprise, then, that after the strike at Leduc, pipeline politics became one of the most controversial matters facing governments and one of the most pressing issues for corporate investment. Pipelines moved both east and west out of the province by the mid-1950s, connecting Alberta oil to the Pacific (via the Trans-Mountain, completed in 1954) and to the Great Lakes transportation system (via the Interprovincial, connected to Superior Wisconsin, on Lake Superior, by 1950). This technological effort was supposed to make oil subject to economic rather than environmental logic: as the boosterish *Western Business and Industry* magazine put it when the Interprovincial finally bypassed Great Lakes ice, "Oil will be flowing at a rate dependent only on market demand and not subject to the vagaries of the weather."[38] But for all the rhetoric of overcoming nature, it was the energy-dense and liquid nature of oil that made such transportation possible.

The Big Four, and their American parents, drove technological innovation up and down the commodity chain. The home-spun "creekology" that dominated exploration in the nineteenth century (so named because of the widespread belief that oil could best be extracted near creek beds) was surpassed by more expert knowledge in the twentieth. Professional geologists moved into the industry itself by the First World War, using a combination of surface indicators, aerial surveys, and core samples to find and assess potential sites. In the 1930s and 1940s, however, seismic mapping allowed the industries to see underground in advance of drilling. The earliest of such geophysical techniques involved measuring the sound waves produced by dynamite explosions to map sub-surface features. "As dramatic as anything that has occurred in the brief history of Alberta is ... the seismographic truck rumbling along dim back trails, their crews unleashing miniature earthquakes," Jack Delong wrote in the *Edmonton Bulletin* in 1947. "Alberta's central and northern hinterlands may well save the day for the greatest industrial civilization in man's history."[39] Refining was another key area of innovation. In 1913, a chemist at Standard-Indiana developed thermal *cracking*, a high-temperature high-pressure form of processing that replaced the simple distillation then common across the industry. In Canada, Imperial contracted for exclusive licence to the procedure in 1914, one of the factors that helped the company maintain its lead over competitors. Continuous flow cracking followed after the First World War, and then, by 1936, catalytic cracking significantly increased outputs. In 1939, Hugh Grant reports, Canadian refineries processed 400 per cent more crude per worker than in 1921.[40] All these developments reflected the increasing presence and influence of professional scientists, particularly in the biggest companies. Imperial hired its first full-time chemist in 1924 and set up a dedicated research facility at Sarnia in 1928. For the most part, however, Canada was R&D poor. Befitting its branch plant economy, most innovations were developed elsewhere and travelled to Canada through continental and global structures of knowledge and capital.[41]

Hydrocarbon society also grew from political choices. State activity stretched right along the commodity chain, smoothing the way for an increasing commitment to petroleum. Governments leased mineral rights (often with exploration-friendly terms), set taxation regimes (ditto), facilitated big projects like pipelines, and brought order to the industry through conservation and pro-rationing of production after 1938. Many policies in these areas, like some more recent (lukewarm) environmental regulations, have been a source of industry complaint, but most governments in oil jurisdictions have been anxious to facilitate development. In Alberta, for example, the conservation regime set up in 1938 was the target of some industry hostility, particularly among the so-called independent operators, but the aim was to ensure rational development and wise use of the resource, not to prevent exploitation.[42] In the post-Leduc years, the province's Arbitration Board served as an example of how even ostensibly neutral procedures aimed to facilitate development. The board settled disputes between oil companies and farmers, but it did so by turning conflicts over land use, access, and environmental harm into financial questions, calculating measurable losses to remove the issue from public contention and to ensure continued development. The board's rulings could not be appealed.[43]

By far the most important state contribution, however, was indirect: road building, which literally and figuratively smoothed the way for increases in petroleum use. Provinces in particular dedicated themselves to so-called blacktop government. Roads were the largest single expenditure for most provinces in the middle decades of the twentieth century, though "blacktop" is something of a misnomer, since hard surfaces were actually a minority of roads in all provinces well into the postwar period. The results were nonetheless impressive. New roads were built, and old roads extended and resurfaced. By 1976, over half of surfaced roads were concrete or asphalt.[44] Municipalities got into the act as well, focused not on megaprojects like the 401 but on a series of more incremental – but highly transformative – schemes of widening, resurfacing, and improving along country roads, urban arterials, and other connectors. Even the 401 system required the improvement of dozens of connecting roads under municipal and county jurisdiction, which is to say that people had to get to the highway to use it. And when they weren't building roads, governments were widening, straightening, and resurfacing them. Blacktop government both literally and figuratively smoothed the way for greater consumption of petroleum.[45]

To undertake these projects, governments developed new fiscal tools. At the provincial level, the gasoline tax became the most important source of road-building funds. The idea emerged in the United States in

1919 and quickly migrated across North America, spreading to every Canadian province between 1922 and 1928. By the 1930s, motor-vehicle revenues (taxes, licence fees, etc.) represented 30 per cent of general net revenues of provinces, the largest single contributor.[46] In the genealogy of fiscal regimes, the gas tax was distinctive in that funds were so often dedicated to road building rather than general revenues – they were, in other words, spent to assist the social groups who paid them. This was sometimes a point of philosophical debate, since the gas tax acted like a user fee but expressed a blacktop consensus that defined road building as a public good. "It seems to me," Walter Gordon told Irwin, a representative of the BC Automobile Association (BCAA) testifying before the Royal Commission on Canada's Economic Prospects in 1955, "that if you really attempted to relate tax revenues to particular sources of expenditure you would get things in an unholy mess." One other commissioner wondered if governments should take "the revenue from liquor taxes and give it back to the liquor consumer." The conversation continued, turning on the fuzzy distinction between user fee and public good:

> IRWIN E.J. [of the BCAA]: It is not the motorist himself who is looking for any personal return; it is a group of motorists who feel that taxwise their money has been collected, or rather that the money which has been collected taxwise has not been used to improve a portion of the country that all motorists should be able to enjoy ...
> GORDON: I have heard people complain about the personal income taxes they have to pay. I have heard corporation executives complain about the taxes their corporations have to pay, and none of them seem to think they get enough back.
> IRWIN: But this is not a personal return.[47]

Personal return or not, road systems were never financially self-sufficient. David Monaghan estimated that between 1919 and 1930, automobile-generated revenue made up just over 40 per cent of the funds spent on roads; in 1953, municipal, provincial, and federal governments spent $480,049,000 on roads but collected only $307,664,000 from drivers.[48] But even if there was typical griping about taxes and levels of spending, no politician was likely to get defeated by promising more and better roads. The blacktop consensus ran across social groups, ideological bounds, partisan divisions, and linguistic differences.

Yet not every province was building a 401. At one level, we can see the broad trajectory of Canadian history was to greater and greater use

– the momentum seemed unstoppable – but a snapshot at each point reveals a much more complex and uneven picture. Levels of gasoline use between provinces (to take one very rudimentary variable) differed quite substantially, as did car ownership. If we consider, for example, a provincial average of one car per household to be a useful working definition of "fully automobilized," then Newfoundland had almost reached that stage in the 1970s, but by that time the average household in more prosperous areas had moved on to two vehicles (see figure 10.4). In absolute terms, Newfoundland had joined the automobile age, but in relative terms it remained backward. Provincial officials well understood their position, making it clear that their own efforts toward auto-modernity were chasing a moving target. "In the next 25 years," one official complained in 1956, "we have to spend on 3500 miles of new roads some $105 million; to improve or rebuild or reconstruct the existing roads, most of which do not deserve the name and would not be called road in other parts of Canada … will cost $30 million …. [W]e will perhaps have reached the level that Nova Scotia has already reached, and Nova Scotia will then be 25 years ahead of us."[49]

How deeply all this penetrated into social history is not clearly known, but it does seem that the practices of ordinary Canadians generally reinforced the momentum of petroleum demand, even as they were structured by it. Although Joshua MacFadyen's chapter in this volume provides a fascinating account of wood, there has been little serious historical research on modern home heating systems – a rather shocking absence for a country with aspirations to nordicity.[50] At the same time, while several historians have taken up aspects of the social life of Canadian cars, we have surprisingly little good research on actual car-buying – on the around-the-kitchen-table trade-offs and in-the-dealership decisions that must have occurred time and time again in the building of an auto-oriented society. There is certainly nothing to compare to Joy Parr's finely nuanced study of domestic technology,[51] though automobiles would surely be conducive to this sort of analysis, given their multi-levelled connections to economic policy, credit, marketing, family, and consumption. But we know enough to say that the social history of gasoline involved necessity and choice, structure and agency. Once the suburban house was bought, bus service reduced, the main street grocery store closed in the face of plaza competition, or a workplace moved to a highway interchange, a car was a virtual necessity. But no one forced middle-class Canadians onto touristy or cottagey jaunts along Highway 401. Still, these decisions were shaped by broader forces, like better roads, cheap

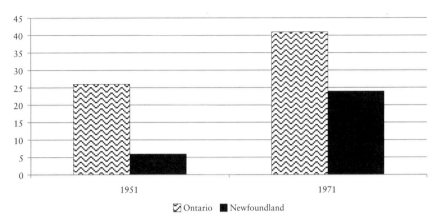

Figure 10.4   Uneven but rising car ownership, Ontario and Newfoundland, 1951 and 1971 (cars per 100 people)

Source: Statistics Canada, Historical Statistics of Canada, series T147–194a, *Motor Vehicle Registrations, 1903–75* (Canada, Newfoundland, Prince Edward Island, Nova Scotia), http://www.statcan.gc.ca/access_acces/archive.action?l=eng&loc=T147_194a-eng.csv; series 147–194b, *Motor Vehicle Registrations by Province 1903–75* (New Brunswick, Quebec, Ontario, Manitoba), http://www.statcan.gc.ca/access_acces/archive.action?l=eng&loc=T147_194b-eng.csv.

oil, growing leisure time, economic affluence, and blacktop government. They were also reflected in statistics, particularly the consistent spike of gasoline sales during summer months. Yet how many were seriously considered in energy terms, rather than as simply going to work or taking a vacation? Only in a few moments was there evidence that consumers seriously measured driving as energy consumption; historians can see these flashes of realization abstracted into miles-per-gallon charts, aggregated into polling data, and mediated by windshield stickers that speak to energy's pocketbook politics. How they were *felt,* and how profoundly they shaped energy consciousness, is more elusive.[52]

Yet underneath all these economic, political, and social dynamics was energy itself. It seems unlikely that petroleum would have become so dominant without, well, petroleum. Early blacktop government was built by a hybrid energy regime that combined human labour, animal work, and some limited mechanization. In Ontario, county governments sometimes reported purchases of machines to the provincial Department of Transport, providing a snapshot of the diversity of energy forms deployed in road construction. In 1918, Welland County bought a three-ton truck, an oil heater, a pressure distributor, and a horse-drawn sweeper. "The motor truck," the report complained, "which was used for hauling stone from the quarry to the road, was not satisfactory, being in the repair shop most of the time. On work of any great extent, one truck is not sufficient if good results are to be obtained; three trucks at least should be employed on hauls up to ten miles." Simcoe County didn't even have machines to complain about: "It would appear necessary before any large programme of construction is undertaken on the country system that suitable machinery equipment be purchased. At the present time the county does not own a steam roller."[53] Both reports were testimony to the growing momentum of fossil fuels in construction – country governments had machines or wanted them – but also to their partial adoption. Blacktop government in the 1920s was built by a combination of steam, ICE, horses, and humans. The Depression actually represented backward step in this technological path, since governments needed to provide relief work to the unemployed. "The work was carried on as an Unemployment Relief measure without the assistance of any efficient machinery now used in road-making," one government engineer noted in 1933. "Every effort was made to use manpower."[54] The war was a turning point. Improvements in military machinery transferred easily into construction equipment, and the era of cheap oil further reduced costs. Between 1920 and 1960, the cost of moving a cubic yard of earth remained basically stable (around forty cents) while the cost of manual labour increased almost 700 per cent. This new economics provided considerable incentives to mechanization, which in turn offered opportunities for bigger and longer projects, like the 401. Road construction continued to require a hybrid energy regime, but the mix of fuel and muscle power had shifted, with machines doing much more, human labour much less, and animal power almost nothing.[55] Here was the classic feedback loop, with petroleum use producing more petroleum use, which, in turn, begat more petroleum use.

To enumerate the increasing use of oil, and the massive economic, social, technological, and labour effort that produced it, is also to speak of environmental consequences. Indeed, we might trace an alternative waste chain that paralleled petroleum's developing commodity chain, running from development (exploding gushers and oil-soaked tailings) through transportation (oozing pipelines and leaky tankers) to use (automobile exhaust pipes and smoke-belching lawn movers). Over time, Canadians have taken account of these issues in different ways. Before the 1960s, the dominant language was conservationist, imagining waste as an expression of irrational development and seeing solutions in limited state action to promote wise use. Alberta's Oil and Natural Gas Conservation Act (1938) set up regulations to prevent "uneconomic" methods and to "encourage the maximum ultimate recovery of petroleum." This involved some government controls, but the ultimate aim was to save the industry from its own wasteful practices.[56] The 1960s wave of environmental politics (and particularly its focus on pollution) raised different issues. Images of oil-covered wildlife after major spills and smog-covered cities on summer days provoked new levels of concern and new attempts at regulation. Such issues were international in scope, although Dimitry Anastakis has argued that new regulations against automobile exhaust actually ran against the continental logic of the industry.[57] Compared to conservation (which placed questions of waste mostly within a debate about the best interests of the industry), environmental politics had more potential to speak of ecologies as intrinsic goods, even if results were often ambivalent and even if older utilitarian ideas survived. By the 1990s, climate change was inserted into this discursive field. Over the 1980s, the growing scientific consensus about the human origins of increasing levels of greenhouse gases (primarily carbon dioxide) entered the public sphere, often through high-profile international meetings like the Toronto Conference of 1988 or the summit at Rio de Janeiro in 1992. This new energy culminated in the Kyoto Protocol in 1997, in which countries committed themselves to mandated reductions in greenhouse gas emissions.[58] Results were uneven. Climate change opened a discussion about the need to consume less, but Canada made few efforts to comply with Kyoto, especially after the Conservative Party came to power in 2006. The new government was more open to the arguments of climate change skeptics and more interested in negotiating industry-specific guidelines that may have been most appealing as stalling tactics. But the problem was really global. Few countries possessed the political will or regulatory mechanisms to even approach Kyoto compliance.

Yet if discourses of environmental consequence have shifted, the waste chain's effects have been real and material. Hydrocarbon society altered ecologies and left observable and measurable traces in earth, air, water, and bodies. In 1967, for example, the National Energy Board's rather bureaucratic table of "leaks, breaks, malfunctions and other pipeline incidents" recorded eighty-two events, a number fairly typical for the period.[59] In 1976, symbolically close to the 401, a large spill on the St Lawrence River flowed over eighty miles downriver, producing an oily sheen on the water surface as far away as Brockville, Ontario.[60] ICEs had dramatic effects on the atmosphere. By 1972, Toronto cars alone spewed over 400,000 tons of carbon monoxide every year; two years earlier, Montreal's Health Department claimed that carbon monoxide levels were severe enough to "cause eye, ear and brain damage in some persons," while researchers at the University of Toronto's Pollution Probe discovered that a single hour of driving in the city could cause measurable increases in carbon monoxide levels in the blood.[61] Cars often drew attention, but exhaust actually had many sources. In one recent study, to take only a small example, Statistics Canada estimated that a single lawn mower could emit as much smog in a year as a car travelling 3,300 kilometres.[62]

Enumerating petroleum's increasing importance also takes account of nature in a different way. Phrases like "environmental consequence" suggest a one-way relationship, with human activity shaping nature and not the other way around. Yet the soul of oil – its foundation and basic essence – remained organic, and even technological systems remained embedded in all manner of natural processes. Petroleum's liquid character made it easy to transport but difficult to contain, flowing through pipelines and into tanks but oozing and spilling out of them as well. Greenhouse gases were also partly a function of the nature of petroleum. Ultimately, carbon dioxide in the atmosphere is the legacy of the carbon-based life forms that, over millennia, became oil; once released from the ground by technological systems and funnelled into the artifacts of machine life, carbon molecules flowed freely into the air as well. Even more basically, moreover, nature sets the ultimate ceiling on exploitation. There is only so much subterranean forest, so all fossil-fuelled societies will eventually confront the limits of supply. It would take four centuries of all the plant and animals life produced on Earth, buried underground for millennia, to generate the fossil fuels that twenty-first century humans now use in a single year.[63] Technological improvement might alter these statistics. Investing in efficiency or in alternative energy systems might slow the depletion of our petroleum inheritance, but so far the globalization of machine life continues to call forth greater demand. At some point in the history of more – though the timing is debated – we will simply run out of nature to exploit.

SIMPLICITY AND COMPLEXITY IN OIL HISTORY

Historians are seldom content with simplicity. Confronted with a linear story, we generally reach for poly-causal and multi-levelled explanations. The ideal history paper probably ends with some version of "the situation was very complex." Yet the raw quantitative growth of consumption shapes so much of Canadian petroleum history. Demand for gasoline, diesel, asphalt, and petrochemicals all surged upwards with relentless tenacity, occasionally sidetracked by some economic or social calamity, but never permanently diverted from their path. Beginning with its natural qualities – an energy-dense liquid stock that could be easily transported and could be processed into many forms of useable fuel – humans built oil into a total system with considerable momentum, touching an astonishing range of human activities. This was not a story of progress, but it was, in many ways, a linear one. To be sure, responsibility for this historical momentum was diffuse, motivations were varied and complex, and many actors and institutions were involved. Historians are also intently interested in context, place, and moment, and petroleum's momentum was always in dialogue with local conditions and environments. At the time construction began on the 401, no doubt average Newfoundlanders would have been surprised to hear that Canada was a hydrocarbon society, or that their province was a blacktop government, even if the future agenda seemed clear. This complexity matters, since our present difficulty of turning *more* into *less* is a direct consequence of petroleum's multi-levelled genealogies. There are too many institutions, too many machines, and too many choices involved to merely turn a valve and wish away the material or cultural power of oil. Still, in the end, it's hard to dispute that the crucial story isn't just *more*, whatever the unevenness and agency we can see along the way. As on the 401, we can switch lanes, pass other cars, vary speeds, and take our own trip, but our collective direction has been, sadly, one way.

NOTES

1 Richardson, "Highway for Today: Ontario's 401."
2 The most through study of the first oil boom – for kerosene lighting – is still Williamson and Daum, *The American Petroleum Industry.* Vol. 1, *The Age of Illumination.*
3 Roberts, *In the Shadow of Detroit: Gordon M. McGregor, Ford of Canada, and Motoropolis*; Davis, "Dependent Motorization: Canada and the Automobile to the 1930s"; Bloomfield, "I Can See a Car in That Crop: Motorization in Saskatchewan," 6.
4 Statistics Canada, Historical Statistics of Canada, series T147–194a, *Motor Vehicle Registrations, 1903–75 (Canada, Newfoundland, Prince Edward Island, Nova Scotia)*, http://www.statcan.gc.ca/access_acces/archive. action?l=eng&loc=T147_194a-eng.csv.
5 Davis, "Dependent Motorization."

6 For a good international study that makes the inseparability of material and cultural factors clear, see Mom, *The Electric Vehicle: Technology and Expectations in the Automobile Age.*

7 Grant, "Canada's Petroleum Industry: An Economic History, 1900–1960," tables 4.1, 5.1, 7.3.

8 To borrow Noam Chomsky's useful academic translation of "I am about to massively simplify."

9 Black, "Oil for Living: Petroleum and American Mass Consumption."

10 Cited in Anastakis, *Autonomous State: The Struggle for a Canadian Auto Industry from OPEC to Free Trade,* 78.

11 Anderson, "Levittown Is Burning! The 1979 Levittown Pennsylvania Gas Line Riot and the Decline of the Blue-Collar American Dream"; Nye, *Consuming Power: A Social History of American Energies,* 217–48. On the Canadian energy crisis, see Anastakis, *Autonomous State*; Sinclair, *Energy in Canada*; Fossum, *Oil, the State and Federalism: The Rise and Demise of Petro-Canada as a Statist Impulse.*

12 From 1980 to 2004, inflation-adjusted gasoline prices dropped 30 per cent. On fuel economy regimes, see Anastakis, *Autonomous State*; Smil, "America's Oil Imports: A Self-Inflicted Burden."

13 Larry McNally estimates a radius of about 100 kilometres in 1921 and almost 600 by 1937, though his chapter is understandably broad, so the point is not very geographically specific. McNally, "Roads, Streets, and Highways."

14 I owe this point to Philip McFee, an MA student at the University of Toronto who is working on the history of trucking.

15 See Hamilton, *Trucking Country: The Road to America's Wal-Mart Economy*; Ahuvia and Izberk-Bilgin, "Limits of the McDonaldization Thesis: eBayization and Ascendant Trends in Post-Industrial Consumer Culture."

16 See chapters 2 and 3 in this volume for more details about the transition away from animal power; as well, Ankli, Helsberg, and Thompson, "The Adoption of the Gasoline Tractor in Western Canada."

17 Stevens, "Roughing It in Comfort: Family Cottaging and Consumer Culture in Postwar Ontario"; Harvey, "Sound Politics: Wilderness, Recreation, and Motors in the Boundary Waters, 1945–64."

18 Rajala, *Clearcutting the Pacific Rain Forest: Production, Science, and Regulation,* 30–40.

19 Canada, Statistics Canada, "Table 1: Lawn Mower Use in Canada and Provinces 2007."

20 See, for example, British Columbia, *Report of the Commissioner, Royal Commission on Gasoline Pricing Structure.*

21 Ross, *Petroleum in Canada,* 100. Italics not in original.

22 Sandwell, "An Introduction to Lighting in Canada."

23 Guillet, *The Story of Canadian Roads,* 203.

24 "Heat with an Oil Right," *Globe* (Toronto), 26 July 1926; "How to Care for Your Oil-Burning Furnace," *Globe and Mail,* 7 October 1954.

25 Figures on household heating can be found in Dominion Bureau of Statistics, *Household Facilities and Equipment.*

26 Bellamy, *Profiting the Crown: Canada's Polymer Corporation, 1942–1990.*

27 Shell Oil Co., Chemical Division, *The Canadian Petrochemical Industry*; Bellamy, *Profiting the Crown*; Purdy, *Petroleum: Prehistoric to Petrochemicals.* This topic is very well served by the Canadian Encyclopedia, which has a good brief outline of the industry. Interestingly, most of the actual power for the petrochemical industry is electricity. Oil and natural gas is "feed stock" – a source of hydrocarbon molecules rather than power.

28 Steward, "Energy Consumption in Canada since Confederation," fig. 1, 239. Unger and Thistle put the date when oil consumption surpassed coal as 1954. *Energy Consumption in Canada,* 130.

29 Smil, "On Energy Transitions."

30 Foley, *The Energy System,* 122.

31 Grant, "Canada's Petroleum Industry."

32 Mackenzie, "Oil for Western Canada," 50

33 Breen, "1947: The Making of Modern Alberta."

34 For a brief survey, see Sinclair, *Energy in Canada.* Dodd, *The Ocean Ranger: Remaking the Promise of Oil,* offers a compelling account of different perceptions of danger in the offshore industry.

35 Lajeunesse, "The New Economics of North American Arctic Oil"; Grant, *Polar Imperative: A History of Arctic Sovereignty in North America.*

36 Nemeth, "Canada Oil and Gas Relations, 1958 to 1974"; Coronil, *The Magical State: Nature, Money and Modernity in Venezuela*; Lizee, "Rhetoric and Reality: Albertans and Their Oil Industry under Peter Lougheed."

37 Holman, "Freedom and Energy Go Together."

38 *Western Business and Industry,* "Petroleum in Western Canada," 58.

39 Breen, "1947: The Making of Modern Alberta," 558.

40 Grant, "Canada's Petroleum Industry," 93.

41 One area where Canadians did perform pioneering research was the tar sands. Karl Clark's experiments set the foundation for later processing of bitumen into useable oil. See Chatsko, *Developing Alberta's Oil Sands:*

From Karl Clark to Kyoto; and Sheppard, *Oil Sands Scientists: The Letters of Karl A. Clark, 1920–1949*.

42 Breen, *Alberta's Petroleum Industry and the Conservation Board*.

43 Lizée, "Betrayed: Leduc, Manning and Surface Rights in Alberta, 1947–55."

44 Monaghan, *Canada's New Main Street: The TransCanada Highway as Idea and Reality, 1912–1956*, 45.

45 Ibid.; Guillet, *The Story of Canadian Roads*.

46 Monaghan, "Canada's New Main Street," 11.

47 Canada, Royal Commission on Canada's Economic Prospects, *Hearings Held at Vancouver, 30 November 1955*, 2972–3.

48 Monaghan, "Canada's New Main Street," 23, 45.

49 Canada, Royal Commission on Canada's Economic Prospects, hearings held at St John's, 18 October 1955, 12–13.

50 In this volume, appendix figure A3.3, "Energy Use in Canadian Homes by Function, Selected Years, 1990–2008 (Petajoules)"; and figure A3.4, "Home Heating Provided by Energy Carriers, Canada, Selected Years, 1941–2008 (%)" provides some data to further this discussion.

51 Parr, *Domestic Goods: The Material, the Moral and the Economic in the Postwar Years*. One excellent recent study of the social and cultural dynamics of automobility is Bradley, *By the Road: Fordism, Automobility, and Landscape Experience in the BC Interior*.

52 One important start is Anderson, "Levittown," but this example requires an anomalously violent episode for narrative and analytic power. Lines were more often boring, suffered in grumpy or stoic silence.

53 Ontario, Department of Public Highways, *Annual Report 1918*, 38, 47.

54 Monaghan, "Canada's New Main Street," 14–15.

55 Richardson, "Highway for Today," 10.

56 Breen, *Alberta's Petroleum Industry and the Conservation Board*, 106–50.

57 Anastakis, "A 'War on Pollution'? Canadian Responses to the Automotive Emissions Problem, 1970–1980." On oil-soaked birds as symbols of pollution, see Morse, "There Will Be Birds: Images of Oil Disasters in the Nineteenth and Twentieth Centuries."

58 Bodansky, "The History of the Global Climate Change Regime."

59 Canada, National Energy Board, *Annual Report* (1967), 6. The figure reports events for both gas and oil pipelines.

60 "St Lawrence Clean Up Starts after Barge Spills Bunker Oil," *Gazette* (Montreal), 25 June 1976.

61 Anastakis, "A War on Pollution?," 107.

62 Canada, Statistics Canada, "Study: Lawns and Gardens and the Environment."

63 Statistics cited in Mitchell, "Carbon Democracy," 402.

11

Manufactured and Natural Gas

Colin A.M. Duncan and R.W. Sandwell

INTRODUCTION

Fossil fuel gases have played an important and varied role in the energy history of Canada. Manufactured gas was first used to provide light, and it did so by means of one of the earliest "modern" industrial systems ever used in Canada. Available in some places from the early 1840s, gas lighting was typically distilled (or "manufactured") from coal in a centralized facility and distributed along a system of underground pipes for municipal, residential, commercial, and industrial consumption. This centralized network system of energy distribution was small by later standards, but although it originally provided lighting to only a few public buildings, the wealthiest of urban households, and some industries, by the early years of the twentieth century hundreds of thousands of customers were using it for cooking and heating, and gas lighting was in decline. Manufactured gas initiated Canada's shift from the "mill model to the network model," from artisanal and vernacular to technocratic and centralized that would eventually define Canadians' relationship to energy.[1]

Natural gas, used by a tiny minority of people in southwestern Ontario from the late nineteenth century and Alberta and Saskatchewan in the early twentieth, was rapidly adopted in the late 1950s after vast supplies of oil and gas were discovered in the Canadian west. By that time the technological means had been developed, mostly in the United States, for extracting gas and transporting it by pipeline over vast distances. By the 1950s, dedicated manufactured gas plants had been phased out in most places, and natural gas provided heat, and electricity provided light. Some coking plants that produced gas as a useful industrial by-product continued to operate. But by 2008, natural gas had become the second-largest

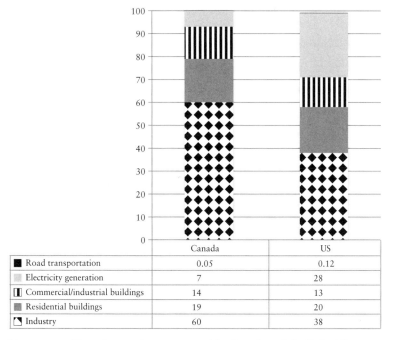

Figure 11.1   Uses of natural gas consumed in Canada and the United States, 2008 (%)

Source: David Suzuki Foundation, Pembina Institute, *Is Natural Gas the Climate Change Solution for Canada?*, 6–7.

source of primary energy consumed in Canada, after oil. As figure 11.1 illustrates, more than 50 per cent of that was for industry (including the fossil fuel sector itself), about 30 per cent to heat buildings, and less than 10 per cent was used to generate electricity. Differences between Canada and the United States can be explained largely by the former's reliance on water power, and not natural gas, to generate electricity.[2]

Not just Canadians consume a lot of Canadian natural gas. Canada is the fifth-largest producer and the fourth largest exporter in the world.[3] About half of Canada's production leaves via pipeline, all to the United States.[4] High production rates are likely to continue, particularly as more elaborate methods of extraction, including hydraulic fracturing and offshore drilling, continue to assure abundant supplies, and demand is almost guaranteed for this relatively "environmentally friendly" (especially because of its low CO_2 emissions) petroleum energy carrier in an energy-hungry continent. Canadian natural gas remains, however, a commodity limited to continental export only. Until the cost of *liquefied* natural gas (achieved by subjecting it to extremely low temperatures) becomes low enough to allow its export beyond North America, Canadian natural gas will not enter international markets.[5] If manufactured gas lighting represented the first network system of the mineral economy, natural gas seems to represent its apogee: cheap for the consumer, profitable for the companies and state revenues, centrally controlled, managed by a phalanx of specialists, invisible, easily transportable, smoke-free, and least productive of CO_2 per unit heat generated by combustion. The extraction, consumption, and export of these flammable gases, usually privately owned and delivered through complex and centralized networks, have had a huge impact on Canadian life and environments, and will continue to shape the country's future.

GAS AT THE HOUSEHOLD SCALE

There have been only two forms of gaseous fuels that did not rely on centralized network distribution systems, but were instead amenable to generation at the household level. In the case of gasoline lamps and stoves, one pumps air into the apparatus so that some of the more volatile *paraffins* present in a gasoline-type fuel (*Gasolene* was a trade name for some time) make a flammable aerosol. These lamps were used widely across rural Canada in the early twentieth century to provide a light that was much brighter, if more dangerous and difficult to use, than the standard kerosene lamps that provided most household light. Scaled up, these pressurized "gas machines" were commonly used until the mid-twentieth century in such isolated sites as "wilderness" resorts, bush camps, or remote military posts, and for heating and cooking as well as lighting. Small versions of these appliances continue to be sold to campers or boaters looking for an efficient portable fuel. In the late nineteenth century, Hamilton inventor Thomas "Carbide" Willson invented a cheap process of producing a bright light by mixing pelletized calcium carbide with water to create acetylene gas. Acetylene's brilliant light was used in carriage and later automobile lights, bicycles, trains, miner's helmets, lighthouses, buoys, and homes, though an explosion in Kingston harbour in the early years of the century killed four people and limited acetylene's appeal; a safer process of creating the gas was, however, soon discovered.[6] In neither case is the gaseous "product" readily susceptible to storage, so it is made, so to speak, only on demand by the immediate user.

Propane and *butane*, the unfortunately named *natural gas liquids*, are extracted in refineries, but they can be purchased, transported, and used in discrete small quantities in ways that resemble coal and fuels of the solar regime, albeit only with specialized containers and burners. Propane gas is naturally rare, but can be liquefied relatively easily on a commercial scale and put into pressurized containers big enough for sustained domestic use and yet small enough to be almost portable. Many propane appliances were created and developed to boost sales for the natural gas industry that was, by the 1960s, in oversupply. It was and is marketed to those living off the electrical or gas grid, providing fuel for heating and cooking stoves, and even refrigeration. Propane tanks big enough for sustained space heating are not exactly portable. Smaller versions, however, now provide portable fuel for campers and boaters, while butane-powered lighters have now largely replaced wooden and cardboard matches as a portable source for lighting combustible materials. In the 1960s, the Arkansas Louisiana Gas Company invented the first propane backyard barbecue, creating an important urban market for the product. Propane and butane need to be properly stored and transported to prevent gas escapes and explosions, but are far less dangerous than gasoline.[7]

MANUFACTURED GAS

The portable gases played a small but significant role in Canadian energy history, but *manufactured gas* (also commonly known at the time as *coal gas* or *town gas*) was the first energy carrier using a network system to make its way into urban life, even preceding water and sewage in some towns (including Toronto). The gas lighting industry had its origins in the new European research in pneumatic chemistry of the 1790s, research stimulated both by scientific curiosity and a growing interest in the applications of scientific knowledge to industrial and commercial enterprises.[8] Declining stocks of wood in eighteenth-century Europe drove the search for a pine tar replacement (mainly for waterproofing ships) and high-energy alternatives to fuelwood, particularly *charcoal*, whose intense heat had been so important to the emerging iron industry. Scientists discovered that important new chemical substances could be "created" by *distillation*, that is, by heating substances like wood or coal, and sorting the resulting distinct gases, liquids, and solids. Manufactured gas lighting emerged as an accidental by-product of a distillation process originally invented to produce tar, coke, and charcoal.[9]

Philippe Lebon began experimenting with extracting gas from wood in the early 1790s, demonstrating the first gas-powered "thermolampe" in Paris in 1801.[10] In England, William Murdoch, the talented chief engineer employed by James Watt (the inventor of the first practical steam engine) at the Boulton and Watt steam engine factory in England, installed gas lighting in his house in 1794. Philippe Lebon and William Murdoch not only discovered that distillation could produce an illuminating gas, but were among the first to realize that the substance had significant commercial potential for lighting. Murdoch calculated that the cost of lighting a mill "with gas (less the sale of the coke by-product) was 600 English pounds, compared to 2,000 pounds for candles producing an equivalent amount of light."[11] By the early years of the nineteenth century gas was lighting mills, factories, and public buildings in England and the eastern United States.[12] The commercial possibilities of scaling up the distribution system from a single mill, using a series of underground pipes to feed multiple lamps across a wide area, was first realized by Frederick Winsor's London and Westminster Gas Light and Coke Company: as early as 1820, London had "122 miles of mains and 30,000 lamps" and thousands of customers in an integrated, necessarily centralized system that was already being imitated across the cities of Europe.[13]

The original gas lighting plants in England were models for later ones elsewhere. They consisted of two or three cast iron ovens, or "retorts," into which coal was loaded and then subjected to distillation.[14] Afterwards, the "lighting gases" were "washed" of unwanted, highly toxic by-products and the obnoxious sulphur smell by moving the gas through water (later limewater), often sending the offensive liquid and solid by-products and waste into nearby watercourses or neighbouring properties.[15] The usable gas, typically composed of *hydrogen, methane, carbon monoxide*, and *ethylene*, was then piped to a gasholder, sometimes called a "gasometer," where it was stored until needed. These ingenious containers were basically huge inverted bell-shaped objects set in a vat of water sitting within a frame. A system of belts and pulleys originally provided the necessary pressurization of the gas: as the gasometer was lowered into the water, gas was forced through the tubing that eventually ended up at a lit burner. In the early nineteenth century some of these pipes were recycled gun barrels in England, or wooden cylinders in the United States, and probably in Canada.[16] The process of "cooking" the coal at temperatures of over 1,000 degrees Fahrenheit turned coal into the more heat-intensive coke, much in demand by the rapidly expanding iron and steel industry. By the late nineteenth century, the tars also generated had become important chemical feedstocks for the dye and chemical industries.[17] Early gas plants experimented with burning a range of products to create illuminating gases, including pine trees and peat.[18] But the most common and longest-used substrate was coal.

Albert Furniss was the first in the colonies to "import the technicians, and – just as important – develop the social forms, the institutions and organizations to operate the technologies" that comprised the first manufactured gas system in Montreal.[19] Municipalities were often early and large customers, using manufactured gas first to light the streets as it provided better and more reliable lighting than oil lanterns.[20] Public demand drove the installation of gas lighting: it was "as desirable an amenity to city dwellers in the mid-nineteenth century as a reliable source of water," for negotiating unpaved roads and poorly maintained sidewalks in the dark could prove unpleasant and dangerous. Street lighting was also widely applauded for lessening crime and vice in rapidly growing towns and cities.[21]

Furniss built his second plant in Toronto in 1841. The gas mains were installed in December, just two years before the roads were dug up again to provide that city with its first piped drinking water.[22] By 1881 there were twenty-seven "gas lighting and heating establishments" across the country, and by 1885, most major towns and cities in Ontario had lighting from manufactured gas plants, as did Quebec, Victoria, Vancouver, Winnipeg, Halifax, St John, Charlottetown, and St John's.[23] By 1923, Canada reportedly had forty-one manufactured gas companies and 375,000 customers, and there were 275,000 gas ranges in use, and 50,000 gas fires, presumably for heating rooms.[24] The gas light industry developed unevenly, however. The logistics of a grid system and economies of scale meant that manufactured gas was generally limited to urban areas with access to cheap coal. The number of gas plants peaked in 1904. Even when the percentage of households with access to a gas plant peaked in 1910, 70 per cent of the population lived outside gas-supplied areas (see figure 11.3).[25]

Manufactured gas companies also had to deal with competition from other new energy carriers: they received a blow in the 1860s and 1870s from the rapid adoption of kerosene lamps, which provided a cheaper and more portable illuminant, and one based on a much simpler infrastructure than gaslight.[26] The competition from electricity, available from the 1880s in the manufactured gas "heartland" of southern Ontario and Quebec, did much less than kerosene, however, to check the growth of the gas industry until the early twentieth century.[27] But the gaslight industry also benefitted from innovations developed elsewhere: it received a great reprieve when, in 1885, the Austrian Karl Auer von Welsbach invented the incandescent (or "Welsbach") gas mantle: a cotton fabric impregnated with thorium and cerium that allowed an intense white light.[28] This invention "improved the efficiency of gas lamps by a factor of seven,"[29] and "notwithstanding the greater yield of light for a lower gas consumption, the output of gas greatly increased."[30]

The significance of manufactured gas in Canadian energy history eventually paled in comparison to its more successful, more enduring, and more massive modern cousins: coal, petroleum, and natural gas. But scaling up a complex, multifaceted system from an individual house or mill to many residences, shops, and mills across an urban area for the first time was "not minor and included many technological and social challenges for the builders of the new network."[31] A quick look at this first modern network system sheds some light on the novel difficulties for those involved. By the time that gas lighting came to Canada in the 1840s, many of the most pressing technical and administrative problems of the new grid system had already been addressed, including the invention of a valve that shut off the gas if the flame went out and of a meter to measure consumption. Municipalities and companies alike struggled, however, with a range of issues. Finding a model for financing such a large, centralized service was in itself an ongoing challenge. By the 1880s, most Canadian cities had municipally owned waterworks, but typically relied on privately owned manufactured gas "local monopolies" that seemed best at meeting the needs of shareholders, "earning high profits from selling poor quality gas at prices that seemed too high."[32] Gas lighting, at least in private homes, was to remain a luxury throughout the so-called gaslight era.

Infrastructure problems with gas production and distribution were legion. They included the difficulties of digging up streets to install and repair underground pipes, particularly in freezing conditions, and meters that failed to work in the cold weather.[33] In the early years it was difficult to find workers with the skills needed to lay the pipes, install the fixtures, and run the plants and the distribution systems. Equipment of all kinds, as well as skilled workers, had to be imported at considerable cost. Production had to be tailored to demand, a daunting task in itself that often pitted producers against consumers. For example, "Gas users tended to view the product as a replacement for candles and oil lamps, one which could be used whenever desired, at any time of day or night. [The gas company] by contrast, wanted to restrict the times at which users could consume gas in order not to exceed the company's generating capacity."[34]

The sheer complexity of a highly integrated system was daunting in itself. As a historian of Consumers' Gas of Toronto complained in 1923,

> Few people have any conception of the vast organization which a gas company must maintain outside of its manufacturing plants in order to render and keep its service efficient and acceptable to the consumer ... The service pipes must be inspected and kept in repair. Leaks and other complaints must be instantly remedied, and gangs of expert workmen have to be kept constantly in readiness day and night to go, in any sort of weather, to the consumers' premises or any other point where trouble occurs. Meters must be kept in good condition and tested periodically, at the expense of the company, and upon demand of any consumer must be moved from house to house.

In addition to meter reading and moving, he notes that books and records had to be kept, "bills printed and made out monthly, delivered and collected and all arrearages [sic] looked after." To do all this, "an office must be kept open, an accounting department maintained and the management provided to supervise the whole organization."[35]

From the gas companies' vantage point, their problems reached beyond those of their own complex networks. Companies realized that their success was "being undermined owning to the installation by consumers of appliances of poor materials, workmanship and construction."[36] After having a good quality gas service installed, most customers were "still using the flat flame burners or mantle lights that they had purchased at bargain sales in department stores."[37] In an important sense, indeed, gas companies' main problem was the consumer. As one manufacturer put it, "The truth is that most of our trouble work [sic] is on account of ignorance on the part of the consumer ... there is the poor light complaint, the burner not giving sufficient light to satisfy the consumer. Then there are petty leaks, a loose cock, a burner orifice plugged etc."[38] Management all too slowly realized that if new customers were to be found and kept for their "product," this was going to involve a commitment to both sales and service of equipment also. Trade magazines fretted endlessly about what comprised good service from a gas company, and how "reliability, availability, courtesy, co-operation, quality, responsiveness, confidence"[39] could be instilled in all employees. As one speaker at the Gas Company Convention of 1913 put it, "No matter how big your corporation, it serves a bigger corporation, the People-Unlimited." That meant being always "polite, genial, careful and smiling" and, as the speaker went on to suggest, "[a] little tactfulness and more sympathy for the man on the other side of the counter who is often wrong – sometimes discourteous, most always misinformed, but always a human being – will work wonders in making the company something of an asset to the community (Applause)."[40]

Consumers had their own extensive critiques of this novel system of delivering energy. As the gas industry was being established in the mid-nineteenth century, the very idea of purchasing any product (let alone an invisible one) that was provided as a centralized service was new and often confusing – or worse – to consumers. From the very first installed service in the 1840s, "consumers of gas and other citizens expressed much dissatisfaction on account of the high price, the uncertainty of the supply, and the poor quality of the gas."[41] They constantly complained about the quality of the lighting (flickering or not bright enough) and its bad smell, factors that related to the quality of the coal and its manufacturing process, and the state of the system of pipes. Some worried about the potentially harmful effects of the fumes; despite assurances from the gas companies of the beneficial "influence of gas in promoting ventilation and circulation of the air and in sterilizing the air through the cremation of bacteria in the bunsen flame," many expressed concern about the supposedly harmful "vitiating effects of gas," as well as the sometimes-excessive heat the burners gave off.[42] They worried about the damaging effect of the gas fumes on their furnishings and wallpaper; as a result, gas lighting was often limited to hallways and kitchens, not sitting rooms or sleeping quarters.[43] Above all, customers objected that they were not getting the amount of gas they were paying for, but had no way of measuring themselves; as a result, they "harboured the darkest suspicions" of the accuracy of the meters, "sustained by the always unexpected quarterly bills and just enough incidents and accidents to justify skepticism."[44] The gas industry itself, tired of the constant complaints and accusations and seeking to reassure their customers, as early as the 1860s was lobbying for limited regulation of meters, and the quality of gas.[45]

While aggregate data document that Toronto's experience was far from typical, we know that Consumers' Gas, the largest gas company in Canada, started as a consumers' co-operative and began with 397 lighting customers in Toronto in 1848. That figure had risen to 1,755 in 1860 and 3,906 in 1880, when the city had a population of just over 86,000. The number of Toronto's gas street lights rose from 977 in 1860 to 2,136 in 1880.[46] Changes in government regulation and gas manufacturing itself transformed the industry in the 1880s. In March 1879, a parliamentary act gave the Consumers' Gas Company permission to "manufacture and sell gas for heating, cooking and other than illuminating purposes," a change related to the switch to a new product made by enriching a manufactured gas by spraying in liquid hydrocarbons derived cheaply from the expanding petroleum sector, and then "cracking" the gas (i.e., exposing it to *thermal cracking*).[47] The price of gas fell almost 40 per cent in Toronto during the 1880s, and the price of gas appliances also fell as the company began to manufacture its own gas stoves and ovens locally in 1881, selling them through its own appliance shop. The number of gas stoves rose in Toronto from under 9,000 in 1905 to almost 110,000 in 1922, with a further 35,000 countertop gas rings.[48] Figure 11.2 documents Toronto's rapid increase in gas consumption after 1881, from about two million cubic feet in 1882 to over 58 million cubic feet in 1892 and 5.4 billion cubic feet in 1923. Gas was clearly being used much more, and in much greater volumes, as it became the fuel of choice for cooking. By 1948, almost half of Canadian homes in metropolitan areas and a third of all households in Quebec and Ontario were cooking with gas, most of them relying on gas manufactured from coal. By that date, gas lighting had been replaced almost entirely by electric in urban Canada, and kerosene lit most rural homes.[49]

While details of the essentially simultaneous decline of gas lighting and rise of gas cooking as well as contemporaneous industrial uses are somewhat opaque, it is clear that by the 1920s the most serious problems of gas lighting – "bad odors, leaks, fires and explosions"[50] – had been solved – by electric lighting. Gas use within the home was mainly for cooking and a small amount of home heating. Few new manufactured gas plants were built after 1890, and after that point plants began to close. But they were massive industries for their time, in capital investment, if not output. In 1906, the thirty-nine "gas lighting & heating establishments" had capital equal to half that of the flour industry, but a total value of product not even half that of the very low-capital harness-and-saddle sector. Only thirty-one gas plants had 5 or more employees (sector total 786).[51] Artificial gas establishments were very few and far between, but were huge in capital, low in value of product, and very low in worker employment. Bear in mind the early 1930s words of Morgan, the author of *American Gas Practice*: "A manufacturer turns his capital over from two to four times a year and a merchant from five to ten times. Utilities on the other hand turn their capital over once in three and a half to five years."[52]

While the first modern network system of gas lighting, cooking, and heating has left surprisingly little impression on the social history of urban

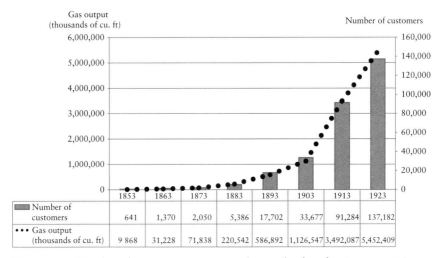

Figure 11.2   Number of customers, gas output (thousands of cu. ft), Consumers' Gas Company of Toronto, 1853–1923

Source: Tucker, *75th Birthday, 1848–1923.*

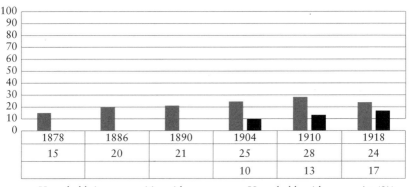

	1878	1886	1890	1904	1910	1918
	15	20	21	25	28	24
				10	13	17

■ Households in communities with   ■ Households with gas service (%)
gas service available (%)

Figure 11.3   Percentage of Canadian households in communities with manufactured gas service available, and with gas service, 1878–1918

Sources: Brown's *American Gas Companies Directories,* 1890, 1904, 1910, 1918; *The American Electrical Directory, 1886; Directory of Gas Light Companies 1878;* Census of Canada, 1951, Volume 3, *Housing and Families* Table 1, Dwellings, households and average number of persons per dwelling and per household, Canada and the provinces, 1881–1951.

Canada, it has nevertheless left an important, if unwanted, historic legacy. The fact that gas lighting created by-products that could be sold to offset the costs of production was an important part of the appeal of manufactured gas. Coke was sold as a fuel, particularly useful in the iron and steel industry, and tar, originally used primarily in shipping, was later sold to tar distillation plants to produce naphthalene, creosote, heavy oil, and tar for roofing and roads.[53] These by-products contributed more than money; the process created substantial noxious and toxic waste. Gasworks were intentionally located near rivers and lakes to make the disposal of fluid waste products easier, particularly after attempts to dump wastes into municipal sewer systems drew unwelcome – and very vocal – complaints from residents suffering from both the fumes and toxic sulphur compounds that made their way into nearby wells. [54] Research has only recently begun to document the long- and short-term environmental consequences of these plants in the history of Canadian cities and towns,[55] but it concurs with that on the English and American gas plants: in addition to producing large volumes of smoke from the burning coal, the "gasification process produced various waste products, the most noxious of which was the spent limewater that also contained sulphuric acid and assorted other substances from the purifiers; the tar and ammonia fluids called ammoniacal liquor."[56] There are few details on the decommissioning of these plants as natural gas replaced manufactured gas across Canada in the 1950s and 1960s. Evidence suggests that former coal gas plants in Canada, as around the world, are significant contemporary sources of "tars, sludges, liquors, and other gas cleaning wastes" that continue to present hazardous pollution of the soil and groundwater today, and are still potentially hazardous to human health.[57]

BIG GAS IN CANADA: NATURAL, PIPED, INCREASINGLY WESTERN, AND VOLATILE

Natural gas, composed mainly of methane and found "naturally" in the ground, abounds in Canada, mainly in Alberta, but also, as figure 11.4 illustrates, in much smaller quantities in British Columbia, Quebec, Nova Scotia, and the Northwest Territories. To extract the gas from a well, a drill (usually set within a large drilling rig) is sunk deep into the ground, the well is encased in steel and concrete, and fluids (mostly water) are typically pumped in to force the gas out. It is captured on escape and put into pipelines for transport. While in the early days gas was often piped directly to the "end user" – such as street lights, or into buildings to

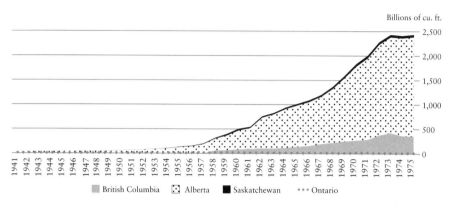

Figure 11.4   Production of natural gas by province, 1941–1975

Source: Statistics Canada, Historical Statistics of Canada, series Q26–30, *Production of Natural Gas by Province, 1941–1975* (millions of cu. ft), converted to billions, http://www.statcan.gc.ca/access_acces/archive.action?l=eng&loc=Q26_30-eng.csv.

provide heat and light – gas is now piped to plants for processing, where propane, butane, and ethane are removed (for sale after liquefaction under pressure), along with "impurities" that otherwise adversely affect its combustion, whether for heating or cooking. Most Canadian natural gas, for example, contains high levels of the poisonous *hydrogen sulphide* that is now extracted and sold as a valuable by-product. While the earliest developments in the industry were rather haphazard and artisanal (the first documented natural gas "plant" in 1827 comprised a fish barrel placed over a gas spring in Westfield on Lake Erie[58]), the natural gas industry of the twentieth century was driven by increasingly specialized technical knowledge, innovation, and a capital-intensive, ever-larger, and more complex physical plant that benefited from innovations in derricks, power transmission, fluid circulation, drill strings, pipeline construction, gas pressure regulation, well casing, and drilling rigs.[59] As the conventional field reserves of natural gas declined, companies looked to new techniques to extract gas from the ground, including horizontal hydraulic fracturing. This process extracts gas that is dispersed through shale by pumping a chemically complex fluid at high pressure into the rock formation to create fractures through which the gas can be captured.

While the origins of the petroleum industry are usually associated with Titusville, Pennsylvania, the first oil well in North America was in fact established in Essex County, Ontario, and "Canadian geologists, land speculators, drillers and investors were among the genuine pioneers in the establishment of an entirely new industry in the 1850s and 1860."[60]

Canadian oilmen began refining and shipping the first commercially viable product from liquid petroleum – kerosene – from hundreds of wells in southwestern Ontario, particularly around Oil Springs and later Petrolia. By the mid-1860s, Canadian oil interests were supplying much of Ontario and Quebec. At the time they also constituted the second-largest supplier (after the United States) of kerosene to Europe (100,000 barrels in 1865).[61] Natural gas is often found along with petroleum, and it initially shared with gasoline an identification as a waste product of the kerosene industry, and a dangerous one at that. While natural gas was originally regarded as "no more than a curiosity and a nuisance for the petroleum industry,"[62] its illuminating and heating qualities had been recognized and even employed locally in at least one New York community since the 1820s. Trois-Rivières was using natural gas for street lighting by 1856.[63] However, natural gas was not developed commercially until the 1880s, as the techniques for capturing, purifying, and particularly transporting gas over long distances emerged only slowly – an interesting contrast to the manufactured gas sector, plants for which in the 1850s "could be built anywhere as long as coal ... was readily available," and that thus acted in many places as a distinct disincentive to the development of a natural gas industry.[64]

A number of discoveries and inventions, many of them in the United States, spurred development of natural gas in the later nineteenth century, as it became an important component in a new growth complex, or energy "development blocks."[65] Pittsburgh began drawing on local reserves of natural gas in 1870, and by the 1880s had become the centre of coal, steel, and natural gas industries. By then, natural gas, becoming better known for its intense and even heat-giving properties, fuelled ten of the city's iron and steel mills, six glass-making factories, a crematorium, and every brewery in the city. Seeking to export natural gas farther afield, entrepreneurs encountered intransigent problems in transporting it: iron pipes were too small to move natural gas over long distances, and the pipe segments were attached by couplings tightened by screws, and always leaked. By the late 1880s engineers had developed a leak-proof rubber ringed-joint that was used by most pipeline companies throughout the world into the 1920s, and gas companies were able for the first time to expand into larger regional areas.[66] By the turn of the century, natural gas had developed in a range of places, from the Appalachians and Ohio to New York and Virginia. Development remained hampered, however, by the rapid depletion of stores of the fuel, often after expensive infrastructure, particularly pipelines, had been built. It was not until massive reserves of natural gas were discovered in the American southwest, and welded pipeline technology was developed in the 1920s, that natural gas began its expansion into the continent-wide energy system that it has become today.[67]

Commercial exploitation of natural gas began in Canada in 1889, when mining engineer Eugene Coste established the Ontario Natural Gas Company and began drilling for gas in Essex County. By 1894, he had almost thirty producing wells in Leamington, Kingsville, and Ruthven, supplying nearby communities with cooking, lighting, and heating fuel.[68] He began exporting gas from a field near Niagara Falls to Buffalo in 1890, and from the Essex fields to Detroit in 1894, where he sold his gas until 1901. In that year, the federal government prohibited the export of natural gas, fearing that exports would threaten domestic access to the rapidly depleting supplies of natural gas in the area.

Hopes of unlimited supplies of gas and oil were frustrated throughout the region, as they were in much of the United States at this time, as local supplies of gas and oil quickly dwindled after discovery and rigorous exploitation. Coste's own Ontario gas fields were depleted by 1904 and abandoned. Further oil discoveries renewed petroleum fever in southwestern Ontario, particularly the abundant Tilbury strike of 1905.[69] Attention was directed at western Canada after natural gas was discovered and exploited as early as 1883, when engineers of the Canadian Pacific Railway, looking for water for their steam engines, found instead a large flow of natural gas. In 1903, a large gas field was discovered in Medicine Hat, and by 1912 the municipality owned six wells that provided natural gas street lighting (finding it cheaper to leave them burning day and night than to turn them on and off), and providing heat and light to 1,900 domestic consumers and sixteen factories.[70] Eugene Coste headed to Alberta in 1906 and found gas on CPR lands near Bow Island. He incorporated the Canadian Western Natural Gas, Light, Heat and Power Company in 1911 and in that year extended a 180-mile pipeline to Lethbridge. The next year, what was rumoured to be "the longest and largest diameter gas pipeline ever built"[71] reached Calgary, where Coste had just bought out the Calgary Coal Gas company, with its 2,200 customers. Coste's company supplied towns and cities with natural gas well into the twentieth century until larger corporations took over.

It was Archibald Wayne Dingman's discovery of what turned out to be relatively large oil and gas fields in the Turner Valley, Alberta, in 1914 that refocused Canada's petroleum industry in the west. Since, as a 1915 report noted, "one of the most important problems which presents itself

... is the provision of an adequate supply of cheap fuel for the population of the Prairie Provinces of Canada,"[72] the large Turner Valley discovery was particularly appreciated. The Canadian Commission of Conservation was already rehearsing the kind of rhetoric that would become familiar to Canadians across the country in the 1960s: "Natural gas is the most perfect fuel with which we are furnished by nature. It is clean, can be readily piped for long distances and has a very high heating power."[73] Natural gas was used in many Prairie cities and towns, and in many rural areas mainly for cooking and heating, and in some industrial processes, including brewing and carbon black production. In 1920, further oil discoveries were made in Norman Wells, Northwest Territories, but these proved too expensive to exploit at the time. In southern regions, though, farmers ran gas lines to heat cattle barns during severe winter weather, for in the 1880s and 1890s some winters had seen massive die-offs of open range cattle. By 1947, Alberta had the highest percentage of Canadian homes with gas ranges.[74] Alberta natural gas had one attributes that was not appreciated, however: hydrogen sulphide. Lethal if inhaled in even tiny concentrations, the gas had an unpleasant rotten egg smell – sometimes known in Alberta as "the smell of money." In 1925 a plant was built in the Turner Valley to remove the poisonous hydrogen sulphide that was characteristic of the new *sour gas* supplies that were becoming available, "sweetening" it by exposing the gas to dissolved ash, in order to make it more widely marketable throughout the province.[75]

Before the much larger strike at Leduc in 1947, natural gas remained a relatively minor energy carrier in Canada as electricity and oil gained ground, and even as natural gas boomed in the United States from the 1920s onward.[76] Problems with the relatively small size of the Canadian fields were compounded by the fact that natural gas compared unfavourably with oil, particularly as the use of internal combustion engines grew. Oil not only fetched a higher price, but it cost about a sixth as much as gas to transport, for oil did not need the "heavy upfront investment in pipelines, the acquisition of technical knowledge, and the negotiation of complex legal agreements with municipalities, corporations, and residential customers" that gas required.[77] As a result, much of the natural gas supply continued to be seen by producers as (and was quite literally called) "waste gas"; many gas reserves that were discovered "were shut-in, flared or consumed within the vicinity of the wells."[78] One estimate suggests that in Ontario alone, nearly two billion cubic feet of gas "was squandered in this manner during these early years."[79] In one extreme case, a bore hole cut by the government officials looking for oil near

Pelican Portage, Alberta, in 1897, found gas instead. The gas accidentally caught fire and was still burning seventeen years later, in 1915, with a "daily escape of 2,900,000 cubic feet per day, 17,994,500,000 cubic feet in 17 years, or $2,951,098."[80]

The attitude to burning gas changed along with Canada's economy after the 1947 oil strike at Leduc, just south of Edmonton.[81] The vast supplies of oil and natural gas discovered met the increased demand for a secure source of continental energy from a United States now focused on the Cold War and at a time when there had been substantial improvements in drilling, pipeline, and processing technologies. Indeed a gas pipeline had been built from Portland, Maine, to Montreal in 1941 in response to to U-boat attacks. The first all-Canadian natural gas pipeline was completed 1958 by Westcoast Energy Inc., extending from northwestern British Columbia's Peace River District to the well-populated lower mainland and on to US markets in the Pacific Northwest. Alberta readily supplied natural gas to Montana, where Cold War fears had boosted production at the Anaconda copper smelter, by means of a new pipeline and little controversy.[82]

Natural gas grew rapidly in the 1950s from a local or regional resource to a continental commodity of considerable national significance. The role of the United States, and the federal government that regulated interprovincial and international trade, was crucial. In 1956, TransCanada PipeLines Ltd was given permission by Parliament to build a pipeline across the prairies and northern Ontario to Montreal. This pipeline was also the longest in the world when completed in 1958 (at a cost over $330 billion). Lateral lines went from Manitoba straight south to the United States, and to Sault Ste Marie, Sarnia, Niagara Falls, and Phillipsburg in Quebec, near the Vermont border at the top of Lake Champlain. The pipeline provided many Canadians with Canadian ("home-grown") natural gas for the first time. It stimulated industrial development along its route, which basically followed the Canadian Pacific Railway, and also considerable national debate about what was emerging on the political stage as an issue of vital, if conflicted, national economic interest.

The Liberal government that had approved the pipeline lost the next election, and the new prime minister, John Diefenbaker, appointed a Royal Commission on Energy to explore the implications of Canada's new energy abundance. The National Energy Board was created to oversee the new interprovincial and international energy trade as a response to the final report in 1958. Particularly contentious in these early oil and gas debates was who should supply the Montreal refineries: the cheaper oil from

Alberta, or the more expensive oil imported from South America and the Middle East. Smaller independent Canadian companies wanted Montreal to refine Alberta oil, whereas the larger multinationals argued that Montreal should refine imported crude. In 1961, Diefenbaker settled the dispute, and the new National Oil Policy declared that west of the Ottawa Valley refineries would use Canadian oil, and to the east, imported oil. The National Energy Board Act of 1959 allowed export of surplus only.

Discussion took place within the context that Canadian demand, however, was much smaller than its supply. Furthermore, the huge expenses of extracting and transporting natural gas to Canadians required "subsidy" from exports. In the late 1950s world oil prices fell, at first in response to Soviet exports. At the same time, the Korean War (1950–53) sparked fears about energy security in both the United States and Canada. In central Canada there was concern that if anything jeopardized exports from southwest Asia, then oil imported from Venezuela and California would go to Europe, and central Canada would then be energy-starved. In the late 1940s Alberta had been reluctant to sell natural gas outside the province, as there was great doubt that supplies could even meet local demand.[83] Now with its new abundance on one hand and Cold War fears about supplies on the other, Canada shifted dramatically away from protectionist policies, encouraging exports to the United States of oil, gas, and hydroelectricity in the early 1960s. The Canadian problem of low population density combined profitably with American thirst for oil and gas, particularly given the increasing insecurity of access to Middle Eastern oil during the Cold War. Since that time, the Canadian gas industry, like its oil industry, has been developed almost exclusively as a US subsidiary, with the United States owning, extracting, and consuming much of Canada's oil and gas supplies as its own dwindled.[84]

In 1968 oil and natural gas found in Alaska sparked a search around the Mackenzie Delta, Beaufort Sea, and Arctic islands, and proposals to pipe Canadian Arctic and Yukon gas and US gas south were made. Oil prices had stayed low until the early 1970s. The United States somewhat protected its own huge oil firms but allowed Canadian imports of natural gas, considering them "domestic." In particular the natural gas price in the United States was kept lower than light and heavy fuel oil, thereby keeping Canadian gas prices low too. Generally the US approach to gas pricing differed from that taken in the rest of the world, as the Americans did not index the price of gas to the price of its competitor, oil, which was the normal practice in Europe from the 1960s and which spread to Asia. So Canadian gas pricing was a function of global oil pricing only indirectly.

In the early 1970s, however, the Canadian Energy Board expressed concerns about its new petroleum future. Foreign (mostly American) ownership in oil and gas operations reached 77.3 per cent by the early 1970s.[85] No new sources of energy had been discovered for almost twenty years, with the exception of the Arctic reserves that were too expensive to exploit in this period of very cheap oil and gas. In 1973, the world experienced its first oil crisis in the wake of the Arab Israeli war, as OPEC quickly raised the price of its oil sixfold and placed an embargo on exports to North America. With no control over its Canadian prices, which were linked to American markets and world prices, and with supplies of foreign oil and gas to eastern Canada threatened, the Liberal government decided it needed to protect the national energy supply. Its attempts to provide low preferential rates for home-grown energy were popular with Canadian voters, but vigorously opposed by the foreign-owned energy companies, and many Albertans more directly reliant on the industry.[86] The government briefly invested directly in the energy market, creating Petro-Canada in 1975, "to supplement regulation" as a means of giving the federal government greater control over oil and gas development.[87] Petro-Canada was required to "act according to government objectives, rather than take the pursuit of of short-term profits as the top priority" on the principle that energy "should not be left entirely to foreign private companies."[88] In 1977 a Yukon plan was approved to send Alaska gas through Yukon to northern BC, on to Alberta, and thence to the United States. There was also a plan for a lateral line to the Mackenzie Delta. But the Polar Gas (pipeline) Project and the Arctic Pilot Project (LNG tankers to be headed for Nova Scotia or Quebec) would be cancelled by 1987 as world prices and supplies changed. By 1981, the total length of the gas pipeline system was 9,344 kilometres, delivering an average eighty-five million cubic metres per day to two million Canadian households. It needs mention that another aspect of the cost of natural gas is "gathering" it for export. In the early 1980s Alberta Gas Trunk Line was operating 10,836 kilometres of pipe inside Alberta to collect from small wells to feed into the large-diameter system owned by TransCanada PipeLines Ltd. But in 1981 Vancouver Island and the Maritimes were still not on the gas pipeline, though the federal government had promised extension to the Maritimes by 1983. This was cancelled, however.

The attempt to develop a coherent energy policy that prioritized energy security for Canadians was short-lived. The National Energy Policy's attempt to provide self-sufficiency in oil and gas was never implemented. There was considerable pressure from oil and gas companies who resented limits on the workings of the increasingly free market that offered such high profits. The conflict was also profoundly regionalized: those in the oil-producing provinces in the west argued that they were being unfairly exploited by oil-consuming provinces in the east. Western governments and corporations provided data suggesting that the only solution to oil and gas problems was to increase their prices to world levels, in order to finance more exploration, research, and development, particularly with finding new sources of oil and gas.[89] In 1984, the new Conservative government dismantled the National Energy Plan with oil price falls as the backdrop. The requirement to preserve a thirty-year natural gas supply for Canadians was reduced to one of ten years. The "new" idea was in fact the old one from over a century ago: sell off fuel resources as fast as possible.[90] Petro-Can was privatized in 1991.

Since 1985, free markets have governed oil and gas production, though with ongoing pressure from the public to regulate the environmental hazards, particularly of transporting oil and gas. The industry has seen massive expansion of natural gas extraction and export since that time. The free trade pressure of the 1980s culminated in 1993 in the NAFTA agreement that "placed Canada in a US dominated security zone." As Peter Sinclair has argued, under terms of the treaty, this meant that Canadian oil and gas became a key contribution to US security. "Contracts with US companies could not be rescinded and supply to the US could only be reduced as a result of a declared shortage under specified dire circumstances, such as armed conflict involving the energy." However, the amount of oil or gas shipped to the United States "could not be less than its proportion of total production in the previous three-year period." Unlike Mexico, also brought into the agreement, Canada is "not allowed to meet its own needs at the expense of shipments that otherwise would go to the US."[91]

As figure 11.5 suggests, natural gas production has increased dramatically since the 1980s, with most of the increase exported to the United States. In the early twenty-first century, natural gas has a reputation as a non-polluting fuel, suggesting that it will "continue to be used primarily for residential and commercial heating, electric power generation, industrial heat processes," and there are indications that its use as a transportation fuel will also grow.[92] In Canada, concerns remain, however, about the extreme price volatility of natural gas, and Americans' increased interest in "home-grown" rather than imported sources of energy.

Environmental issues are also a concern. While the environmental impacts of natural gas have been less damaging than those of other fossil

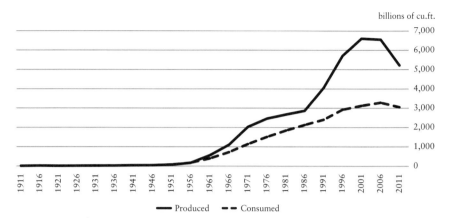

Figure 11.5   Canadian natural gas production and consumption, selected years, 1911–2009 (billions of cu. ft)

Sources: Historical Statistics of Canada, series Q31–37, *Production and Trade in Natural Gas, 1892 to 1976* (millions of cu. ft) (converted to billions), http://www.statcan.gc.ca/pub/11-516-x/sectionq/4057756-eng.htm#Q31_37; Index Mundi, "Canadian Consumption of Canadian Gas, 1980–2012."

fuels, they have not been negligible in this natural gas-exporting nation. Hydrogen sulphide in much of Alberta's sour gas was used untreated in homes and street lights before the first "sweetening" plants were created in the 1920s. A market for sulphur was found in fertilizer, mineral refining, and pulp and paper industries, and sulphuric acid was used to extract uranium from ore in mines in the Northwest Territories and Saskatchewan and Ontario. Not all toxins were removed, however; for many decades "sweetening" involved simply burning the remaining hydrogen sulphide emissions, and dispersing the sulphur dioxide thus created into the air from the smokestacks. Sulphur dioxide is "less toxic than hydrogen sulphide but as it oxidizes it can form very fine suspended particles (aerosol sulphate) that penetrate deep into the lungs potentially causing respiratory distress,"[93] and in higher volumes can destroy local environments. It was not until the 1960s, long after communities complained about the degradation of local environments and their own health problems, that companies were finally required to desist from disposing of the poisonous compounds.[94] While celebrated as an environmentally friendly fuel in comparison with its low CO_2 greenhouse gas emissions, some scientists have recently argued that natural gas is not, in fact, the climate-change solution for Canada, as the new processes of extracting natural gas from shale have significantly increased these emissions. [95] While concerns continue about other airborne pollutants such as nitrogen and sulphur oxide, new so-called fracking practices needed to extract shale gas have also been associated with extensive and serious groundwater pollution and landscape degradation. Many worry about the impact of gas extraction on the next frontier of development, sensitive Arctic environments.[96]

CONCLUSION

In the nineteenth and twentieth centuries, manufactured gas remained a local fuel until displaced by a variety of other national energy carriers, including oil, electricity, and natural gas. In the early days of lighting with manufactured gas, many urban dwellers welcomed a new source of light, but household demand was constrained by people's unfamiliarity with gases, as well as cost and supply problems. It is striking that the adoption of natural gas occurred much later.[97] Natural gas sales were constrained for some time by technical difficulties of transportation that affected both its cost and availability, many of which were related to problems of scale. Canada's modest population dispersed over vast distances and severe climate ranges presented an unattractive "market," yet vast amounts of natural gas have to be exploited in order for it to be profitable. Transcontinental pipelines built in response to American Cold War concerns resulted in Canada becoming a continent-wide trader in this energy carrier.

Many questions remain about the impact of both manufactured and natural gas for Canadians. We need a better understanding of how many Canadians used manufactured gas, and why. We need a better understanding of the transition to natural gas. The experience in other countries suggests that natural gas, which provided a much hotter flame than manufactured gas, was probably supplied through the same infrastructure of underground pipes that had supplied manufactured gas, but the actual burners and sometimes the entire appliance had to be replaced. In most industrial countries, the transition was expensive, technically difficult, and inconvenient to distributors and consumers. In the United States, "the conversion took place over approximately about thirty years, with the change-over of the largest utilities occurring in the 1940s and 1950s ... whereas the British shifted entirely to natural gas in a single operation." Carried out over about ten years in that tiny country, involving about thirteen million customers, and "estimated to cost one thousand million pounds," the British conversion from manufactured to natural gas was "perhaps the greatest peacetime operation in the nation's history."[98] We have only anecdotal accounts to chart the transformation in individual towns and cities in Canada.[99] Did most Canadians make the transition from manufactured to natural gas? Or did most move directly from wood or oil? Did Canadians need substantial persuasion to adopt this new energy carrier, as they did with manufactured gas and electricity?[100] And when they were convinced, how did the exigencies of natural gas production, transportation, centralized distribution, and unintended consequences of its use affect their lives? These and other urgent questions await historians in this and upcoming generations of scholars.

NOTES

1 On the role of petroleum in defining the second industrial revolution, see Tomory, "Building the First Gas Network: 1812–1820"; Hurley, "Creating Ecological Wastelands: Oil Pollution in New York City, 1870–1900."

2 David Suzuki Foundation, Pembina Institute, *Is Natural Gas the Climate Change Solution for Canada?*, 6–7.

3 About 40 per cent of Canada's energy needs come from each of oil and gas. Ibid., table 2, 7. In 2013, Canada was the world's fifth-largest producer of natural gas. In 2004, it was third. Canada, Statistics Canada, *Human Activity and the Environment: Annual Statistics 2004*, table 1.1.

4 Canada actually exported 59 per cent of its natural gas in 2008, but also some imported some back for a net total of 50 per cent exports. David Suzuki Foundation, Pembina Institute, *Is Natural Gas the Solution to Climate Change?*, 9. For Canada's natural gas export status internationally, see US Energy Information Administration, *Report Canada*.

5 Many attempts have been made to propose export facilities, but as of 2014, there is only one, established in Moncton, New Brunswick, in 2009, a mere terminal used only for trans-shipment of overseas sources of liquefied natural gas (LNG) to New England. The very explosive nature of the liquefied gas, the huge expenses in liquefying and de-liquefying it, and the fact that other countries are already cheaply supplying international markets for the product are limiting factors.

6 The process was later made safer by an automated feed. Black, *Canadian Scientists and Inventors: Biographies of People Who Shaped Our World*, 123–8. In one of the many positive feedback loops so characteristic of the mineral energy regime, oxyacetylene torches began to be used in the 1920s to weld joints on gas long-distance pipelines, transforming the industry. Castaneda, "Natural Gas," 167.

7 For a explanation of the properties of these gases, see appendix 1.

8 Tomory, *Progressive Enlightenment: The Origins of the Gaslight Industry, 1780–1820*, 6, chap. 2; Russell, *A Heritage of Light: Lamps and Lighting in the Early Canadian Home*, 288.

9 Murdoch was explicit in his 1791 patent that he was looking for tar for shipping, while Lebon's research, and his legacy, demonstrate his interest in industrial distillation. Tomory, *Progressive Englightenment*, 10, 74. The two almost certainly did not know each other.

10 Ibid., 6, chap. 2; Russell, *Heritage of Light*, 288.

11 Castaneda, *Invisible Fuel: Manufactured and Natural Gas in America, 1800–2000*, 6.

12 Tomory, *Progressive Enlightenment*; Castaneda, *Invisible Fuel*, 6.

13 Tomory, "Building the First Gas Network," 101. Frederic Winsor was the first to patent coal gas lighting, in 1804, and received the charter for London and Westminster Gas Light and Coke Company (GLCC) in 1812, making him the first to develop and implement the novel concept of an *integrated network* of underground pipes from a centralized plant. Tomory, *Progressive Enlightenment*, 7, 11, chap. 2; Tucker, *75 Years 1848–1923: The Consumers Gas Company of Toronto*, 35.

14 This description is paraphrased from Tomory, "Building the First Gas Network," 79.

15 The polluting aspects of the gas industry will be discussed below. See Tomory, "The Environmental History of the Early British Gas Industry, 1812–1830"; Tarr, "Toxic Legacy: The Environmental Impact of the Gas Industry in the United States."

16 American cities were using gas lighting from 1817 (Baltimore), but the industry did not really take off until coal supplies became more available on the East Coast, when it began to manufacture its own cast iron pipes in the 1830s, and no longer had to rely on British imports. Castaneda, *Invisible Fuel*, 28. The first natural gas pipelines in the United States, to Rochester, New York, in 1870 used Canadian Pine, and further research will likely demonstrate that many of the early Canadian pipes were wood. Castaneda, "Natural Gas," 165.

17 Tarr, "Toxic Legacy," 112.

18 The first gasworks in the United States, in Baltimore and Philadelphia, relied on distilled gas from pine tar until bituminous coal could be transported by canal and rail to the northeastern United States from the mines of Virginia in the late 1840s. Tucker, *75 Years*, 35; Castaneda, *Invisible Fuel*, 17; Jones, "A Landscape of Energy Abundance: Anthracite Coal Canals and the Roots of American Fossil Fuel Dependence, 1820–1860."

19 Armstrong and Nelles, *Monopoly's Moment: The Organization and Regulation of Canadian Utilities, 1830–1930,* 12.
20 On the importance of street lighting in the early development of gas in Toronto and Montreal, see ibid., 13, 20–33.
21 Ibid., 20.
22 Tucker, *75 Years,* 55.
23 Russell, *Heritage of Light,* 291; Canada, Department of Agriculture, "Manufactures of Provinces, Compared by Establishments and Products for 1881, 1891 and 1901," table 15, *Census of Canada, 1901,* vol. 3, *Manufactures.* See appendix 4, "Tentative List of Gas Plants," for details of where the plants were installed.
24 Tucker, *75 Years,* 51.
25 Armstrong and Nelles, *Monopoly's Moment,* 20.
26 Sandwell, "An Introduction to Lighting in Canada, 1840–1920." For an excellent history of kerosene in the United States, see Williamson and Daum, *The Age of Illumination;* Wlasiuk, "Refining Nature: Standard Oil and the Limits of Efficiency, 1863–1920."
27 E.J. Tucker, quoting from Consumers' Gas, *Annual Report,* 1862, stating, "The use of coal oil for illuminating purposes ... has in some measure interfered with the consumption of gas." Tucker, *75 Years,* 65. The number of Consumers' Gas customers fell throughout the 1860s, before rising in 1880. Armstrong and Nelles, *Monopoly's Moment,* table 3, 4, 27–8.
28 Russell, *Heritage of Light,* 297.
29 Nordhaus, "Do Real-Output and Real-Wage Measures Capture Reality? The History of Lighting Suggests Not," 38.
30 Tucker, *75 Years,* 47–8. The mantle, in spite of its illuminating virtues, was slow to catch on in both Canada and Europe before the 1890s, possibly because of price and problems with its "form and manufacture." O'Dea, *The Social History of Lighting,* 61; Russell, *Heritage of Light,* 297.
31 See Tomory, "Building the First Gas Network," for a delightful and detailed account of the administrative, technical and "people" (i.e., customer) problems created by "scaling up."
32 Armstrong and Nelles, *Monopoly's Moment,* 30.
33 When metering was invented by Samuel Clegg in 1815, it measured gas by moving it through water and so froze in the Canadian winters. Armstrong and Nelles, *Monopoly's Moment,* 27.
34 Tomory, "Building the First Gas Networks," 78. Further research is needed in the Canadian context. For a further discussion of the "domestication" of the British gas industry for home consumption, see Clendinning, *Demons of Domesticity: Women and the English Gas Industry, 1889–1939.*

35 Tucker, *75 Years*, 85.

36 Hanlan, "Special Campaigns," 77. Tomory, in "Building the First Gas Works," has interesting discussions of this point in the English context.

37 J.D.C. Clark, letter to the editor, *Gas Industry: Heat, Light, Power*, 77.

38 *Gas Industry: Heat, Light, Power*, 62.

39 See, for example, ibid.

40 E. St Elmo Lewis, "Efficiency in the Advertising and New Business Dept."

41 Tucker, *75 Years*, 57.

42 Pierce, "The Sale of Gas for Illumination," 64–5. The gas companies blamed the competing electricity industry for spreading false health rumours about the dangers of gas.

43 On some of the difficulties of convincing people to consume manufactured gas, see Graeme Gooday, *Domesticating Electricity: Technology, Uncertainty and Gender, 1880–1914*, 26–30.

44 Armstrong and Nelles, *Monopoly's Moment*, 27.

45 Ibid., 25–7.

46 Tucker, *75 Years*, 57; Krywulak, *Fuelling Progress: One Hundred Years of the Canadian Gas Association, 1907–2007*, 9.

47 The Lowe System of manufacturing gas, or *water gas* as it was sometimes called, installed in Toronto in 1878 and throughout the United States, revolutionized the industry. See appendix 1, "Primer on Terminology"; also Armstrong and Nelles, *Monopoly's Moment*, 27–30; and Tucker, *75 Years*, 67. One of the fullest descriptions is by Tarr, "Transforming an Energy System: The Evolution of the Manufactured Gas Industry and the Transition to Natural Gas in the United States, 1870–1954," 22–4.

48 Armstrong and Nelles, *Monopoly's Moment*, 78; Tucker, *75 Years*, 67.

49 Table 3, "Estimates of Types of Cooking Equipment Used in Metropolitan, Other Non-Farm and Farm Households in Canada," in Canada, Dominion Bureau of Statistics, *Household Equipment November 1948*, 4. Of all homes in Canada in 1948, about 25 per cent cooked with gas, about 20 per cent cooked with electricity, and 50 per cent cooked with wood. In 1941, just under 4 per cent of Canadian homes relied on gas for heating and 79 per cent of farm homes used kerosene for lighting. Canada, Dominion Bureau of Statistics, *Eighth Census of Canada, 1941*, vol. 9, *Housing*, tables 11, 13.

50 Castaneda, *Invisible Fuel*, 60. Forty-four per cent of all Canadian homes had electricity as early as 1921. Sandwell, "Mapping Fuel Use," 248.

51 Canada, Census and Statistics Office, *Census of the Manufactures of Canada, Bulletin 2, Canada 1906*.

52 Morgan, *A Textbook of American Gas Practice*, 930.

53 Castaneda, *Invisible Fuel*; Russell, *Heritage of Light*, 290.

54 Tomory, "Environmental History," 32; Tarr, "Toxic Legacy," esp. 124.

55 See Bonnell, *Reclaiming the Don: An Environmental History of Toronto's Don River Valley*, esp. 31–8; Cruikshank and Bouchier, "Blighted Areas and Obnoxious Industries: Constructing Environmental Inequality on an Industrial Waterfront, Hamilton, Ontario, 1890–1960"; Hatheway, *Remediation of Former Manufactured Gas Plants and Other Coal-Tar Sites*; Tarr, "Toxic Legacy" 107–8.

56 Tomory, "Environmental History," 32–3.

57 Several compounds in the tars have been documented as cancer causing in Ontario towns. Further problems remain in the iron oxide, sulphur, cyanide, and ammonia compounds remaining. Intera Technologies, *Inventory of Coal Gasification Plants in Ontario*, 1:17–20.

58 Castaneda, "Natural Gas," 164.

59 Gow, *Roughnecks, Rock Bits and Rigs: The Evolution of Oil Well Drilling Technology in Alberta, 1883–1970*.

60 Laxer, *Oil and Gas*, 5.

61 Kander, Malanima, and Warde, *Power to the People: Energy in Europe over the Last Five Centuries*, 261. For a history of Ontario's oil industry, see Cobban, *Cities of Oil: Municipalities and Petroleum Manufacturing in Southern Ontario, 1860–1960*; and Morritt, *Rivers of Oil*.

62 Drummond, *Progress without Planning: The Economic History of Ontario from Confederation to the Second World War*, 97.

63 The earliest was Freedonia, New York, where gas was used for lighting in the mid-1820s. Castaneda, "Natural Gas." For Trois-Rivières, see Krywulak, *Fuelling Progress*, 12.

64 Castaneda, "Natural Gas," 164.

65 Kander, Malanima, and Warde, *Power to the People*, 5; or what Wrigley calls "positive feedback loops."

66 Castaneda, "Natural Gas," 165–6.

67 Changes in field methods of ditch-digging and the use of protective coatings were other important innovations. Tarr, "Transforming an Energy System," 27.

68 Krywulak, *Fuelling Progress*, 14.

69 Ibid., 25.

70 De Mille, *Oil in Canada West: The Early Years*, 66.

71 Bott, *Evolution of Canada's Oil and Gas Industry*, 15.

72 Adams, "Committee on Minerals: Our Mineral Resources and the Problem of the Proper Conservation," 12.

73 Ibid., 13.

74 At 41 per cent, compared to 31 per cent in Quebec and 32 per cent in Ontario. "Table 1, Estimated Number of Households Using the Different Types of Cooking Equipment by Province," Canada, Dominion Bureau of Statistics, *Household Equipment November 1948*, 3.

75 Bott, *Evolution of Canada's Oil and Gas*, 26.

76 From the mid-1920s, natural gas from the southwest boosted dwindling supplies in the northeast, and long-distance pipelines, mainly financed by coal gas distribution companies, brought southwestern gas to Los Angeles, then San Francisco in 1929, and Chicago in 1931. Castaneda, "Natural Gas," 168–9.

77 Canada, Energy Mines and Resources, *Energy in Canada: A Background Paper*, 26.

78 Sinclair, *Energy in Canada*, 23.

79 Krywulak, *Fuelling Progress*, 26. A Federal Trade Commission Report to the Senate on Public Utility Corporations listed the billions of cubic feet of natural gas wasted in the United States between 1919 and 1930, concluding that in most years the amount flared off or vented exceeded the amount consumed. The annual amounts wasted grew from 213 billion cubic feet in 1919 to a peak of 585 in 1929. Castaneda, "Natural Gas," 168.

80 Adams, *Committee on Minerals*, 19.

81 Imperial Oil had previously drilled 133 "dry holes" in the region. Bott, *Evolution*, 18.

82 Ibid., 25–7.

83 Sinclair, *Energy in Canada*, 26.

84 Ibid.

85 Laxer, *Oil and Gas*, 7–8; Sinclair, *Energy in Canada*, 25.

86 Laxer, *Canada's Energy Crisis*.

87 Sinclair, *Energy in Canada*, 28.

88 Ibid.

89 Laxer, *Canada's Energy Crisis*, provides convincing evidence that there was neither an oil and gas shortage, nor the perception of one; political pressure forced the changes. Sinclair is a little less direct. For a discussion of the contemporary chaos of energy in one province, Nova Scotia, see Richard Starr, *Power Failure?*

90 For a detailed description and critique of this approach, see Sinclair, *Energy In Canada*.

91 Ibid., 31.

92 Castaneda, "Natural Gas," 173.

93 The University of Calgary Environmental Science Program, "Impacts of Airborne Pollution from the Sour Gas Industry on Southern Alberta."

94 Bott, *Evolution of Canada's Oil and Gas Industry*, 27.

95 David Suzuki Foundation, Pembina Institute, *Is Natural Gas the Climate Change Solution for Canada?*, 12–14.

96 Ibid., 15–19.

97 Fouquet, "The Slow Search for Solutions: Lessons from Historical Energy Transitions by Sector and Service," 6592.

98 Tarr, "Transforming an Energy System," 31.

99 See, for example, BC Hydro Power Pioneers, *Gaslights to Gigawatts: A Human History of BC Hydro and Its Predecessors by the BC Hydro Power Pioneers*, 127–32.

100 Sandwell, "The Pedagogies of the Unimpressed: Re-Educating Ontario Women for the Mineral Economy, 1900–1940."

12

Nuclear Power

Laurel Sefton MacDowell

Nuclear power, like the most of the energy provided by fossil fuels and hydroelectricity, requires highly complex, integrated, centralized, and expensive systems of extraction, transportation, processing, distribution, and waste disposal. While nuclear power shares the economies of scale and technological complexity of oil, gas, and electricity, the way that nuclear power is generated is significantly different from these, and indeed all of the other sources of energy used by people on the planet. We get energy directly from digesting our food because the *chemical reactions* involved in the process release usable energy. Similarly, energy is produced from the chemical changes involved in the *combustion* (burning) of organic materials like wood and the super-condensed forms of organic matter that comprise fossil fuels. Human beings also deliberately harness kinetic energy from wind and water in order to move things, including geared machinery and ships and turbines that in turn move a conductor in relation to magnets in a way that transforms kinetic energy to electrical by stimulating the movement of electrons. *Nuclear energy* does not come from kinetic, magnetic/electrical, or chemical processes; it is produced when the atomic structure of particular kinds of matter is transformed by dividing the nucleus – "splitting the atom" – to create a new kind of matter. Unlike digestion, combustion, and electrical energy, "nuclear power" refers not to energy released by rearrangements amongst atoms; instead, massive amounts of energy are released by changing the atoms.

In spite of its unique source of generation, nuclear energy is used in Canada for one energy-creating purpose only: to generate electricity via steam power. The energy released by the nuclear reaction acts as a

substitute for the energy released by burning of coal or oil in other thermal electrical generating stations, providing the energy needed to heat
water to create steam, which moves turbines to generate electricity. In the
case of hydroelectric generating plants, the electricity-generating turbines
are turned by moving water. In spite of its dramatic novelty and complexity as a source of energy, therefore, nuclear power is used only to generate
electricity, and in a familiar way: by heating water to create steam.

Although the energy released by "splitting the atom" is now used in
Canada to heat water, its origins, and indeed much of its contemporary
use, has been tied to military purposes since it was first harnessed in
nuclear reactors in the 1940s. First and foremost the energy released by
nuclear fission was used to create bombs, those dropped on Hiroshima
and Nagasaki in August 1945, which ended the Second World War.
Nuclear reactors continue to be employed for this purpose, although not
in Canada, but military uses of nuclear energy are not limited to bomb-
making. The military also has been interested in using nuclear power to
generate electricity, particularly to fuel military submarines, where small
nuclear reactors promise months of power without the necessity of either
refuelling or resurfacing.

There are two broadly different processes, worldwide, for creating the
"peaceful uses" of nuclear power to generate electricity, one developed
and used in Canada, and the other used in the rest of the world. The more
technical aspects of these two processes are outlined in appendix 1. To
put it as simply as possible here: the CANDU reactor developed in Canada
is the only nuclear reactor that does not use enriched (weapons-grade)
uranium to create nuclear power. Instead CANDU reactors rely on a process involving *heavy water* (*deuterium oxide*), a difference that not only
makes the process of making power safer – "meltdowns" are not possible
in heavy water reactors, scientists argue – but its non-enriched fuel cannot be used to create bombs.

When the CANDU reactors were first developed, Canadians champi-
oned this uniquely safe and peaceful feature of its nuclear industry. As we
will see below, however, Canada's claims of superiority were difficult to
maintain through the worldwide protests against the potential dangers of
nuclear power in the 1970s and 1980s. Canada was never able to sustain
fully the unique safety claims of the CANDU reactors, particularly in light
of revelations about the ever-growing problems of disposing of the
nuclear waste – the radioactive by-products of power generation.
Furthermore, the allegedly peaceful nature of nuclear power from CANDU
was seriously challenged on a number of fronts. Not only was Canada

responsible for extracting, processing, and exporting much of the weapons-grade plutonium that *was* used in the world's other nuclear reactors and bombs, but the world learned that at least one of the CANDU reactors sold on international markets in the 1970s had been adapted for the creation of a nuclear bomb.

In spite of ongoing debates about the safety of nuclear power and discussions of Canada's role in creating weapons-grade plutonium for export, nuclear power became a significant source of power, particularly in Canada's most industrialized province, Ontario. Between 1970, when nuclear power was first used commercially, and 1976, the proportion of total electricity provided by the nuclear industry in Canada rose from 1 to 6 per cent. By 2007, that figure had increased to over 14 per cent of all electrical generation, and about 40 per cent of all thermal power.[1] Many critics of nuclear power have changed their ideas about its relative harm in light of concerns about climate change: nuclear power provides much less carbon dioxide per kilowatt hour of power than thermal power from coal or oil; but environmentalists point out that the nuclear industry does produce carbon and pollution in mining and processing the uranium needed to produce nuclear power. Nevertheless, the government of Canada is backing away from supporting nuclear power generally and the CANDU reactor in particular, presumably because of concerns about the huge cost of this form of electrical power, and in response to ongoing protests about safety and Canada's role in supplying the armaments industry. The Ontario government recently committed to refurbishing some nuclear plants (Darlington and Bruce Power), but it is not building new ones.[2]

MINING FOR NUCLEAR POWER

While oil and gas were among the important energy sources supplementing wood, water, and coal in the twentieth century, some scientists have suggested that "harnessing nuclear energy for war and peace was arguably the most momentous event in the twentieth century."[3] On one level, nuclear power developed in Canada during the Second World War simply because uranium was available. Uranium extraction began with the founding of Eldorado Mining Company in 1925 by mining entrepreneur Gilbert LaBine and the discovery, in 1930, of high-grade *pitchblende* (uranium ore) on the eastern shore of Great Bear Lake in the Northwest Territories. A Department of Mines staff member considered the find "the most important mineral discovery in many years,"[4] highlighting the earliest function of nuclear energy: its potential for military use as an extremely powerful weapon. In 1933 the Eldorado mine began production, and company employees lived in the new company town site of Port Radium. The mine closed briefly in 1940, reopened in 1942, and was nationalized in 1944 as its uranium was used in the American atomic bomb project.[5]

Prospectors and geologists found other deposits of commercial grade uranium between Great Bear and Great Slave Lakes and in northwestern Saskatchewan north of Lake Athabasca. With several ore bodies in the far north, distance and climate partly determined how the mines developed. Men and supplies were flown in; heavy equipment was shipped up the Athabasca and Mackenzie Rivers on barges, and camps mushroomed. The industry invaded the Great Bear Lake region, clearing the edges of northern lakes, extracting ore, and moving it south. As historians Arn Keeling and John Sandlos have described, the company relied largely on local Indigenous people "to perform unskilled work at the site and along the ore transportation route, including the loading and unloading of uranium ore in burlap sacks for barge transport along northern waterways to a railhead in Alberta."[6]

Farther south on the north shore of Lake Huron, in 1949, prospector-geologist Franc Joubin discovered uranium and sold some claims to mining companies for large sums of money. Several mining companies, among them Rio Algom Mining Ltd and Denison Uranium Mining Ltd, developed land bordering the Serpent River First Nation, an Ojibway people living by fishing and hunting on the Serpent River watershed between Sault Ste Marie and Sudbury. They developed mines and mills in the new town of Elliot Lake. There, between 1954 and 1958, eleven uranium mines sank shafts, and over ten thousand miners worked the vast uniform and continuous uranium deposits.[7] The industry employed three types of mining methods – underground stope, underground room and pillar, and open pit – to take out the ore by drilling and blasting – using cold weather methods to cope with dampness and permafrost. After uranium ore was mined, it was milled to separate the uranium from other elements, and the ore was crushed into a powder called "yellowcake." The yellowcake product was shipped to refineries, converted into uranium fuel pellets and then fuel bundles to power CANDU reactors. The second form of demand influencing the production of uranium was, therefore, using it as a fuel for heating vast amounts of water, which converted into steam to generate electricity.

Elliot Lake became the main uranium-mining centre in Canada in the 1950s and 1960s. A boom-and-bust community, it started with temporary quarters, mushrooming trailer camps, small service businesses in plywood buildings, and dusty dirt roads. By 1959, its peak year, the mines had eleven treatment plants operating and produced 74 per cent of Canada's uranium oxide. When the world market for uranium collapsed in 1963 as the result of competition from US deposits, the Elliot Lake mining industry partially shut down. When the 1970s oil crisis renewed demand for Canadian-made fuels and energy, uranium revived along with the growth of domestic nuclear power generation and a stronger relationship with Ontario Hydro. The number of mines increased to thirteen, by then all owned and operated by Dennison Mines and Rio Algom Mines. The last mine closed in 1996, as the companies moved to mine higher-grade ore in the Athabasca Basin on Dene lands.[8]

Uranium production in northern Saskatchewan expanded between 1952 and 1955, as the American market for uranium stimulated the construction of the Beaverlodge Mine, a mill, a transportation system, and Uranium City, located on the north shore of Lake Athabasca, the third-largest and most southerly of the northern great lakes. Discoveries of uranium in the Athabasca Basin began around 1968 with Rabbit Lake, just outside the Basin, and the Key Lake deposit in 1975. Cigar Lake and McArthur River, the largest known deposits, comprised a second phase of development from 1977 to 1992.[9] In the mid-1970s, the early Saskatchewan discoveries at Rabbit Lake, Cluff Lake, and Key Lake were of sufficient size and quality to pre-empt most other known sources of uranium, including the already-depleting, costly areas around Elliot Lake and Uranium City.[10] The low-cost production in Saskatchewan contributed to lower prices for uranium. Throughout the 1990s, Cameco's (formerly Eldorado Nuclear's) Key Lake was the world's largest high-grade uranium mine, supplying 15 per cent of the world's uranium mine production in 1997. Of four new uranium projects in northern Saskatchewan, three used the treatment plant at McClean Lake. In 1999, the McArthur River mine operated by Cameco produced high-grade ore, and further expansion in the area has continued to the present. These mines trucked the ore for treatment to mills seventy to eighty kilometres away.[11] Despite an economic downturn in 1981, by the 1990s Saskatchewan was known as the uranium capital of the world, with the richest uranium ores in Canada. The low-cost production in Saskatchewan contributed to lower prices for uranium. Production continues and is expected to peak around 2020.

Table 12.1   Annual uranium production, 2000–2011 (tonnes U_3O_8)

	2000	2001	2002	2003	2004	2005	2006	2007	2008	2009	2010	2011
McArthur River	6,877	8,491	8,491	8,492	8,492	7,528	8,654	9,029	9,064	8,868	9,135	8,675
Cigar Lake	–	–	–	–	–	–	–	–	–	–	–	156
McClean Lake	2,734	2,724	2,490	814	867	1,476	1,637	785	–	–	–	51
Rabbit Lake	2,690	2,462	2,732	2,326	1,821	1,613	1,706	1,726	1,721	1,744	1,872	1,889
Cluff Lake	32	–	–	–	–	–	–	–	–	–	–	–
Total	12,333	13,676	13,713	11,632	11,180	10,617	11,997	11,540	10,785	10,612	11.007	10,771
World	41,998	47,430	49,052	46,499	48,680	51,611	59,772	63,285	63,085	68,805	70,015	66,297

Source: World Nuclear Association, "Uranium in Canada."

Canada is rich in uranium resources with a long history of exploration, mining, and generation of nuclear power, and for many years it was the world's largest uranium producer. Until 2008, more uranium had been mined in Canada than in any other country – 428,000 tonnes which was 18 per cent of the world's total. Today, Canada is the second-highest producer globally; most production is from the McArthur mine in northern Saskatchewan, with smaller developments in Labrador, Nunavut, and Quebec[12] (see tables 12.1 and 12.2).

Though six uranium companies mined profitably in the 1970s, uranium deposits discovered in British Columbia, Nova Scotia, and Labrador were not developed, because of citizens' opposition.[13] Almost from the beginning, the nuclear industry generated controversy, partly because in the early years, nuclear energy was used exclusively for military purposes. Canadians became aware that the Canadian nuclear industry had sent large amounts of its uranium ore to the United States to its wartime Manhattan Project, which made the first atomic bomb. The shock of Hiroshima and ongoing tests of hydrogen bombs in the context of the Cold War created fear, and a peace movement emerged in Canada, as elsewhere. From the Ban the Bomb movement in the 1950s, the peace movement continuously renewed itself as an anti-nuclear movement that continues to the present. The power of the new technology and its potential for destruction contributed to a cultural flood of information that frequently unnerved the public.[14]

The nuclear industry was characterized by instability, with its rapid development in the early years, and the boom-and-bust towns that sprang up around the uranium mines in Canada. And it has been characterized

Table 12.2 Annual uranium production, 2002–2014 (tonnes U) (as opposed to U_3O_8)

	2002	2003	2004	2005	2006	2007	2008	2009	2010	2011	2012	2013	2014
McArthur River	7,199	5,831	7,200	7,200	7,200	7,199	6,383	7,339	7,656	7,686	7,520	7,744	7,356
Cigar Lake	–	–	–	–	–	–	–	–	–	–	–	–	132
McClean Lake	2,342	2,318	2,310	2,112	690	734	1,249	1,388	666	–	–	–	43
Rabbit Lake	440	2,281	2,087	2,316	1,972	1,544	1,368	1,447	1,464	1,459	1,479	1,587	1,602
Cluff Lake	1,626	27	–	–	–	–	–	–	–	–	–	–	–
Total	11,607	10,458	11,597	11,628	9,863	9,477	9,000	10,173	9,786	9,145	8,999	9,331	9,134
World	36,063	35,613	40,219	41,595	39,429	41,279	43,764	50,684	56,663	53,494	58,344	59,372	56,217

Source: World Nuclear Association, "Uranium in Canada."

by both an innovative technology on the world stage and the widespread public concern and protest that it generated. The generation of nuclear power stimulated anxieties about war and peace and citizens were increasingly concerned about both the disastrous potential of accidents in the peaceful use of nuclear power, and by the 1970s about the environmental degradation and health hazards attendant on its extraction.[15]

ESTABLISHING THE NUCLEAR INFRASTRUCTURE

In the nuclear industry's early stages, British and American military interests led Canada to establish the infrastructure necessary for its nuclear industry. Canada was the first country to mine uranium on a large scale. During the Second World War, it developed "a network of mines, processing plants, and transportation that made it a leading exporter of uranium for atomic bombs."[16] As mentioned above, initially nuclear energy was used for military purposes; the nuclear industry sent large amounts of Canadian uranium ore to the United States to its wartime Manhattan Project, which made the first atomic bomb. The Eldorado (later Comeco) refinery in Port Hope, located on Lake Ontario in central Canada, had good rail connections to the north and was crucial to the bomb project, processing 2,032 tonnes of uranium oxide from 1942 to 1946. As a result of collaboration between British and Canadian nuclear researchers in Montreal, in 1944 the National Research Council of Canada (NRC) opened the Chalk River labs, which in 1945 developed the first operational nuclear reactor outside the United States. Chalk River had an experimental heavy water reactor to control a nuclear chain reaction, and this new technology became the basis of the Canadian nuclear production, which was exported for military purposes and used for civilian nuclear energy programs. These nuclear reactors became known as CANDU reactors. Chalk River labs received government approval in 1955 to develop nuclear energy for civilian use. Eldorado initially was involved in selling radium to treat cancer, but its use declined gradually as Chalk River produced cobalt-60 to replace it. Today, Canada produces 85 per cent of the world's supply of cobalt-60 in the aging Chalk River reactor, which continues to be used for scientific research. A second research reactor, the National Research Universal (NRU), which began operations at Chalk River in 1957, produces more than half of the world's medical *isotopes* as well as 80 per cent of cobalt-60 used in cancer therapy.[17]

From the beginning, the nuclear industry was regulated in the wartime context because of the nature of its production. It shared with other high-energy forms of power such as oil, gas, and hydroelectricity a tendency to have close ties to the Canadian and provincial governments, and these characterized the industry's culture and decision-making. The Canadian government originally bought shares in Eldorado but in 1944 converted the company to a Crown corporation. It established the Atomic Energy Control Board (AECB) in 1946 to regulate atomic energy in Canada. In 1952 it set up Atomic Energy of Canada Ltd (AECL) as a Crown corporation, to design, build, market, and manage reactors, do nuclear research, oversee nuclear waste management, and develop nuclear energy for peaceful purposes. AECL also took over the operation of the Chalk River labs engaged in experimenting with nuclear energy and testing nuclear reactors.

After the war, private companies had greater access to the nuclear infrastructure than during the war; the privately owned uranium mines operated under special agreements and shipped ore to Eldorado, which retained a monopoly on processing. Canada supplied the American government with 30 per cent of the ore it needed for weapons production. Until 1963, its nuclear scientists refined plutonium-making techniques and sold plutonium to the Americans. Keeling writes of Uranium City, but his comment applies in general to all the uranium mining communities: "Uranium development stitched together: distant markets; extensive transportation networks and processing facilities; the extensive and voracious resource demands of an industrial mining operation; regional, national and international political economies; and regional populations, ecologies and landscapes."[18]

In the 1950s, consumer demand for energy exploded along with economic growth. Governments in Canada confirmed their support for large-scale, high-tech projects, which included the nuclear industry. Notwithstanding the rapid growth of hydroelectrical power generation, nuclear energy was seen as an important addition for developing the country's industrial potential, particularly in the coal-poor industrial heartland of Ontario. Canada's position as a nuclear power contributed to its international status, notwithstanding the fact that from 1949, it chose to pursue "the peaceful applications of nuclear energy"[19] with the uranium it mined. As a result of this policy direction, the relationship between the AECB and Ontario Hydro commenced in 1951 and expanded. In 1954 Ontario Hydro's annual report first mentioned the nuclear energy partnership and a plan to build a nuclear power plant near Chalk River.

In 1959 the United States did not renew its contracts to import uranium from Canada, choosing instead to rely on its own increasing production. When the world market for uranium collapsed in 1963, the federal and Ontario governments supported the shrinking uranium mining industry by negotiating stretched-out contracts with the Americans so the industry could stay afloat. The Ontario government stockpiled uranium ore and encouraged the expansion of nuclear energy to power electrical plants. Ontario Hydro started to produce nuclear energy in 1962 at the country's first nuclear power plant complete with a CANDU reactor, in Rolphton, Ontario. A similar private/public relationship applied to the western region, where the Saskatchewan Mining Development Corporation, a provincial Crown corporation, took "a 20 percent interest in the Cluff Lake development and a 50 percent interest in Key Lake." In 1988 it merged with Eldorado Nuclear Ltd, which had closed the Beaverlodge Mine in 1981, to form Cameco Corporation, now the world's largest uranium producer.[20]

The growing postwar energy demand expanded the market for nuclear power to support the production of electricity. As a domestic nuclear energy industry expanded in Canada, the conditions for production, extraction, and transportation were in place, and the industry developed closer ties with hydroelectric companies, particularly in Ontario.

In Ontario, with no remaining untapped hydroelectricity sites and no oil, gas, or coal but plenty of uranium, the provincial government developed its nuclear energy capacity to service growing communities. Taxpayers subsidized the massive costs, and the government's partnership with the nuclear industry provided Ontario Hydro's personnel with training and status. In 1968, Ontario Hydro and AECL collaborated to build another CANDU reactor at Kincardine, Ontario, which operated until 1984. The Pickering and Bruce atomic complexes started up in 1973, a 200-megawatt nuclear generating station was built at Douglas Point on the eastern shore of Lake Huron, as was a four-unit station in 1984 at Darlington. Quebec purchased two reactors, and New Brunswick ordered one reactor from AECL after 1963, but Ontario expanded to nineteen nuclear reactors for a total of twenty-two in Canada. Canada's nuclear energy industry got a boost from the oil crisis in the early 1970s so that a portion of the federal government's National Energy Program (1980) involved nuclear energy.

THE PRODUCTION AND CONSUMPTION OF URANIUM AND NUCLEAR ENERGY

In 1976, 5 per cent of Canada's electricity came from nuclear power. By 1992, nuclear energy produced 48 per cent of Ontario's electricity and 15 per cent of Canada's and was increasing. "Canada generated 636 billion kWh in 2011, of which about 14.3% was from nuclear generation, compared with 59% from hydro, 13% from coal and 8.4% from gas. Annual electricity use is about 14,000 kWh per person, one of the highest levels in the world." The huge cost of nuclear power was facilitated by the continuing government partnership with the nuclear industry.[21] But the Canadian nuclear industry also had a substantial commercial export component. It sold uranium concentrates to Britain as early as 1957 and later to Germany, Switzerland, and other countries. Briefly from 1972 to 1975, it formed a world cartel with other nuclear nations to force uranium prices up, and after 1973 when oil prices quadrupled, the demand and price for uranium rose.[22] The federal government spent billions promoting nuclear energy in Canada and selling CANDU reactors abroad for nuclear energy use to India, Pakistan, Argentina, South Korea, and Romania. India misused its Canadian nuclear reactor to develop a bomb, which it detonated in 1974, despite safeguards in the contract. Canada renegotiated contracts, which met its new, tougher, but not necessarily effective non-proliferation requirements.[23] Sales dropped off after 1978, but in the 1990s, as the number of nuclear nations increased, AECL used aggressive marketing to introduce the next generation of CANDUS. It delivered nuclear plants to South Korea and China, in 2003 sold a CANDU-6 reactor to China, and produced the compact ACR-700s, which use light-water cooling, increase power and safety, but reduce cost. Changes in the international political climate resulting in the creation and strengthening of the international non-proliferation regime and more recently greater awareness of the importance of environmental protection have affected CANDU exports.[24]

In mid-2011 AECL sold its reactor division to SNC-Lavalin's Candu Energy subsidiary for $15 million; the Canadian government retained intellectual property rights for the CANDU reactors, in the hope of future royalties from new-build and life-extension projects "while reducing taxpayers' exposure to nuclear commercial risks." As CANDU energy sells new third-generation reactors, the government will contribute $75 million of support money. It will complete refurbishment projects at four plants and absorb 1,200 employees. CANDU power reactors also produce almost all the world's supply of the cobalt-60 radioisotope for medical and sterilization use. Two fuel fabrication plants (Blind River and Port Hope) in Ontario process some "1900 tonnes of uranium per year to UO_2 fuel pellets in a two stage process, mainly for domestic CANDU reactors but it is also enriched outside Canada for use in light water reactors. Between 15 and 20 percent of Canada's uranium production is consumed domestically"[25] (see table 12.3).

Table 12.3   Canadian uranium production, use and exports, 2005–2010 (tonnes)

	2005	2006	2007	2008	2009	2010
Canadian production	11,628	9,863	9,477	9,000	10,173	9,786
Less: domestic use	1,607	1,620	1,661[a]	1,670[a]	1,845[a]	1,675[a]
Canadian export	10,021	8,243	7,816	7,330	8,328	8,111

[a] Figures from World Nuclear Association Market Report
Source: World Nuclear Association, "Uranium in Canada."

WASTE PRODUCTS IN THE NUCLEAR INDUSTRY AND CONCERNS ABOUT THE EFFECTS ON THE ENVIRONMENT, ECONOMY, AND SOCIETY

The nuclear industry has embodied several elements: it represented Canada's emergence as a modern industrial nation and an atomic power; it expressed Canada's desire to develop resources in "the new North" and settle its new resource towns with largely imported labour forces from the south; and it reflected older northern colonization notions that sought to assimilate Aboriginal peoples, which disadvantaged them in the new economy. Such notions reflected class, gender, and racial biases.[26] Invariably the industry with government support was able to generate great wealth for corporate owners and shareholders, but its detractors note the destructive social effects on workers (occupational health and safety impacts) and Indigenous peoples (radiological contamination impacts). The refining process produced mine tailings in the form of fine yellow gravel, and mill wastes from acid leaching processes that contained elements such as thorium. Thorium remains in the environment for 250,000 years or more and has health impacts that are not well understood.[27]

The case of the Serpent River First Nation (SRFN) is instructive. In 1955 the federal government reactivated an old company lease and in return for jobs, the SRFN agreed to a sulphuric acid plant on its reserve to produce acid for the uranium mills upstream. They were told the land would be returned to them in its original condition. When the plant was abandoned in 1962, residents discovered that the land was contaminated.[28] The degrading environmental effects of the nuclear industry involved tons of tailings, lakes destroyed by wastes, overuse of forest resources in the construction boom, loss of wildlife from urban development and pollution, and costly remediation projects "in perpetuity" operated at public expense.

In the course of producing and processing uranium up to the 1970s, the mining industry processed uranium ore as it would any other mineral, using similar methods of mining and waste disposal, and similar standards of employees' occupational health and safety. Evidence suggests the industry was, however, well aware that radiation posed special problems. The International Commission on Radiological Protection had set levels of maximum exposure, which it lowered in 1959 as it learned more. In 1960 the federal Atomic Energy Control Board (AECB) enacted its own radiological regulations, which the province of Ontario was initially going to enforce. Instead it allowed the industry to self-regulate.[29] Governments imposed only limited regulatory approaches, allowing radiation to damage environmental resources such as waterways and the health of persons in contact with it. Uranium mining waste from the tailings was significant (over fifty million tons a year between the start of operations and 1964) and radiation pollution in the waters from the tailings and carried through the air by wind was dangerous to miners and to those living near the mines.[30] As protesters and critics argued, these actions could be understood as governments placing the needs of the industry before the public interests in environmental or human health. As Robynne Mellor put it in her comparative study of uranium mine and mill tailings regulation in Canada and the United States, "From the end of World War II until the beginnings of regulation in the 1970s, the effluents from uranium mines and mills in Elliot Lake polluted landscapes. The structures engineered to contain these tailings routinely failed, and as more mines shut down, there was no one to take responsibility for this problem. The AECB unsuccessfully regulated the industry's destruction of the landscape."[31]

As with other mining, milling, and processing industries, great damage was done to the environment. The nuclear industry's approach to employee safety and environmental protection was neglected. The serious problem that resulted by the 1970s led to a transition from a hands-off policy by government to an era of more regulation. The adverse occupational health and safety (OHS) effects on the uranium mines' employees, exposed to high levels of silica dust and radiation, resulted in soaring cancer rates three times the expected rate and high levels of silicosis, and developed into a political crisis in Ontario. In 1974, unionized Elliot Lake uranium miners went on a wildcat strike to draw public attention to their plight. Under political pressure, the province appointed the Ontario Royal Commission on the Health and Safety of Workers in the Mines (called the Ham Commission in the media), which wrote a scathing report about the treatment of miners and recommended a new OHS system for Ontario; the recommendation resulted in the Occupation Health and Safety Act (1978).[32]

Ironically, even though the issues in the uranium mines sparked the Ham Commission, uranium mining was a federal matter. Ontario uranium miners were not covered by the new provincial legislation. In 1974 the AECB had its own Mine Safety Advisory Committee investigate health and safety matters. In 1980, the United Steelworkers' Union demanded the same protection for uranium miners as for other workers under the Ontario act. Not until 1984, however, did the federal board pass the Uranium (Ontario) Occupational Health and Safety Regulations, which stated that uranium mines must comply with the Ontario Occupational Health and Safety Act.

Meanwhile in the Port Hope refinery, established by Eldorado, employees were exposed to hazardous radiation levels, which with greater knowledge, employees' training, and money gradually were reduced. As late as 1979, the AECB approved Eldorado's plan to expand the uranium refinery, but even then the plan made no provisions for waste disposal or for a study of health effects. Doctors reported an unusually high incidence of cancer in the area, and concerned residents pressed for a community health study to track increasing rates of cancer, asthma, emphysema, thyroid problems, learning disabilities, and ailments possibly linked to radioactive contamination.[33]

Despite its earlier casual approach to occupational and environmental health and other regulations, in 1978 the AECB organized an Advisory Panel on Tailings, which acknowledged the tailings problem and recommended the remediation and review of the problem. Uranium-mining operations underwent more stringent AECB on-site inspections, and companies had to report accidents to the board. In 1983 the AECB passed regulations about transporting and packaging radioactive substances.[34]

Also beginning in the 1970s, the Ontario government grew concerned about the environmental health of the uranium mining and processing communities, such as Elliot Lake and Port Hope, Ontario. This was partly as a result of reports by the Ontario Water Resources Commission, which were based on its testing of waters, studying the reduction of fish populations and aquatic biota, and researching the effects of the pollution on people living and working in the region. The government was also affected by the emergence of an OHS movement spurred by unions like the United Steelworkers, by the emerging environmental movement focused on pollution, and by an anti-nuclear movement that was ongoing from the 1945 American atomic bombing of Japan.[35]

Some legislators and members of the public questioned the viability of the industry after uncovering "unmarked radioactive nuclear dumps, unsafe uranium mining practices, unacknowledged radioactive spills, and a general failure to prepare for a serious accident." The money spent on three heavy-water plants in Cape Breton was widely dismissed as wasted, and the public had serious concerns about "the over-designed and frequently malfunctioning Douglas Point reactor."[36] The anti-nuclear movement grew in Canada and organized protests, including one in 1979 at Darlington. Canadian activists were influenced by the Three Mile Island meltdown that year, which was "the most serious nuclear accident" in American history. The story of the suspicious death of whistle-blower Karen Silkwood, after she told AEC officials in Washington DC that her employer, the Kerr-McGee plutonium processing plant, was violating health and safety regulations, was in 1983 made into a popular film *Silkwood* and gained wide exposure. The Chernobyl disaster in Russia in 1986 released a hundred times more radioactivity than the two bombs dropped on Japan and demonstrated the devastating effects of even the peaceful use of nuclear energy. After its radiation wafted over Europe, international efforts resulted in two treaties to maintain uniform safety standards. As the nuclear industry in the United States contracted in the 1980s, nuclear research budgets in Canada and the international market for reactors also shrank.[37]

The anti-nuclear movement in the 1970s sought more public input into the regulatory process. Although the AECB had five members and increasing numbers of staff, it had no members from environmental, labour, or Aboriginal groups. In 1977, the Hare Commission investigating the safety of nuclear reactors, and nuclear waste management recommended that the AECB substitute its secrecy for more openness in its deliberations and that the membership and scope of the board be broadened, as it was too close to Ontario Hydro. In the 1980s the AECB established an office of public information and guidelines to consult the public, but it still could issue a licence to a nuclear facility without a hearing, and the environmental assessment process was permissive, not mandatory.

The dominance of a hybrid techno-political regime of engineers, politicians, and businessmen associated with the nuclear industry was centralized and secretive in its nuclear decision-making structure. It was predominantly male and did not include some stakeholders, including Aboriginal peoples affected by the industry and environmentalists critical of it. Business and government collaboration resulted in a lack of accountability and gave priority to nuclear technology over health and

environmental protection in Canadian society. As a result they created a
tailings pollution problem, which the Canadian government has entrusted
private companies to remedy.[38]

Canadian public support for the anti-nuclear movement increased; by
1983 the majority favoured phasing out nuclear power completely. Lower
oil prices in the 1980s made gas and oil more popular. Bad publicity for
nuclear power from mishaps contributed to public anxiety; at the aging
Pickering reactor complex, a discovery in 1992 of a crack in its cooling
device, led to an accidental spill of 3,000 litres of radioactive water into
Lake Ontario. Citizens also were perturbed by the industry's huge capital
expenses, cost over-runs leading to enormous public debts, and by the
unsolved problem of what to do with nuclear waste.

The generation of nuclear power peaked in the 1990s and then declined
as one-third of Ontario's nuclear generators shut down. In 1990, Premier
Bob Rae announced a moratorium on nuclear energy facilities in response
to public safety concerns, escalating costs, and environmental and waste
disposal issues. In 1993 Ontario Hydro rates rose 7.9 per cent, and 80 per
cent of the rise in costs was to repair the Darlington nuclear plant.
Nevertheless, Rae's successor, Premier Mike Harris, reversed the nuclear
moratorium, privatized Ontario Hydro, and relied on coal stations to
generate enough electricity.[39] In 2000, the federal government overhauled
its nuclear regulatory system, replacing the old AECB, which never held
public hearings, with a more transparent Canadian Nuclear Safety
Commission, also considered too close to the hydro industry.[40] Public
safety remains a concern with nuclear energy, particularly after the 2011
Fukushima accident in Japan.[41]

Nuclear waste is a problem to which the industry has not found a solution, either in Canada or in the international community, and perhaps it
cannot solve. About a million bundles of nuclear fuel waste from the
twenty-two reactors and AECL's prototype and research reactors are
stored on local sites and increase by about sixty thousand bundles annually. The local system can last a few more decades, but requires constant
monitoring for security. The life of a CANDU reactor is about forty years
or less, but then it has to be replaced. There has long been a need for a
more permanent national solution for its safe disposal, but after a 1977
report, *The Management of Canada's Nuclear Wastes*, recommended
burial in the Canadian Shield, the public reacted fearfully. The study estimated that the process would cost $10–30 billion over 70 to 100 years,
and the concept development itself cost $700 million of mostly federal
money. In 1998 as a result of the Policy Framework for Radioactive

Waste (1996), an environmental assessment was completed, but hearings revealed insufficient public acceptance so no action was taken. Though technical experts claimed that the "geological solution" of burying the waste in the Canadian Shield is viable, the government recognizes that if it is implemented and then becomes a problem, a way to reverse the process needs to be available. Water seepage is possible, and such storage leaves a dubious legacy with financial and environmental liabilities for future generations.[42]

The problem is particularly pressing because the nuclear industry produces deadly waste products at every stage, from mining to decommissioning. In Canada uranium mining produced millions of tonnes of tailings and "continues to have one of the world's largest inventories of low-level radioactive waste, scattered across sites remote from public scrutiny and awareness, and for the most part abandoned and unregulated."[43] Eldorado mine dumped two million tonnes of radioactive tailings into Great Bear Lake. "Although it was decommissioned and remediated upon closure in 1982, tailings piles, the Gunnar open pit, the Lorado mill site and dozens of other satellite mines in the Beaverlodge region pose physical and chemical hazards only now undergoing remediation."[44] The interior of the Port Radium mines was hazardous to miners' health despite ventilation improvements made after the war. Northern uranium mining communities were "disaster zones where tons of radioactive mine tailings scar the landscape. These are the leftover piles of radioactive sand remaining after rocks containing uranium are brought to the surface and crushed." Underground in rock form, the ore was not hazardous, but on the surface, the wastes could blow and "land on distant vegetation, enter the food chain, and contaminate distant rivers and lakes."[45]

Eldorado also was careless with its radioactive waste from the Port Hope refinery. The Ganaraska River contained low levels of radioactivity after the company in the 1930s dumped waste into the town's harbour. The highly radioactive sediment was not cleaned up because a federal government report did not consider it dangerous. The town and a 1950s dump just outside of Port Hope became heavily polluted with radioactive and toxic wastes, such as arsenic and lead. Nearby farmers sued Eldorado when their cows got sick and died, but settled out of court. In 1976 it was learned that during the war, an Eldorado contractor had dumped 200,000 tonnes of radioactive waste all over the town. Some of it was used as landfill in over 100 other sites, including two schools, so that some buildings emitted radon gas and had unacceptably high radiation levels. The AECB at first took a hands-off approach as its regulations were designed to encourage prospecting and mining with a minimum of controls. But after an investigation, St Mary's school was evacuated in 1975 because of very high levels of radon gas; in 1979 the AECB ordered a seven million dollar cleanup of Port Hope. Environmental pollution persisted, but an AECL health study was not done. When in 1997 an independent study showed there was cause for medical concern, Health Canada responded that its health reports showed no such need. In 2007, the Ontario Environmental Department urged Eldorado Nuclear Ltd to complete the cleanup in the next few years. Today a conflict in the town between citizens concerned about public health and those who are pro-nuclear and worried about property values divides the town.[46]

In the 1990s, when the mining industry in Elliot Lake closed, it left 160 million tonnes of radioactive, acidic uranium mill tailings stored behind poorly constructed earthen dams, which between 1955 and 1990 leaked and spilled waste into nearby valleys, filling lakes, rivers, and ponds and contaminating the Serpent River watershed, whose source is above the city of Elliot Lake and extends to its entry into Lake Huron.[47] After the last mine in Elliot Lake closed in 1996, the federal government appointed a decommission panel; its report concluded, "The tailings of the Elliot Lake uranium mines present a perpetual environmental hazard," programs to maintain the sites would be "in perpetuity," and the tailings hazard created uncertainty about the surrounding complex ecological systems. It recommended a permanent endowment fund to support research on the mines' waste facilities. The government left the cleanup to the companies, which in the 1990s implemented a decommissioning and cleanup plan, declared it a success, enthusiastically supported the town's plan to attract seniors to enjoy recreation and leisurely retirement, and then set up a company that specialized in decommissioning mines and introducing environmental monitoring systems for profit.[48]

Another concern with nuclear plants is their emission of small quantities of radioactive effluent into the atmosphere and adjoining water bodies. An Ontario Hydro / Ministry of the Environment study between 1986 to 1989 found samples of *tritium*, a carcinogen, in grass near the Pickering plants and in nearby fruits and vegetables. The AECB sets maximum limits for radioactive emissions and monitors them to keep them low. But the "hidden" costs of environmental degradation and damage to health from both radiation and radioactive waste-disposal sites remain an issue. Environmental lawyers David Estrin and John Swaigen note, "Until the modern nuclear power era, the costs of exposure to radiation were paid by a limited number of people, such as the victims of nuclear warfare. Now, in the context of worldwide power expansion, radiation exposure is a potential threat to a large part of the population; as radio-active materials proliferate, so do the dangers associated with them."[49] Neither scientists nor doctors know the threshold exposure level below which there is no risk for any exposed individual. They do know that exposure to a high level of radiation causes death, and that low-level exposure over time may lead to cancer, leukemia, and genetic mutilation. A pattern has emerged that as scientists learn more about radiation, they recommend lower permissible levels of radiation exposure.

Critics argue that Canadian nuclear industry's production process and waste products are inimical to the environment and to people exposed to them. Historically the industry's lackadaisical approach to OHS and environmental issues and the government's collusion have made the situation worse than it might have been. As a result, the public remains ambivalent about an industry the Ontario government supports. The federal government also supports the nuclear industry but has sought to privatize parts of its infrastructure.

CONCLUSIONS

As John Cadham, summarized it, "Canada was one of the first countries to adopt nuclear energy. It is the world's largest supplier of natural uranium and a supplier of nuclear technology and expertise."[50] Recent announcements such as the Ontario government's plans to postpone indefinitely the construction of two nuclear reactors at its Darlington facility, the federal government's intention to privatize Atomic Energy of Canada Limited (AECL), and the stunning accident in Fukushima Japan in 2011 raise questions about the future of nuclear energy in Canada.

Nuclear power remains expensive, controversial, and dangerous in an era of increased global terrorism. It is difficult to imagine how it can be sustainable without a solution to the nuclear waste problem, and as with fossil fuels, its supply is ultimately limited. There is only so much uranium in the ground. As Van Jones notes, "Mining uranium is messy, destructive and potentially hazardous for neighbouring communities. Highly toxic nuclear waste poses new dangers today; fanatics would love to nuke a Western city. And despite clever attempts at repackaging, nuclear power is not even a good 'low-carbon' solution for global warming. Constructing a single plant requires gigantic amounts of concrete – the creation of which spews tons of carbon into the atmosphere."[51]

American journalist Paul Roberts concluded that "nuclear energy has so many technical, economic and political problems that its future is in doubt, while fusion energy – the so-called good nuclear power – is by most accounts probably a century away from being feasible on a large scale." Since the Fukushima disaster, politics, particularly in Europe, have shifted against nuclear power.[52]

However, despite its record of laxity in health, safety, and environmental standards, the nuclear energy industry in Canada has argued that in an era of global climate change and excessive emissions from fossil fuels, it is a clean "green" industry because its production does not emit greenhouse gases so that full use of its twenty-two reactors could reduce such emissions by 15 to 20 per cent. Despite this claim that it is clean because it does not produce carbon in the last stage, the processing from uranium mining to producing pellets does produce carbon, and its environmental effects and its inability to deal with the waste products have been negative.

The effects of uranium mining on employees and local populations and the exploitation of First Nations over many years have been "pervasive and insidious," with the result that many Natives have been activists in the antinuclear movement. The impacts on the health of people and on the environment in company towns such as Uranium City, Elliot Lake, Port Hope, and Chalk River have been very damaging. The industry's radioactive contamination and radioactive wastes remain an unsolved and perhaps insolvable problem. The nuclear industry in Canada has always been highly subsidized by governments and it is expensive. Politicians' persistent support for nuclear energy works against a new sustainable energy regime. "The amount of capital apportioned to nuclear power represents an obstacle to the development of new energy alternatives." For both environmental and cost reasons, it cannot be part of a future sustainable energy policy.[53]

Thus far in Canada, decision-makers remain conventional in their assumptions and seem unwilling or unable to deal with the repercussions of climate change. Some see nuclear energy as a solution, including scientist James Lovelock and Patrick Moore, the past head of Greenpeace, but Lovelock's shift results from his pessimism about people's ability to deal with climate change, and for Moore it is a matter of opportunism and financial gain, as he receives consulting fees from large companies that are polluters.[54] The fact remains that the nuclear industry in Canada, the United States, and elsewhere has had a dubious environmental record, and globally there remains no viable solution to nuclear waste disposal. Unlike many Canadian politicians, the Canadian public's support for nuclear power remains hesitant and lower than in the United States.

Alternative energy sources seem infinitely easier and less expensive to develop so long as there is political will, and they are much less damaging to the environment, because they are readily available. These include wind power, solar energy, geothermal energy from the heat of the earth below the surface, and tidal energy, a form of hydropower that converts the energy from tides (in Canada's three oceans) into useful forms of power – mainly electricity.

NOTES

1 Canada, Statistics Canada, *Electric Power Generation, Transmission and Distribution*; Canada, Historical Statistics Canada, Series Q85-91, *Electrical Generation by Utilities and Industrial Establishments, by Type of Prime Mover, 1919 to 1976.*

2 CBC *News Toronto*, "Darlington Nuclear Plant Gets Renewed Licence."

3 Engels, *Rosalie Bertell: Scientist, Eco-Feminist, Visionary*, 2.

4 Bothwell, *Eldorado: Canada's National Uranium Company*, 30.

5 Keeling and Sandlos, "Environmental Justice Goes Underground? Historical Notes from Canada's Northern Mining Frontier," 117.

6 Ibid., 117; Van Wyck, "The Highway of the Atom: Recollections along a Route."

7 World Nuclear Association, "Brief History of Uranium Mining in Canada."

8 Tataryn, *Dying for a Living*, 64; Stanley, "Citizenship and the Production of Landscape and Knowledge in Contemporary Canadian Nuclear Fuel Waste Management," 67.

9 Macdonald, "Rocks to Reactors: Uranium Exploration and the Market," 1.

10 Ibid., 4.

11 Ibid.; World Nuclear Association, "Brief History of Uranium Mining in Canada."

12 World Nuclear Association, "Uranium in Canada."

13 Canadian Coalition for Nuclear Responsibility (CCNR), "Uranium: A Discussion Guide," 8.

14 Hersey, *Hiroshima*. This novella appeared in the *New Yorker* in segments during 1945 and then was published as a small book. Simone de Beauvoir's *Mandarins* (1954) about intellectuals in Paris, in the war years to the mid-1950s, conveyed the shock the first nuclear bombs had on them and their world. Subsequently other writers and scientists influenced public opinion and increased the public's unease about nuclear power. Several scientists at Los Alamos objected to the United States developing more atomic weapons that were even more powerful than the bombs that devastated Japan, and they had an impact. But the Soviet Union had an atomic bomb test in 1949 and President Truman launched a program to develop the hydrogen

(fusion) bomb. In 1952 the new bomb was tested in the Marshall Islands. After a further test in 1954, which created nuclear fallout that contaminated 18,000 square kilometres and the Marshall Islands population, it provoked peace movements to push for a ban on atmospheric nuclear tests. In 1963, all nuclear and many non-nuclear states signed a Limited Test Ban Treaty, pledging to refrain from testing nuclear weapons in the atmosphere, underwater, or in outer space. The treaty permitted underground tests. Thereafter occasional nuclear accidents contributed to the public's ongoing unease about the nuclear industry.

15 Keeling and Sandlos, "Environmental Justice Goes Underground," 117.

16 Engels, *Rosalie Bertell*, 109.

17 Babin, *The Nuclear Power Game*, 46, 64.

18 Keeling and Sandlos, "Environmental Justice Goes Underground," 231.

19 Babin, *The Nuclear Power Game*, 38.

20 World Nuclear Association, "Brief History of Uranium Mining in Canada."

21 Eggleston, *National Research in Canada: The NRC 1916–66*, 317; World Nuclear Association, "Nuclear Power in Canada."

22 Babin, *Nuclear Power Game*, 61, 66; it was not the first Canadian cartel in the industry. In 1938, the Belgians and Canadians formed a cartel to control global radium prices, but radium use declined with high-profile cases about its horrifying side effects to radium dial painters, for example. Clark, *Radium Girls: Women and Industrial Health Reform, 1910–1935*. The loss of markets led to the closure of Port Radium mine in the Northwest Territories, and in 1941 the cartel ended.

23 Giangrande, *The Nuclear North: The People, the Regions and the Arms Race*, 92–9, 142.

24 Bratt, *The Poliltics of CANDU Exports*.

25 World Nuclear Association, "Nuclear Power in Canada."

26 For a gender perspective, Stanley, "Labours of Land: Domesticity, Wilderness and Dispossession in the Development of Canadian Uranium Markets."

27 World Nuclear Association, "Thorium," on the half life of thorium; United States Environmental Protection Association, "Radiation Protection," on health effects.

28 Stanley, "Risk, Scale and Exclusion in Canadian Nuclear Fuel Waste Management," 208–9; Leddy, "Poisoning the Serpent: Uranium Exploitation and the Serpent River First Nation, 1953–1988."

29 "Part 1: Accidents and Related Representations 1959," vol. 71, *Report of Special Committee on Mining Practices at Elliot Lake* (hereafter *1959 Report*), Union of Steel Workers Papers, Library and Archives Canada;

published in Ontario, Department of Mines, *Bulletin 155, Being the Report of the Special Committee on Mining Practices at Elliot Lake,* pt 2, *Accident Review, Ventilation, Ground Control and Related Subjects,* 53–5.

30 Lepperd, "MPC (Maximum Permissible Concentration) Objectives for Drinking Water Contaminated by Certain Uranium-Thorium Daughter Mixtures," 94–5.

31 Mellor, "A Comparative Case Study of Uranium Mine and Mill Tailings Regulation in Canada and the United States," 38.

32 MacDowell, "The Elliot Lake Uranium Miners' Battle to Gain Occupational Health and Safety Improvements in the Post-War Period."

33 Giangrande, *Nuclear North,* 100–4; Bothwell, *Eldorado,* 206, 347; Engels, *Rosalie Bertell,* 105.

34 Sims, *A History of the Atomic Energy Control Board,* 88; Griffith, *The Uranium Industry: Its History, Technology and Prospects,* 39.

35 Ontario Water Resources Commission, "Water Pollution from the Uranium Mining Industry in the Elliot Lake and Bancroft Areas"; see summary report, https://archive.org/stream/waterpollutionfro1onta/ VVATERPOLLUTIONFR_00_SNSN_08772_djvu.txt; Storey, "From the Environment to the Workplace … and Back Again? Occupational Health and Safety Activism, 1970s–2000+."

36 Dewar, "Nuclear Reaction," 71.

37 Tataryn, *Dying for a Living*; Weyler, *Greenpeace,* chaps 1–4; Krensky, *Four against the Odds: The Struggle to Save Our Environment,* 53; Engels, *Rosalie Bertell,* 86–93, 128, 141–52.

38 Bothwell, *Eldorado,* 112, 115; Bothwell, *Nucleus: A History of Atomic Energy of Canada,* 423; Doern, Dorman, and Morrison, "Canadian Nuclear Energy Policy"; Doern, Dorman, Morrison, *Canadian Nuclear Energy Policy: Changing Ideas, Institutions, and Interests,* 35; Engels, *Rosalie Bertell,* 104; Hecht, *The Radiance of France: Nuclear Power and National Identity after World War II.*

39 Estrin and Swaigen, *Environment on Trial: A Guide to Ontario Environmental Law and Policy,* 674–5; Dewar, "Nuclear Reaction," 69–84.

40 Bothwell, *Nucleus,* 449; Dewar, "Nuclear Reaction," 69–84.

41 Pritchard, "Japan Forum: An Envirotechnical Disaster: Nature, Technology and Politics at Fukushima."

42 Durant, "Burying Globally, Acting Locally: Control and Co-option in Nuclear Waste Management"; Bothwell, *Nucleus,* chap. 4; Doern, Dorman, and Morrison, *Canadian Nuclear Energy Policy,* 24, 44, 114, 122, 124, 203.

43 Engels, *Rosalie Bertell*, 111. In 2007, 441 nuclear plants in the world annually produced "almost 13,000 tons of high-level nuclear scrap." Weisman, *The World without Us*, 268.

44 Keeling and Sandlos, "Environmental Justice Goes Underground," 246; Saskatchewan Research Council, *Annual Report, 2009*, 27.

45 Engels, *Rosalie Bertell*, 109.

46 Edwards, "Uranium: The Deadliest Metal," 3; Harries, "Nuclear Reaction: Accusations of Nuclear Fallout Divide a Small Ontario Town."

47 Stanley, "Citizenship and the Production of Landscape," 167.

48 Canadian Environmental Assessment Agency, *Decommissioning of Uranium Mine Tailings Management Areas in the Elliot Lake Area*, 1. See Denison Environmental Services, "Closed Mines."

49 Estrin and Swaigen, *Environment on Trial*, 674.

50 Cadham, "The Canadian Nuclear Industry: Status and Prospects."

51 Jones, *The Green Collar Economy*, 7; Uekoetter, "Fukushima, Europe, and the Authoritarian Nature of Nuclear Technology."

52 Jones, *Green Collar Economy*, 7; Roberts, *The End of Oil: On the Edge of a Perilous New World*, 190.

53 Babin, *The Nuclear Power Game*, 14–16, 24; Rekmans, Lewis, and Dwyer, *This Is My Homeland: Stories of the Effects of Nuclear Industries by People of the Serpent River First Nation*.

54 On Patrick Moore, see Sinclair, "Who Founded Greenpeace? Not Patrick Moore"; Centre for Media and Democracy, "Patrick Moore: Media Coverage That Doesn't Disclose Moore's Nuclear Consultancy Work." Lovelock, *The Revenge of Gaia*.

13

Reflections, Questions,
and Tentative Conclusions

R.W. Sandwell

Changes in the kind and amount of energy people consumed transformed just about every aspect of Canadian life between 1800 and the 1980s. The augmentation and eventual near-substitution of seasonal flows of energy from organic sources with centralized stocks of dense, highly efficient energy in fossil fuels, electricity, and nuclear power (even if the last has been used only to heat water and make bombs) have changed society in foundational ways. It is not environmental determinism to conclude that changes in energy carriers have dramatically shaped the choices people can make as a society, and that these choices come with material, as well as political, social, cultural, and economic opportunities, constraints, and consequences. Chapters in this volume have explored energy carriers, old and new, revealing how producing and consuming energy continues to bind people intimately to their environments, profoundly influencing society and culture in the process. This is the case even if those environments are now far away, if people no longer touch the energy carrier they use, or have little understanding of the science behind splitting an atom or oil refining. Escalating population growth, climate change, and rates of obesity, as well as continued economic growth are reminders of these connections.

   The rich and detailed studies of energy production, use, and consumption presented here are meant as a foundation for – or as a fertile ground upon which to grow – a broad history of energy in Canada. The chapters highlight trends and patterns, continuities and ruptures as Canadians moved from a way of life built around tapping flows of energy from organic sources, to drawing on vast stores of energy from fossil fuels and readily available electricity. They reveal themes in Canada's great energy transition: its uneven nature, notwithstanding Canada's heavy energy demands; the interrelationships between old and new forms of energy; the irregular distribution of the benefits of the mineral energy regime; and disturbing consequences for people and environments. These essays hint at the broader range of tangled issues relating to changing human energy use, including cheapness and familiarity, adaptability to local customs and ways of life, as well as advertising, propaganda, and fear.

Particularly striking is the contrast between the huge amounts of energy used by Canadians by the late twentieth century, particularly for transportation, heating, and entertainment, and the invisibility of its supply. Few people are aware of the complex networks of finance, technology, and the material world that are invoked by turning on the stove, or driving down the road, and even fewer could understand them. The obscurity and complexity of contemporary energy carriers also stands in stark contrast with people's historical relationship to energy, which in the organic economy used to be direct, visceral, and almost constant. When energy had to be wrested from flows of energy in the local environment, the labour involved – the energy they used in order to produce and consume energy – occupied most of people's days, whether planting or harvesting, tending to animals, cutting wood, cooking food, or simply moving from one place to another. The chapters here confirm that Canadians have made choices about energy use, but they often did so on the basis of terms and conditions that were not visible or clear to them. While few miss the labour expended supporting life in the organic regime and "saved" in the mineral, or the periodic famine that characterized the organic economy, the spectre of climate change does make it seem more possible to look with nostalgia at a world where organic limits forced constraints on human uses and abuses of the environment. Energy history allows us a deeper understanding of the nature of the changes involved, as well as a better evaluation of their consequences for people and the environments in which they live.

These essays confirm what comparative international statistical data (available only after these chapters were first written and discussed) suggest: Canadians have long been high per-capita energy consumers and were considerably later than those in other industrial countries at making the transition to the mineral economy. These are rich areas for historians to explore. What is the significance of the apparent fact that Canada's political economy, or important components of it, remained rooted in the organic economy long into the twentieth century? While Unger and Thistle suggest in their brief statistical overview that both trends can be explained as a supply-side issues – Canadians generally had abundant supplies of wood and needed them – recent historical studies hint at more complex explanations. As I noted in the introduction, most Canadians before the Second World War, unlike Europeans, lived in rural and small communities, where they owned or had free access to trees. In addition, many owned or had access to abundant land and the animals and plants that they yielded, often "for free." This surely made industrialization – the shift away from organic flows of energy to the mineral regime – a very different process than it was in England or Wales, where people's access to the bounty of the land was strictly limited, and they had little choice but to work for wages in order to obtain the energy they required to sustain life. The distinctive political economy of rural Canada, where households were able to interrupt flows of energy to provide for their own support, even as they engaged with global economies through trade and their seasonal work in resource extraction, may have more to tell us about the slow nature of Canadians' energy transitions than previously believed. It is our hope that this volume will clarify some of the material contexts of energy use, and stimulate new questions and research about the foundational relationships that the production and consumption of energy forges amongst people, their environment, and society.

Full of information about the contexts and consequences of energy use, these chapters also provide ways of approaching the huge questions about the changing and rapidly increasing role of energy in Canadian society. But many pressing energy-related issues are not addressed at all here, or not enough. Choosing to focus largely on kinds of energy production, rather than more precisely on energy functions (heat, light, and work), or regions or historical periods or kinds of people (by race, class, and gender, for example) has meant that many of the connections amongst people, time, place, and energy use appear, inaccurately, to be of secondary concern. More discussion is badly needed as well about energy periodization to articulate the turning points in energy production and consumption across the country. Excellent new studies of large-scale modern energy networks and their consequences are appearing, but we still await mature evaluations of many energy-related themes and events, including the energy crisis of 1973 and the development of the National Energy Policy and its failure. We would benefit from much more work that, following H.V. Nelles, Chris Armstrong, Joy Parr, Tina Loo, Timothy Mitchell, and authors in this collection, articulates the deep connections between modern power networks and politics. We badly need more histories of how energy was produced and consumed by industry, commerce, and people's bodies and homes. These and other issues are going to have to wait, however, for other volumes.

In the end, the authors hope that this unique collection meets our three goals: to bring together the "back story" of changing power, fuel, and energy use in Canada in one volume; to demonstrate that each and all of the energy carriers explored do indeed have a history; and to suggest that their history is central to the history of Canada.

APPENDICES

Primer on Terminology

Colin A.M. Duncan

PURPOSE AND ORGANIZATION OF THIS PRIMER

The contributors to this first general book on Canadian energy history found that the organization of sub-topics adopted for the symposium in Bloomfield worked well enough at the conference table. Any structure would have generated its own pattern of cross-references and would have involved going back and forth in time. We changed the original sequence of presentation and have designed this book's index specifically to facilitate making connections and/or comparisons. Unprecedented massive changes in the sourcing of energy carriers started to accumulate a bit more than a century ago, mostly in response to the sudden resort to fuels extracted from underground that had lain undisturbed for hundreds of millions of years. Using them (and hydroelectricity) to perform work on other stuff or inside machines, not just to heat things or light spaces, allowed unprecedented further changes in all sectors of what came to be denoted as "industry." The energy carriers presented complicated physico-chemical aspects, including invisible ones, and most require special apparatus to be useable at all. Everyone now uses these sources of heat, light, and work directly and/or indirectly all the time, but most people have little if any idea where things come from or how they work. They take them for granted. It was not so straightforward in the past. Energy sourcing and processing not only migrated from the province of the vernacular to highly commodified trade regimes, but also from being matters of ordinary unaided experience to arcana deep in the technosphere. There has been a great disconnect between people and environment. Never have so many used up so much with so little understanding of the processes involved (let alone their consequences). The significances of this great divide (scarcely specific to Canada) drove us to break our Bloomfield mould and reallocate the chapters. To make it easier for the reader to grasp the implications of shifting so much to energies from underground or sources otherwise less visible, and to compare all the chapters on specifics, and thus build a cumulative understanding, it seemed useful to have a section of the book devoted to technical concepts lurking behind units used for comparing forms of energy, especially because most of the key issues are stuck in terminology that was developed essentially on the fly, and that continues to baffle most people. Industrialization extended itself so fast and so broadly that language could not keep up. So reading right through this primer's semantic analyses may give the reader a vicarious and/or virtual historical experience. Because so many sources of heat, light, and work were multipurpose and appeared in a rapid sequence, especially over the last century and a half, ordinary people (generally with very limited budgets) were trying to make complex comparisons on the basis of shifting and almost always inadequate knowledge. Folk had to make discriminations that were often just as fine as the technical analyses of investments being made inside the energy sector. Since the purpose of this book is to provide a comprehensive introduction for the non-specialist, this primer sketches the bases for past judgments made by ordinary people, and for the analyses made by engineers, technicians, and research scientists.

No single scholarly source, secondary or primary, is as comprehensive as this historical book on Canada, so some information for this primer had to come from considering non-Canadian cases. By the same token it should be useful to anyone seeking to understand the current general human predicament.

To give the reader a foretaste, admittedly not very appetizing, of the complexity to come, I present here a list adapted from a table in Morgan's two-volume 1930s compendium *American Gas Practice* comparing sources of heat. Each successive entry of cents and/or dollars (very carefully not rounded up to whole pennies in the first column) is the invoiced cost of the stuff as an industrial source of one million British thermal units of heat energy. The price listed at the end of each line shows how it was normally packaged for purchase back then. The alternative prices given for some categories give insight into the considerable normal spread at that time in America. The grid had not yet manifested itself fully, to put it mildly. In the accompanying text it was pointed out that the comparisons are unfair to gas as compared to coal or oil, because the cost "at the burner" would be the tougher comparison inasmuch as labour costs for handling gas fuel on site are relatively minuscule.

If you were a budding captain of industry about to set up a factory requiring substantial heat to be routinely applied, which option(s) would

Table A1.1   Compared costs of heat sources, early 1930s

Invoiced cost per million Btu	Heat source	Cost as packaged for purchase
$0.172	Bituminous coal rated at 14,500 Btu/lb.	$5.00/ton
$0.200	Heavy fuel oil rated at 150,000 Btu/gal.	$0.03/gal.
$0.241	Bituminous coal rated at 14,500 Btu/lb.	$7.00/ton
$0.250	Natural gas rated at 1,000 Btu/cu. ft	$0.025/million cu. ft
$0.350	Fuel oil rated at 143,000 Btu/gal.	$0.05/gal.
$0.417	Anthracite coal rated at 12,000 Btu/lb.	$10.00/ton
$0.500	Natural gas rated at 1,000 Btu/cu. ft	$0.50/million cu. ft
$0.507	Furnace oil rated at 138,000 Btu/gal.	$0.07/gal.
$1.000	Manufactured gas rated at 500 Btu/cu. ft	$0.50/million cu. ft
$2.000	Manufactured gas rated at 500 Btu/cu. ft	$1.00/million cu. ft
$2.174	Gasoline rated at 11,500 Btu/gal.	$0.25 /gal.
$2.932	Electricity rated at 3,411 Btu/kWh	$0.01/kWh
$8.795	Electricity rated at 3,411 Btu/kWh	$0.03/kwh

Source: Adapted from Morgan's *American Gas Practice*.

you have preferred? Wood fuel is not even given a chance. Hydroelectricity probably lurks behind at least one of the very different electricity prices, because no specific location is presumed in the comparisons. In fact the items listed vary too much in many important respects for price alone to settle the issue. The table presents comparisons with a degree of abstraction that is practically useless.

This primer aims to help introduce the kinds of considerations involved and is broken into three further sections, one on units, one on hydrocarbon chemistry, and one on the very concept of fuel. They must be read in that sequence, as various points build on each other. Though connected in countless ways to material in the other chapters, this primer was written so that it can be profitably read through in isolation, ideally indeed prior to the rest of the book! Where a word is in *italics* in this primer it is most succinctly defined, or at least explained! Such words are also in *italics* where first used here and in the rest of this volume.

ON UNITS COMMON IN THE LITERATURE (PAST AND PRESENT) AND ON KEY CONCEPTS BEHIND THEM

Many statistics concern things that children can count (trees, horses, farms, ships, dollars even, certainly coins), but with energy we are often dealing with forces and other invisibles, with quanta that, in many cases, are dissipating even while we are trying to count them, and that are gone forever once used! Here follow some rough equivalences for reading historical statistics concerning energy, whether in heat, light, or work. They are tiresome, but you need to know about them.

First we present two common measures: one of volume and the other of mass. In the oil sector 1 *barrel* is 42 US gallons or 160 litres. A *tonne* (a.k.a. *metric ton*) is 110% of a normal ("short") ton, 2,200 pounds, or 1,000 kilograms.

A series of comparative examples will take us into the energy past: one barrel of oil per day is equal to 50 tonnes of oil per year, a tonne of oil being the energy equivalent of 1.5 tonnes of coal, or 12,000 kilowatt hours of electricity, and one cubic metre of natural gas being equivalent to ~10.5 kilowatt hours, one watt being a unit of work considered over time, 746 watts being equivalent to one *horsepower*, which was defined in the 19th century as the work done by a horse pulling a load of 180 pounds at 180 feet per minute (walking in a circle), which work was said to be equivalent to raising (against gravity) a weight of 33,000 (~180 x ~180) pounds up one foot into the air, in one minute. It takes three normal people working hard enough to be exhausted after half an hour to match what a horse can do. And horses can work hard for much longer periods than can humans.

Tiresomely, there has been added by the authority of scientists' SI (the universal French acronym for Système Internationale) yet another abstract, anti-vernacular unit for energy and hence for work, one watt now defined as one joule per second. The joule is the new unit for energy in the sense favoured by scientists. Each human heartbeat consumes about one joule, and 15,000 are needed to make a small cup of coffee. Joules have to be integrated with yet another unit, the newton, because a joule is defined as the force needed to push against one newton for one metre. A small apple hanging from a tree experiences a force of about one newton (due to gravity). Holding a heavy physics textbook gives you an idea what 10 newtons feels like. More on pressure later, but we can say here that 100,000 newtons per square metre equals the normal pressure in the atmosphere near sea level. The reader will already appreciate that with some abstract terms, sample equivalences are more helpful than terse definitions. Dictionary definitions of SI units have been tending over the last half century to total unintelligibility, so I will not repeat many here.[1] Only special apparatus carefully used reveals their real meaning. This primer is focused on words ordinary people actually use(d).

One still familiar unit for energy, the *calorie*, was long defined as the heat needed to raise the temperature of one gram of water one degree

Celsius, before it was fully noticed that that relationship is not linear across temperatures. The old calorie unit (stably assessed around 15 degrees Celsius) is now said to equal ~ 4.2 joules. In very wide use in our sources is the British thermal unit (a.k.a. *Btu*), which was the heat required to raise one pound of water one degree Fahrenheit (= about 1,055 joules or 252 calories). One hundred thousand Btu is a *therm*.

The old measure of luminescence (a.k.a. light) was candlepowers. One million lumen-hours, the main unit now used for measuring light, is the same as 100 hours of a 100-watt incandescent light bulb. Completely unlike other energy uses, in the case of light, "consumption" necessarily exactly equals production, but if this sounds reassuring, it misleads. For the United Kingdom it has been calculated that an average family now uses 200 times as much artificial light as their equivalent did two centuries ago, but the British economy as a whole increased its use over that period by a factor of 25,000.[2] Clearly, immense quantities are wasted.

One other difficulty we must confront constantly in trying to compare the measures of help people got from other bits of Nature, both living and inorganic bits, is the difference between potential and actual work/energy. We will here discuss just the units most historical Canadians had to deal with. To assess the power latent in a (healthy, adult, trained, and unexhausted) horse, Watt had had to measure how much work some actual horse actually did in a definite amount of time (all described above). But when we burn a fuel to get some work done (e.g., moving a 5,000 pound shore-bound wooden boat one foot so it clears a land storage cradle and can be swung out over the water to be launched), we cannot so easily measure the contribution of the fuel itself as such, because its performance crucially depends on the machine used – on how efficiently it uses the fuel put into it. Unfortunately two completely separate issues arise: how efficiently the fuel is combusted and made into power, and how efficiently the rest of the machine harnesses and transmits the power awaiting use, a.k.a. dissipation. In the case just described, it was a 25-ton crane equipped with caterpillar treads and a system of rusty cables and pulleys towering in a steel frame above a diesel engine rated at 85 horsepower (most of this gear a bit over half a century old). The crane's owner (even older) avers that fuel costs are too low to worry about, but the crane is worse than useless without its fuel!

We considered the measure of horsepower above as an example of work, but the calorie unit was devised to reduce the complexity of measuring the energy latent in any fuel by giving it a simple job: just increase the temperature of a bit of water. Many discussions refer to the "heat content" of various fuels using various measures, but because they are considering mere potential, not actual transformations, the discussants leave the machinery and time out of the assessment. So in this kind of study we are dealing not just with apples and oranges, but sometimes apples over time as compared to oranges abstracted from time. Sometimes it feels like we are being forced to compare apples and orangutans!

One complication about gas in particular is that it must be measured not just in quantity (like any fuel), but any actual manifestation of a gas is at some pressure or other, unless it is escaping somewhere. The basic physical explanation of gas behaviour is given below. In most writings about gas, whether contemporary or historical, little or no effort is made even to ballpark the pressures involved, even though variability in gas pressure was a subject of almost constant complaint by consumers, as well as of worry to producers. At first, it seems gas pressure was just too hard to measure and/or modify. Calculating just the volume of a gas passing through a meter was easier, and volume for gases is a significant if counterintuitive measure, adequate for many purposes, as explained below. The first suppliers of manufactured gas did not even try to quantify what they were selling. For them the cost of the gas-making equipment and delivery pipes so vastly overshadowed the price of the fuel itself that gas charges were set by measures of time, not quantity of the stuff. After one's building was connected to the supply pipes, one simply rented the use for a period set in a contract of a gas supply (of unknown and variable quality and quantity and pressure). Whether the customer used none, or had it burning all day every day, made no difference to anybody. A similar thing happened initially with hydroelectricity in many places (including Canada). The later (and bogus) atomic energy promise, "too cheap to meter," actually did apply a long time ago, sort of!

If we look at how pressure is defined for arithmetic purposes, we see readily why measuring it is tricky. *Pressure* is defined as force per unit area. *Force* is mass multiplied by acceleration. *Acceleration* is velocity per unit time. *Velocity* is length (travelled) per unit time. That amounts to too many denominators by far for most people! And just what exactly is pushing gases is far from obvious, because the whole phenomenon is invisible. But pressure matters, for depending on its pressure, a gas may or may not be easy and/or safe to capture and/or use. Understanding how pressure is measured can be made easier by some connection to the vernacular, so we will start there. When the needle on your marine or household barometer is vertical, typically a "fair" day, is deemed normal air pressure. The pressure your body deals with on that day coming from the weight of the air in our planet's atmosphere pressing down on, and thus at you from the side too, is said to be normal, and one reasonable name for this standard unit of pressure is one *atmosphere*. That quantity of pressure is what the *P* stands for in the scientists' phrase, "STP" (standard temperature and pressure). We come to temperature later. As we have seen already there are, sadly, many other units for pressure in use, but we will consider one that also connects to the vernacular. One atmosphere equals the push of almost 15 pounds per square inch, or if you prefer, a kilogram per square centimetre. We are unaware of normal atmospheric pressure because our body does not signal to our brain that anything odd is happening. In extremely low atmospheric pressure, as in the immediate run-up to a hurricane, some premature births in late-pregnancy mammals are induced naturally. Regardless of the weather, pressure also decreases as we go higher on Earth, because there is less air above us, so at sea level is generally where it presses hardest. At a height of two miles, twice as high as Calgary, the air pressure is less than three-quarters, and most people find that hard to take. At the top of Mount Everest pressure is less than a third what it is at sea level, and people cannot endure that. They can look around for a few minutes and then they have to leave, in a hurry. Most scientists are like most people in one respect: they live and work at or very near sea level.

Atmospheric pressure used to be measured mostly using a glass tube set in a container of mercury so that as the air weighed down on the surface of the mercury some of it rose up the tube. The level of the mercury would thus go up and down in the tube as air pressure rose and fell, and 29.92 inches of mercury on a standard tube was deemed standard *P*. Generally pressure varies only from about 29.0 inches to about 31.0 inches over the course of normal weather changes, or less than 10% in that system's units. But such changes matter a lot, because pressure differences cause wind, and wind brings us our weather and also used to do a lot of work for us (as detailed in this volume). Another unit widely used on barometers is the millibar, with 1,010 of them equalling one atmosphere. By the way, 29.92 inches is equivalent to no fewer than 760 millimetres. Yet another unit, the SI one, is expressed as 101,325 pascals for one atmosphere of pressure. One pascal of pressure is a force of one newton over a square metre.

Underwater, pressure is much greater, and it increases very fast with depth. Every 33 feet down adds the equivalent of an atmosphere. The most an unprotected human can cope with is about three atmospheres, so a bit over 100 feet down is our limit without special equipment. Whereas when we hold a large onion in our palm we feel about one-tenth of a pound per square inch, at half a mile down in the sea the pressure is half a ton per square inch.

Avogadro was the chemist whose strange but insightful concept helps us understand gases (and thus more about pressure). He made a seemingly odd suggestion that has been found basically true and makes the world a lot easier to analyze. All gases, being composed of molecules separate and actively moving all the time, behave, if contained, the same way against the walls of their containers (with which the molecules collide repeatedly). The collisions with their cage occur because every single molecule behaves chaotically, and that is why seventeenth-century scientists decided the Greek word *chaos*, which came to be pronounced *gas* was the appropriate general name. But because it is a matter only of chaotic impacts of very small things, the same numbers of molecules of any chemical all exert the same force as gases. So equal volumes of gases (at the same temperature and pressure) contain the same number of molecules. Therefore a cubic foot of natural gas at some pressure and temperature is a definite number of molecules, and compressing them does not change how many there are, just changes how much space they occupy. Heating them makes them more active so the pressure goes up if the volume available to them is kept the same. Likewise the total *heat content* of that set of molecules (the latent heat in them released only if and when they all combust) also does not change just because the pressure did. So when workers at a wellhead let a volume of natural gas into a container and seal it, they have captured a definite quantity of potential heat.

When measured in quantity, natural gas is by (US) stipulative definition a heat source containing 1,075 British thermal units at the wellhead, and 1,055 at the point of delivery, both per cubic foot. These numbers are arbitrary and have no statistical basis, in the sense they are not averages, weighted or otherwise. The hydrocarbon chemistry behind this messy state of affairs is explained in the next section. Different countries use different temperatures and pressures and also tolerate different admixtures of water vapour in their different "standard" measures of gas heat content, but generally nowadays one cubic foot of gas has 1,000 Btus. Other non-jurisdictional complications with this measure are explained below when we consider liquids other than water taken from samples of gases that are "contaminated" with those liquids. Caught between the British Empire and the United States, Canada adopted the scientists' "international" metric system several decades ago, but retains English spelling. There are 35.31 cubic feet in a cubic metre. At the foot of its highly schematic 2013 poster map of "Canada's Energy Production and Transmission Power Ways" *Canadian Geographic*[3] tells us that one cubic metre of gas is enough to heat a single family house for three hours. But the label on your standard gas barbecue cylinder includes a Btu number, thanks to the influence of the recreational market in the United States! The normal container size (five US gallons) is about two-thirds of a cubic foot in volume (or about 19 litres, or almost a fiftieth of a cubic metre). Because corners are structurally weak points in metal things, gas containers are made almost spherical, so they are not subject to simple measurements and calculations, the way rectangular and truly spherical containers are. Although gas molecules individually move randomly, as a group they will move directionally, but only towards an opening. Inherently hyperactive, gaseous fuels therefore must be absolutely contained except when being burned. For reasons clarified below, a gas that is only "relatively" contained gas is very willing to explode. So the opening of a gas container must be finely controllable. Precision manufacturing is essential. Very fortunately, not all gases are flammable.

Things commonly gaseous on Earth include nitrogen (formula N_2, as nitrogen atoms stably exist only outside of chemical compounds in pairs, being *diatomic*, as they say). Nitrogen is about 78% of our atmosphere by volume. The heavier element oxygen (O_2) is also common, around 20+% of the atmosphere, but since oxygen is highly reactive, it is constantly getting entangled with other elements (on which more below). Oxygen atoms locked inside various compounds constitute almost half of the Earth's crust by weight. Conceptually familiar as a product of every breath and every combustion act, and as the feedstock or starting molecule for all photosynthesis, but actually very rare in the atmosphere, is carbon dioxide (CO_2). It should be only 0.03% of our atmosphere but reached 0.04 in the first week of May 2013, just prior to our symposium. Also gaseous normally on Earth and more or less useful are hydrogen (H_2), helium (He), and carbon monoxide (CO), as well as argon, neon, xenon, and krypton. These last four elements are extremely non-reactive, so much so they were called *inert* (a bit prematurely we now realize). Their general non-reactivity made and makes them ideal for use inside electric light bulbs. Argon is much more common than CO_2, but the other three are vastly less abundant. They are abstracted in useful quantities by a variety of special processes called the liquefaction of air, which was perfected by a man named Linde.

Although gaseous substances are hard to grasp, electricity is in many respects the real odd man out in our story, because it is measured in so many different respects, using many different units, including watts (in a special non-horsepower sense), but also ohms, amperes, volts, and even hertz. Thankfully, there has been consistency in using only the one set of words for electricity, and most of them are used for nothing else. This extreme semantic narrowness is due to there being absolutely nothing vernacular about electricity historically.

Amongst our "energy carriers" (a deliberately vague term), *electricity* most exemplifies the physicists' sense of energy as an abstraction. Grasping how gases move is tricky enough, but explaining how electricity moves is beyond most people. Metaphysicians today use the word *stuff* as a term of art, but since electricity is clearly no kind of "stuff," we might usefully think of electricity as more like a state of affairs between things. Perhaps electricity should be likened to political power, as it exists only in, and as, a set of relations amongst entities. In the case of electricity, the entities in question are generally more or less contiguous metal objects, but depending on circumstances, human bodies can get also involved in the relationships (often fatally). Again like political power, electricity presents complex problems of delivery, as it cannot be discretely transported (in units of any kind) and also cannot itself be stored (except temporarily with heavy and non-durable apparatus, a.k.a. "batteries," cheap ones weighing in at around 25 kilograms per kilowatt hour of deliverable power). So although electricity conveniently can be made from myriad sources, it has to be made always anew to be usable at all, because neither the site of generation nor the site of use can serve as a site of storage of itself.

Electricity in bulk thus needed an extended and contiguous material grid for delivery. The huge systems of metal wires, like the pipelines for gas and oil, were normally built only with government help. More on the socio-economic implications of grids elsewhere in this volume. Here we focus on technical issues. Pressure does fall seriously with distance in a very long gas pipeline, but that has been easily solved by installing compressors here and there. Whereas systems of pipes are comparatively intuitive, electric grids present arcane problems in physics in the form of very severe practical trade-offs that can only be sketched here. Direct current (DC) is easier to use, and safer in general, but at low (safe) voltages could not be moved more than a mile or so before it dissipated beyond usefulness. Voltage and current being inversely related, as transmission losses are proportional to the square of the current, doubling the voltage (for example) reduces dissipation losses fourfold. High-voltage transmission over long distances marked by many use-sites along the way (whether huge factories or urban areas) generally had to be done with alternating current (AC), raising a need for specialized extra apparatus at both ends of any journey for conversions between direct and alternating to occur. But in some special cases, especially northern Canadian ones, very long distances between the site of generation and the first site of use made direct current (high voltage direct current or HVDC lines) the relatively attractive choice. Contemporary proponents of building mass solar farms far from where people live (e.g., in the Sahara) argue about transmission schemes more or less constantly.

Because they are DC, batteries are fine over short distances. Computers, when not relying on their batteries, require that AC from the wall socket be reduced in power and transformed to DC, whereas old-fashioned electric lights plugged into the wall need neither operation. What actually happens inside a battery or transformer is beyond the scope of this primer because there is no agreement on how to describe electricity in terms drawn from analogies with ordinary experience. Most introductory texts favour the physicists' "electron" theory, but in 1991 Clearwater published Kenn Amdahl's remarkably accessible, practical, and poetical introduction to electricity provocatively called *There Are No Electrons*.[4]

ON THE CHEMICAL SEMANTICS OF HYDROCARBONS AND RELATED STUFF

Beyond (or underlying) the problems with units we face a multitude of problems with words in our topic area. Some words now in common use are old trade names, while others have gone out of service. Some actually make no sense but are used anyway. This "bad language" problem particularly bedevils any attempt at quantitative comparison of different sources of heat, light, and help at work. Because the ordinary language for our subject is used so unsystematically, we must first consider some semantic points and then some key facts from chemistry. So the italicized words here include ones that are (or were) used in both chemistry and daily life, as well as some common umbrella terms. For technical reasons the close etymological discussion of the word *fuel* is delayed to the final part of this primer (under the nuclear rubric), but we may say here the word's root used to mean a more or less contained fire. *Fossil fuels* were so called because they came from operations involving digging into the Earth because the Latin *fossa* means "trench."

The equivalent term *mineral fuels* likewise betrays its origin to be the source, so to speak, because *mineral* is a medieval Latin noun for ore, in

the most general sense of dug-up-stuff, what has to be mined, as we now
say. The chemical compounds referred to by these synonymous phrases
(concerning which the chemical basics are given below) are *hydrocarbons*, which are molecules consisting of only, or mostly, hydrogen atoms
and carbon atoms. In general the compounds are modified remnants of
either marine animals or terrestrial plants, respectively alive half a billion
and only one-third of a billion years ago in many parts of the Earth's
moving surface plates. Both sorts thus present to us in geological formations, inside rocks. The marine remnants are mostly liquid (a.k.a. oil or
petroleum from Latin words for rock and for oil/fat). The land plant
remains are mostly solid (a.k.a. *coal*, which is just an old German word
for coal). But the rock formations in both cases generally contain a mix
of hydrocarbons in various states – gaseous, liquid and solid – the first
sort fairly demanding their freedom, the last being generally suspended
inside the second. But readers may as well be warned here that, for scientists, the word *mineral* has another, essentially opposite sense. Most
"mineralogists" stipulate that what they study is non-organic rocks, ones
that have a specifiable internal crystalline structure. A century ago this
mattered a lot at law in Ontario. In 1912 the Judicial Committee of the
Privy Council in London (then the British Empire's highest court) ruled
that, unlike ores and oily liquids, the natural gas in the ground below the
surface belongs to the landowner, who in Canada was generally a farm
family at that time, thanks to the decades-long policy of mass land settlement. Wanting the revenues from natural gases, an English land settlement company that had sold land to farmers near Lake Erie claimed
retention of the mineral rights, in accord with the norm separating mineral rights from rights to use soil and cut timber. The court found for the
famers, following the testimony of geologists that, to count as a mineral,
a stuff has to have its chemical constituents in a definite chemical structure, a stipulation applicable to hydrocarbon gases only down around
-200 degrees Celsius!

Liquid oil too has no definite three-dimensional structure, and getting
it out of rock has been likened to squeezing molasses out of a brick – not
easy. Sometimes pressures built up in the rock formations as the result of
tectonic and more direct thermal forces, making it needless to do anything like squeezing the rock. *Gushers* occurred when a puncture of a
rock formation containing non-solid hydrocarbons that were naturally
under pressure resulted in an immediate and rapid release of liquid oil.
Similar bursts sometimes happened with gases, but the *plumes* were not
visible (except when ignited). By far the most common sort of coal, called

bituminous for no very good reason (see below on *bitumen*), burns with a lot of sooty smoke, but where a seam of it was bent by geological forces the resulting (mechanical) heat drove off the more volatile hydrocarbons, leaving a harder, denser, clean-burning *stone* coal. Understandably it has long commanded a premium and has also been called *anthracite,* which comes from the ancient Greek word for coal, *anthrax* – which went on to refer to blackish pustules on sheep and humans. *Seacoals* is just a very old English term for coal imported by boat, e.g., from the Newcastle region down the east coast to wood-starved London.

For reasons unclear to this agricultural historian, sites of extraction of fossil fuels have been called fields, even though these fields are always invisible and vastly more complicatedly three dimensional, and in many cases far underwater! Equally disturbing environmentally is the fact that the extraction of fossil fuels has been termed production, though it bears no resemblance to either farm or factory production, both of which involve renewable and/or recyclable materials being made into discrete objects for consumers. Extraction would have been a more honest general term, and some people do use it. Canadian geologists in 1851 referred to the *gumbeds* of the most southerly part of Ontario,[5] which later turned out to have oil and gas in or beneath them. It seems to me a great shame that word went out of fashion.

Paraffin was a general name for all hydrocarbons that are maximally *hydrogenated* (all classes of molecule in which each carbon atom is normally bonded only to hydrogen atoms or to other carbon atoms, thus giving the very simple general *chemical formula* C_nH_{2n+2}, because hydrogen atoms can form only one bond each, whereas carbon atoms can bond with as many as four entities). Because they are so *saturated* with hydrogen, paraffins are chemically strikingly indifferent (and not very reactive), the empirical point clarified in the etymology: Latin *para* for "barely" plus *affinis* for "related." So people have problems with these chemicals, not because they are unusually reactive (quite the contrary); rather, the trouble is many paraffins are quite *volatile* (ready to evaporate) and thus hard to contain, and once escaped into air they are completely surrounded by lots of oxygen, which is all they need to burn. So these chemicals are both safe and not safe. The other general name for paraffins is *alkanes*, a word like *algebra, alfalfa,* and *alkali,* coming from the Arabic article *al* plus other Arabic words, in the last two examples connoting green fodder and the product of combustion we call ashes (very roughly translated so as to include ashes with metals in them, such as *potash* which contains and gave us the name for potassium, via "modern Latin," which was invented by scientists to share their insights across the Babel of ordinary languages). The paraffin (a.k.a. alkane) group was first figured out by Reichenbach in 1830. At STP (recall "standard temperature and pressure") only the first four chemically distinct in the group of alkanes are gases (C_1H_4 to C_4H_{10}): methane, ethane, propane, and butane (a.k.a. quartane). Being paraffins these gases will not interact with each other and so can cohabit spaces happily enough. That is what they do in natural gas formations (as explained below), but if separated they contain respectively per cubic foot: 995, 1730, 2,480, and 3,215 Btus.

On human uses of these valuable gases see the dedicated gas chapter. The less hydrogen-saturated simple hydrocarbon compounds include *alkenes* (which have "double bonds" between the carbons, the simplest being *ethylene* (C_2H_4), a.k.a. ethene, a key *substrate* or raw material for the chemical industry). There are also *alkynes* (which have carbon-to-carbon triple bonds, the simplest being ethyne [C_2H_2], a.k.a. *acetylene*, a flammable gas). It is worth noting here that the manufacture of acetylene (like ethylene, an essential in the plastics industry) first required the use of high-temperature electric furnaces to make useful quantities of "calcium carbide" (one of two key raw materials) out of limestone and *coke* (which is bituminous coal heated so as to volatilize, and thus purge, the lighter alkanes naturally present in it, the solid residue becoming something like good anthracite, better if done very carefully). Two centuries ago the rather helpful alternative spelling *coak* was in use. Some enthusiasts called it *charcoal of coal* in contradistinction to "charcoal of wood" previously the most favoured fuel for fine work. On account of starting as a late product of photosynthesis, wood charcoal is defined later, in the fuel section of this primer. But we can note here that because anthracite coal was so clean and reliable, it was for a time also called *natural charcoal*.

Technical cross-linkages amongst materials and fuels abound in our general subject area, and by the same token resist systematic presentation. Acetylene can also be made from methane more directly but only profitably because the by-products – carbon monoxide and hydrogen – can now also be packaged and sold. More has to be said about the term *carbide*, developed shortly after the start of the oil industry and applied to essentially any compound containing carbon atoms. For a time it was used in the names of companies that made gas available in small communities by placing calcium carbide in water, a process discovered inadvertently by a Canadian to give off acetylene gas. Water in Canada is not rare, and calcium carbide could be had and moved fairly cheaply. Many

Prairie towns far from either coal or natural gas used acetylene plants (primarily to heat public buildings).

Here we go back to the alkane gases, because they are best at introducing the arcane subject of *stereochemistry*. The term borrows the Greek word *stereo*, meaning "solid," and it names the study of the three-dimensional shape (a.k.a., "structure," more or less), of chemicals, a feature that affects not just their reactivity, but also their melting and boiling temperatures, and hence their ranges of specific usefulness (by convention defined as at STP). The alkane called methane (one carbon evenly surrounded by hydrogens) can take only one shape in reality. The necessarily linear molecules ethane and propane can be twisted torsionally, collisions nudging them back and forth amongst slightly different conformations. Butane, the final alkane gas, can take either of two very different shapes, linear or branched, and these change their state at slightly different temperatures. Chemically the same in terms of reactivity, the different shapes taken by a chemical of a particular formula are said to be *isomers* of each other. The branched (hence rounder) version of C_4H_{10}, called "isobutane," is gassier, happy to take flight at -12 degrees Celsius, whereas the linear isomer waits for zero Celsius (which is the T in STP) to go from liquid to gas. This difference it seems explains the problems unevenly experienced by people in central Canada wanting to use their otherwise very fuel-efficient marine gas barbecues in January. All samples of hydrocarbons encountered outside the laboratory contain many different kinds of molecules, many in several shapes, and, depending on the source, these all may be in very different proportions. This is part of why they are not quite as substitutable as one might hope or suppose (on which more below). The first alkane not normally a gas at STP, pentane, can take three shapes, with boiling points of 9.5 degrees Celsius for the most compact symmetrically branched version, 28 degrees for the other branched shape, and 36 degrees for the linear one. Just to underscore the complexity of the (simple formula) alkanes alone, consider the fact that the 20 carbon 42 hydrogen alkane, eicosane, has 366,319 possible shapes, changing state at different temperatures.[6] I doubt all of these have been experimentally verified, so we may presume some pairs (trios?) of shapes may by fluke share the same temperature for state-changing, a complication happily of no significance.

The non-gaseous (at STP) members of the alkane family, likewise mutually tolerant, are naturally found not just each in many possible shapes, but also in various relative proportions together in liquid or solid clumps (of oil or coal). They thus require more or less separation from each other before they can serve highly varied human purposes. Every distinct alkane has a particular temperature at which it becomes more individually capturable, so *distilling* (a.k.a., *refining*, in cases where the starting stuff is liquid) consists in heating a solid fossil fuel "chunk" or container's worth of liquid fossil fuel, in such a graduated way that the different constituents as they rise out sequentially (from lightest to heaviest) can be made to separate and then be kept apart, that is, contained separately (after condensation in the case of fractions temporarily gasified in the process). Many paraffins could also be distilled apart, starting from other base materials: wood, peat, and wax, but are not now.

It cannot be over-emphasized that alkanes can be separated so easily precisely because of their low affinities for each other. Generally the higher the number of carbon atoms, the higher the temperature at which the chemical becomes a gas, i.e., can be distilled away, to elsewhere be condensed again to liquid state. For instance, above 40 degrees Celsius much of what is in what we call gasoline takes to the air, but the temperature at which what we call kerosene becomes gaseous starts only at 175 degrees Celsius. These physical differences amongst the constituent chemicals are why different fuels are differently susceptible to explosion, can thus be more or less easily stored, can differentially dissolve different ranges of other compounds for cleaning purposes, etc. The especially vague word *naphtha* (said by the *Oxford English Dictionary* to be of "Oriental origin") referred millennia ago to volatile liquids coming out of the ground here and there, and later was applied to such liquids obtained by distilling coal, petroleum, or shale. I cannot discover why.

Gasoline is another, newer, horribly complex word, with a similar (overlapping) range of chemical denotation to naphtha. It should not be confused with *Gasolene*, which it seems started just as a variant spelling but then became a (now obsolete) proprietary name for a relatively small-scale technique for making a gaseous fuel out of fairly volatile liquid hydrocarbons (briefly considered alongside acetylene gas in the gases chapter). What we now call gasoline obtained normally just by distilling naturally found petroleum at an oil refinery, includes alkanes C_5H_{12} through $C_{10}H_{22}$ as well as a number of near-alkane relatives that are not linear but rather circular (formulas highly various, detailed structures even more so). In connecting back to itself, a ring-shaped hydrocarbon necessarily has two fewer hydrogen atoms, but the circular shape gives it other special properties, many of great usefulness, not just of great interest to stereo-chemists.

Benzene is an extreme, strong-smelling (thus *aromatic*) hydrocarbon relative of the alkanes with the strangely lean formula C_6H_6 intuited by Kekulé in 1865 to be circular as he dreamed one night dozing over his research notes about a snake swallowing its own tail. Having three "double bonds" as well as three single bonds linking the carbons turned out to be the key difference resolving the apparent riddle presented by the consistently sparing results of analysis. Often benzene is added to natural gasoline (along with many other special hydrocarbons) to give what we call high test gasoline. Benzene, which itself also serves as raw material for many plastics, is mostly made, mostly as a by-product of coking, two pounds resulting per ton of starting coal. Chemists say that may sound paltry but not to worry, so much coal gets coked worldwide we are unlikely ever to lack for benzene. Adding ethanol to ordinary gasoline is a way to achieve a performance effect somewhat opposite to adding benzene. As every drinker should know, Scotch whisky is flavour, plus ethanol, plus water, plus caramel for colour, but Scotch is a smooth mixture only because ethanol and water have an extremely high affinity for each other, and that in turn means that in any application where water vapour is liable to enter and then condense inside a gasoline container, what is effectively a non-fuel may dominate for a while in a small-diameter fuel hose. Rarely used fuel-efficient outboard boat motors come to mind.

Somewhat oddly, *benzina* gave the Italians their general name for what in North America we call "gasoline" and British people call "*petrol.*" It is unclear why the word "gas" is so oxymoronically inside *gasoline*, now a general name for a liquid, a word constructed and found in a publication first in 1864.[7] Likely it is just that the hydrocarbon fuels in gasoline are all volatilizing at normal operating conditions inside the equipment burning them. At any rate, the name Gasolene preceded the development of *reliable internal combustion engines* (ICE), which by definition burn gas mixtures in metal containers. The first such engines used actual gases as the starting fuel. In later ones, fuel gasification also occurs in the container. For a long time, gas engines and oil engines competed, both meriting articles in the *Encyclopedia Britannica*.[8] Soon, oil engines took over and now dominate in automobile design, but the gas (at STP) called propane has made a significant comeback as a fuel for internal combustion, in taxis notably. The key point is that mixing two gases, namely the gassy hydrocarbons and the atmosphere, leads to (explosively) faster combustion, with a concomitant release of pressurized force. This is because gases, being quintessentially invasive, fully interpenetrate each other's spaces when in a container, in a way a gas plus a solid or liquid will not.

The internal combustion fire being thus more effective is why the engine can be smaller (and lighter). We will never see chainsaws or lawnmowers, let alone planes, powered directly by coal. Coal, like nuclear power, does most of its work heating water (though coal has been used historically to heat a variety of other substances for distillation, refining, metal forging and simply cooking food), but any coal-fuelled apparatus has vastly higher mass per unit work performed than the ICE. In a stationary engine the weight may not matter at all (*weight*, recall, is mass only pushing down, due to gravity). But we all want cars, and trains, to move (mostly sideways). Planes we also want to have go sideways for long distances, but only after laboriously going up. Planes, of course, are only too willing to descend and can do so without power-assist. On external combustion engines, more below in the section on the fuel concept.

Depending where petroleums are found, they naturally contain different sets in different proportions of alkanes and other hydrocarbons. And we can change some of them. The useful alkene, ethylene, can be obtained from petroleum by *cracking*, which is the term for breaking alkanes by applying heat as they are passed through a chamber. If a gas has undergone cracking, it is said to have been *reformed*. The term *still gas* is short for *distilled gas* and denotes the gases produced as by-products at oil refineries. It may be emphasized here that the fossil fuel sector has complicated pricing because at least two commodities are made simultaneously as the result of technical considerations. The notorious problems of monopoly and oligopoly pricing in the sector may have been partly an historical function of that complication, and thus not just ascribable to greed. Hugely complicating the pricing has been the perceived need by governments to tolerate and/or approve high prices in order to encourage exploration for new sources. The sources have tended to become harder to exploit and thus inherently more costly. This rising *get ratio* refers to the unit cost in a fuel of getting another unit of the fuel.

The solvent *turpentine*, distilled from trees, overlaps with the gasoline family's general hydrocarbon size range and so is similarly volatile (i.e., near-gaseous at STP), but contains no alkanes at all. *Kerosene* is the common (originally Canadian proprietary) name for coal- and/or petroleum-sourced alkanes $C_{12}H_{26}$ through $C_{18}H_{38}$ usually found mixed with other more or less aromatic hydrocarbons. The first fraction from petroleum used on a mass scale was kerosene, and for almost half a century the use was almost exclusively for lighting. It was later adopted for tractors (and later still, jet engines). *Gas oil* is the name for a mix containing $C_{12}H_{26}$ and higher such numbers in general, up to but not including alkanes so heavy as to serve only as lubricants or which, even grosser yet, are actually solids at STP. Gas oil is also called *diesel* fuel, so named because a man named Diesel designed the relevant kind of engine, harder to start in cold weather than a gasoline engine because its fuel is less volatile, but in other ways easier to manage (more reliable). The chemists' term *gas oil* also includes *furnace oil*, the components of which overlap with the set ascribed to diesel but include heavier molecules too.

Since *lubricants* are required to deal with friction and thus heat, their melting points are critical and so, although far from tending to be gaseous, they nonetheless have to be more or less liquid at STP for ease of pre-use penetration into axles, between ball bearings, etc. Lubricants have structures mixing circles and chains. For a long time the very best and most durable *machine oils* could be gotten cheaply only by distilling oils taken from dead whales. *Asphalt*, also called *petroleum coke*, is now the name for the solid stuff left after all else is distilled away from heated petroleum, but asphalt had earlier senses (on which see below on bitumen). So petroleum is the main source for useful alkanes, but petroleum, like coal, contains many (essentially countless different kinds of) hydrocarbons with various structures (some of the more solid ones incorporate many ring-shaped sub-parts).

Natural gas (so called because it comes out of the ground on its own, as it were, and willing to burn, as is) is a mixture of gases in all non-laboratory cases, but always overwhelmingly methane (~90% on average), which, being the simplest organic chemical, is the one whose combustion gives the greatest quantity of energy per unit carbon dioxide created. This is because the big energy comes from breaking the bonds between hydrogen and carbon (on *combustion*, more in the final section of this primer). This matters because (along with water vapour) the chief chemical by-product of all combustion is carbon dioxide, a local and a global poison! In fact, as a global poison, methane itself is in action ten times worse, though vastly less long-lived in the atmosphere. Since we cannot know the pace of future climate change, there can be no way to reduce these two issues of toxicity and durability to one another.

So-called *sour gas* sometimes came from the same places as natural gas but is not a hydrocarbon. It is *hydrogen sulphide* (a.k.a. *sulphuretted hydrogen*) and though nasty and nasty-smelling is colourless and flammable. It is a gas above -60 degrees Celsius.

Where oil and gas come from the same field, the gas is called *associated gas*. If geologically on its own, it is *unassociated gas*.

Loose terms also abound in the discourse on fossil fuels. *Sweet* is a word applied to the *light* unrefined (a.k.a. *crude*) oils that are low in the sour component and thus smell sweeter, but the real point is that light crude has more of its molecules in the gasoline range of hydrocarbons per unit than does *heavy crude*. The obviously variable safety implications in transportation were until very recently ignored, with many catastrophic results in Canada and the United States over the last few years.

Coal gas and *coal oil* were respectively, somewhat and rather more, similar chemically to natural gas and kerosene, but were distilled from coal, not simply dug up, or otherwise liberated from the ground. Many types of gas were made by distilling and were called generically *manufactured gas*, or less often, *artificial gas*. Of course the coal distilled was not anthracite, but bituminous, the volatile components in which range from 15% to 50%, depending on source bed. Two centuries ago it was strenuously argued that one resulting by-product, coke, was alone so valuable that no form of lighting other than coal gas could ever make sense.[9] At that time much gas made in coking was simply wasted, burned off, or allowed to disperse. Ordinary coal is much better as fuel if split into two parts, gas and coke, as both can be burned more fully and more precisely than can coal itself. Moreover tar and some other useful liquids may be simultaneously had thereby. To get the gas, the coal has to be heated in a sealed retort with a pipe to convey the gases and liquids to various cooling containers, each called a "refrigeratory" about two centuries ago. Remarkably the coke (the solid residue) taken from the retort afterwards constituted a fuel sufficient to repeat the process. We glimpse here some of the deep origins of the eventually widespread and wildly insouciant belief in the inexhaustibility of natural resources that so warped the development of economic analysis, especially entrenched after David Ricardo launched his vicious, uncomprehending attack on Adam Smith's views about the natural world and what it does for us.[10]

Coal gas was commercially licensed in London as early as 1812 but had been used in Europe decades earlier. Called *town gas* (or *city gas*) at times because only distributed in towns, if at all, people were very happy eventually to replace it with "natural gas" for several reasons. The urban gas gave only half the calories per cubic unit. Especially if it was mostly *water gas* (made by blasting air and steam alternately over glowing coke or coal, breaking even some of the water molecules apart), it was liable to contain a high proportion of hydrogen gas, as well as the chemically very simple poison carbon monoxide. Carbon monoxide in high concentrations in an airspace can kill because it has an affinity for the hemoglobin molecules in blood that is 300-fold stronger than that of oxygen, which we absolutely need, and so quite quickly it can crowd it out. With gas supplies containing a lot of carbon monoxide, the need for reliable and automatic shut-off valves was thus doubly imperative.

Water gas was sometimes disparagingly called *blue water gas,* as its flame was blue (and so useless for illumination, on which more in the fuel section). In some places over two centuries ago, people seeking a better flame for lighting took water gas and enriched its carbon content by *carburetting* it, that is by passing it through a brick latticework structure into which oil was injected and vaporized. A carburettor is generically an apparatus for mixing air and hydrocarbons so as to add carbon to the air, and so also later came to be the name for a key part in small internal combustion engines.

What was called *producer gas* was made by a process related to but not quite as analytical as laboratory distilling (steam was partially substituted for air in the process), and this also gave a mixture of gases, mostly not hydrocarbons indeed, and since the main hydrocarbon fraction was overwhelmingly carbon monoxide and the overall calorific value was only a fifth of that of town gas, it soon lost market share once natural gas was more widely distributed (well after the Second World War). For awhile in some places, e.g., the United States, a mixture of manufactured and natural gas was sold. Industry had tended to stick to more directly using coal, e.g., in 1932 buying only 8% of the manufactured gas sold in the United States. To see things from an industrial perspective, consider again the example of systematic comparison of fuels included near the start of this primer.

Gas manufactured in such a way as to more closely match natural gas in heat content is called *synthetic natural gas* or *substitute natural gas.* For awhile, starting in the 1960s in the United Kingdom, companies made consumer gas in special high-temperature, high-pressure plants near oil refineries, using fractions of the imported petroleum as the feedstock (raw material). Oddly not called "oil gas" by simple analogy with the very familiar coal gas, but rather "gas oil" (see above, around diesel on the first kind of thing given that name), it was cheaper than coal gas. At about that time the United Kingdom also started importing by ship liquefied natural gas (explained below), first from Louisiana as an experiment, then from Algeria. For household consumers this was reformed down to be equivalent to the quality of town gas (~500 Btus). In 1965 the United Kingdom found a lot of natural gas offshore under the North Sea, but using that for household consumption grew only later, in the 1970s, because it required either destructively reforming the already much-processed gas or converting all the gas appliances installed in homes. Only the rising demand, especially for central heating, decided the state to pick the latter approach. The colossal expense of converting, for example, 200 million home stoves was committed to by a long series of government ministries who also ensured the fuel's price was lower than its competitors. Later the gas industry was denationalized and the commercialized remnant of that is a firm called National Grid. In recent decades British governments encouraged a unique "dash for gas" that, among other things, pushed coal (and coal-mining unions) ever more firmly into the background.

The substance called *liquefied natural gas* is made by people looking to transport natural gas between continents, and for a few other reasons explained later. The procedure for liquefying natural gas is daunting. Propane liquefies at -45 Celsius at one atmosphere of pressure, a common state of affairs at the poles during the season when the sun is "away." So it is easy to liquefy propane at more normal terrestrial temperatures by using pressure instead. Butane also is easy to liquefy with pressure, as it changes state at just below zero Centigrade. Ethane goes liquid at -89, tiresome! But natural gas is mostly methane, and methane cannot be liquefied using pressure alone. If pressurized to 45 atmospheres, methane will liquefy at -82.5 Celsius (the "normal" extreme any of us ever encounter is in -80 degree freezers in research facilities for long-term storage of biological samples). But transoceanic shipping of stuff so massively pressurized is unattractive. It was decided to liquefy methane by using temperature alone at normal atmospheric pressure, implying -161 degrees Celsius (a rather unearthly temperature, even as a wind chill factor). Nevertheless as a consequence, ships starting generally from destinations near the Equator, and thus quite hot, must keep their cargo that cold. The small amount of cargo that warms and volatilizes on the way is used directly on board to fuel the refrigeration. This fuel liquefied thus requires only 1/600th the storage volume it needs as a gas. So on land people in some places liquefy natural gas in order to store it more easily against large fluctuations in demand for the gas. Some gas is stored as LNG in special tanks, though some can instead be stored in the ground in some geological formations. It seems that both space and pressure come at premiums exceeding those of refrigeration, even though the laws of thermodynamics tell us that cooling something always takes more energy than heating it, degree for degree! The whole business of LNG makes for a remarkably bad fossil fuel get-ratio. Also everything to do with liquefied natural gas has to be planned to the utmost extent, as nobody wants to gamble about safety or finances in such a business. During the early years with supertankers (for oil), several of them exploded, as it took awhile to learn how to use them safely, and the oil crisis starting in 1973 made people do things hurriedly.[11] Today LNG is shipped around the world, but only *from* places with lots of gas relative to local demand (and too poor to build pipelines for their own slightly more distant customers), and only *to* rich places very short on oil.

Remarkably enough, the first American liquefaction of natural gas on a large scale was done back in 1917. But the object then was not a fuel – quite the opposite! Helium occurs amongst natural gases in some sites, and the Americans liquefied the alkanes in order to end up with pure helium (which stays gaseous down to -269 degrees Celsius). This they sought as a means to keep airships aloft, as pure helium is lighter than normal air (our atmosphere is a mix of many gases, several very heavy). That helium is also flammable and did burn many airships is a complication beyond our scope, but a looming general point about incommensurability of purposes amongst energy carriers is worth developing further here. Makers of soldered metal toys in England two centuries ago knew that the flame from burning gas is the best heat source for fine work. It remains much more quickly controlled than electric equivalents. Even so, a major use today for natural gas in Canada is just to boil water, with the steam being used either to make electricity by driving turbines or for injection into tar sands in pursuit of heavier alkanes. That kind of pushing action (literal work) was done by natural gas itself as early as 1890 in the United States, where petroleum seekers pumped escaping natural gas back into oilfields to induce the oils to emerge. Another non-fuel use of natural gas, which has been of even greater importance and also for some time, is as a substrate for the artificial synthesis of ammonia for soil fertilizers. But the variegated relationship between natural gas and human purposes is nowhere more starkly revealed than in the controversial practice of flaring. Whenever and wherever natural gas is a nuisance, its flammability has encouraged people to burn it to no purpose other than riddance. Many flares burned for decades, and flaring has been revived in contemporary Montana and the Dakotas. But back around 1980 when most of the oil industry's infrastructure was still in the United States (508,000 of the world's 582,000 non-Communist operating wells), American flaring had essentially ceased, while elsewhere almost all gas was flared off, annually over three million million cubic feet, the equivalent then of all commercial energy consumption in Africa or the Indian subcontinent. The chief difficulty is the expense (and danger) inherent in storing gas anywhere. At times people have stored it in rock formations. Keeping a length of pipeline full of gas not going anywhere for a while is called *linepacking*. Perhaps the extravagance of LNG shipping is no worse than any other way of treating gas that is in the wrong place, but it must be recalled that shipping gas by pipeline costs several times as much as shipping oil by pipeline, and that LNG can be moved only at sea. It is always liquefied at the coast for export and deliquefied at the point of import before being put in a pipeline (again).

Gases are essentially dry in the sense that, like solids, they are *not* liquid. But it is not quite so simple semantically. There are two versions of natural gas: *dry gas* and *wet gas*, and different geological formations give usually one or the other. Dry gas just is the alkanes gaseous at STP. Wet gas also contains the small alkanes nearest the true gas family in size that come along with the natural gas as a dampish contaminant. A "wet" natural gas sample might contain, say, 80%–90% methane, 5%–10% ethane, 3–5% propane, 1%–2% isobutane and butane, and 1%–2% pentanes and upwards. The heavier, less gaseous alkanes are easily made definitely liquid by lowering the temperature and/or increasing the pressure conditions. Amongst those so-called *natural gas liquids* is a substance ludicrously called *natural gasoline* at a time (around the First World War) when it was heavily sought as aviation fuel for what then counted as high-altitude flying. In 1918 7.5% of US gasoline production was from wet gas. After the Second World War it became normal for natural gas to be processed so as to get the natural gas liquids, some two-thirds of natural gas receiving the treatment. The product was equivalent in quantity to about a tenth of US crude production in 1955 (which also equalled crude imports that year). The somewhat more helpful because less vague general name *liquefied petroleum gases* also refers to these sorts of stuffs, mostly propane and butane, modified and packaged by natural gas plants for sale. Propane, which volatilizes on its own at minus 42 degrees Celsius and thus very suitable for the general public as a portable high-power fuel, was long known as *bottled gas*. My eight- kilogram gas barbecue cylinder says on it "LPG 1075." For a long time, plants had wasted all these alkane components.

Dry gas is defined to contain less than a tenth of a gallon of natural gas liquids per thousand cubic feet of natural gas. Because natural gases are defined for measurement to contain a definite number of British thermal units per cubic foot, gas reserve statements on wet gases understate the heat content, because once liquefied and removed, the natural gas liquids appear as an extra in the heat content accounting.

Pintsch gas was a proprietary gas made by refining naphtha into a high-grade oil distillate, which, unlike coal gas, could be compressed and transported without losing its power to illuminate (and so was much used by railways to light passenger cars).

Biogas, which comes from microbial digestion of organic wastes, is a bit more than half composed of methane, but the rest is not alkanes. So, though cheap to make, it is not commercially very valuable. However, it has been a great boon in warm and damp poor countries, as it can be made at the household level, but also reduces the infectious health risks from organic waste stuffs, and also performs a global service, as the people burn that greenhouse gas methane that would inexorably come from unsupervised putrescence anyway. In some cities attempts were made a century ago or so to capture the gases escaping from enormous municipal garbage dumps, but details about the practice are hard to come by. At present in the United States, mounds of refuse are fitted with vents, which are monitored, but the gases are not used. It may need noting here that over the last two centuries the composition and volume of stuff thrown out in urban areas has varied immensely, so using it is very difficult, regardless whether putrescence or recycling is the chosen method.

We must now consider *tar sands*. After highly discreditable public relations considerations in international negotiations on carbon emissions, this stuff was renamed by a Canadian government *bitumen*, a conveniently vague term from antiquity originally denoting a kind of semisolid pitch. This material was also called "asphalt" by the ancient Greeks, who borrowed the word used back then to denote some stuff found floating on water near Babylon. In antiquity bitumen was used in, or as, mortar in buildings. The word later expanded by apparently random adoptions to refer to liquids and solids of many kinds. Tar sands are in fact what the name suggests – sand grains coated with tar – and can give products similar to what can be gotten from oil. But on average about two tonnes of surface material has to be mined per barrel of oil equivalent eventually obtained after processing. Tar sands are far more abundant in the Athabaska region than anywhere else in the world. People have long pondered what to do with them. At one point the idea was discussed in Canadian and American government circles of using underground nuclear explosions to both liquefy the oil and create a large storage reservoir. This proposed killing of two birds with one stone was turned down. Perhaps it was not such a comparatively bad idea. For well over a decade, James Lovelock, the eminent British geochemist (originator with Lyn Margulis of the Gaia hypothesis) has been strenuously urging the fossil fuel–addicted rich countries to switch to nuclear power as the only conceivable way they can avoid having their civilizations crash-land under conditions of massive climate shift.[12]

Oil and/or gas deep underground in some sedimentary rocks such as shale (abundant in parts of the United States) can be acquired by fracturing the rock-bound deposits with explosives and pressurized liquids containing minuscule particles able to keep tiny cracks open so the hydrocarbons can get out. The scale of associated underground pollution is hard to grasp. Moreover, the forcing underground of the proprietary materials has been credibly though not conclusively implicated in earthquake causation and/ or exacerbation. This elaborate method of getting oil substances from shales, though costly enough and prone to failure wherever associated clays abound (as in China, Australia, and Poland) is cheaper than mining the shales (digging them up) for processing. To get the same energy quantum would require on average five times as much mining in the case of oil shales, as compared to mining coal.

Now, however, most American coal is extracted not by mining anymore in the sense of digging. It is gotten hold of by exploding the entire hill bearing the coal, from the top down. The resulting fragments are loaded onto railcars. There is an immense amount of automation used in the transportation phase, many of the trains having no crew whatsoever. This new kind of mining was first applied to metal ores, notably very low-grade copper, and has recently been frankly and aptly (because most literally) termed "mass destruction," a phrase adopted by historian Timothy LeCain as the title for his book on the subject.

Presumably it is because the very latest fossil fuel type, called shale gas, is much easier to extract than shale oil that it threatens those heavily invested in liquefied natural gas, which is also very costly. China is currently exploring shale gas very quickly, presumably as a way to reduce dependence on coal as soon as possible.

It best be noted last but not least that many early uses of hydrocarbons in many cultures had nothing to do whatsoever with energy in any sense. Oils were used as ointments, called embrocations if medical use was the object. Much of the money funding Mr Drake's famous 1859 Pennsylvania oil find (generally, though as outlined in our "Manufactured and Natural Gas" chapter, mistakenly counted as the beginnings of the oil industry) came from the Seneca Oil Co., I believe, but have not yet been able to verify that this is the origin of the bizarre American phrase *snake oil*.

SOME REFLECTIONS ON THE CONCEPT OF "FUEL," ESPECIALLY AS EXEMPLIFIED IN "NUCLEAR (OR ATOMIC) FUELS"

It is now merely an enlightening historical curiosity that the very useful 1911 *Encyclopædia Britannica* distinguished *fuel* as something put in a furnace, from an *illuminant* as something burnt in a lamp. In fact the word *fuel* comes from a medieval Latin word *focalia*, which was a kind of right to firewood. It was constructed from the Latin *focus*, which denoted a domestic fire, that is, a contained indoor fire, a.k.a. a hearth. We now routinely use fuel for many non-domestic combustibles, but the common concept seems to be "containment," itself a word used none too coherently. So *fuel* is the name we use for stuff consumed inside an apparatus: originally just a fireplace with chimney, or a stove, but nowadays we usually have in mind a much more complex container – an engine in an airplane, car, or powerboat. These are properly considered kinds of internal combustion engines, unlike, for example, steam engines, in which the bulk of the engine is the boiler containing water that is becoming steam. The steam-engine fire, the heat source, is located below, external to the boiler. In a complex machine such as a railway steam locomotive of course the burning wood and/or coal was in a metal container – the firebox – under the boiler, but that does not count as internal combustion, because the heat from the burning is not contained (as in a piston) but rather touches the outside surface of the boiler. It is not clear whether the use of fuel in a rocket is properly seen as internal combustion, as the desired thrust is outwards (backwards, indeed), but since very few people use rockets directly, we will leave them to one side. On flame behaviour, more a bit later.

Another, slightly less uncommon use of the word fuel also might seem to complicate, even contradict the point about containability. Ecologists are extending the concept of fuel when they warn about the "natural fuel-load" in a heavily wooded (or grassland) area subject to lightning strikes during a drought. Forest fires start as uncontained, often are indeed utterly uncontainable, despite mass effort and expense. The environmental historian Stephen Pyne has told us a great deal about such fires in many books, and other historians have written about unintended conflagrations in urban places as well as about deliberately caused firestorms during war.[13] The fact that fire and its control have always presented as a difficult pair is illustrated in the origin of the word *curfew*. It came from the French *couvre-feu*, which means "fire cover" and is associated with a municipal requirement that an outdoor fire be smothered after dark when fewer people would be out and thus less likely to see and respond to an uncontained outdoor fire that escaped.

At any rate, fuel is now a multi-purpose word for a source or form of energy. Indeed we learned in the last half-century to speak of "nuclear fuel" (a.k.a. "atomic fuel"), but in that case, although the need for containment is absolute, the word fuel has been extended beyond reasonable analogy. The concept behind this new kind of fuel needs considerable explication. Unlike with all other fuels, the technical details are purely scientific, so there has never been a vernacular use of these fuels, nor could there be. It is important to keep in mind that the only way people can use nuclear fuel as a fuel is to help them heat something, which is always just water. The point of heating the water has always been to make electricity, not tea or soup. Nor could anybody directly use nuclear fuel to heat a house. Electricity presents as a power, not a fuel, and that must be why energy coming from atomic fuel is in fact called "power" and so measured in watts. We speak of a "nuclear power plant." In Canada we could just as well call them "atomic power plants," but nowadays few do. It seems some like to reserve "atomic" mostly for bombs, but we also speak of "nuclear weapons," so the semantic drift here is thoroughly unedifying. In this text we will let the terms *nuclear* and *atomic* oscillate a bit, but it is reasonable to ask why there is any flipping at all between the two words. It seems best to approach that question by setting up the contrast with the case of chemistry.

By definition, chemical change happens amongst atoms that are connected to each other or not, as they change which ones they are connected to. The atoms themselves are unaffected by the chemical changes they get involved in. They go in and out of molecules. A *molecule* is a set of atoms chemically connected. But every *atom* has inside it a *nucleus*, and the key stuff that happens in a nuclear power plant is in fact stuff happening not just inside and to an atom, but inside and to its nucleus. In *atomic energy* and *nuclear energy* it is unfortunate that two terms are used where one would do. That the words they come from – atom and nucleus – have different and nested meanings has confused many people.

The atomic sector is not just semantically challenged. Nuclear fuel is different from all other fuels in at least three other broad respects: ease of access in the field, mode of consumption or use, and the behaviour of by-products. To go conceptually from grasping something, to using it, and on to dealing with waste is simply to mirror the actual chronology of real life, but considering what actually happens when a normal fuel is consumed is perhaps the easiest way to start the comparisons. Food provides power for those doing muscular work by *chemical reactions* inside the worker's body, whether ox, horse, human, or mule. When wood or coal, or a liquid fraction of petroleum, or a gas is burned, it is also a chemical reaction that occurs. In both those cases, food eaten and combustibles burnt, the energy is released in the course of one set of chemicals becoming another set by means of the atoms shuffling their partners. Shuffling requires breaking some bonds, and then maybe forming other ones. As noted earlier, for the most part when we burn a fuel (in our body or in an engine) the desired energy (heat content) is released when chemical bonds between carbon and hydrogen are broken. The biochemical details in a body are more complicated than what happens in an engine, but in both cases the carbon and hydrogen (and any other) atoms in the molecules at the start are all still intact afterwards, just in different combinations with each other, in different molecules. The atoms in molecules that are gaseous at STP may wander off, but they all still exist as the same atoms. No atom of any element has disappeared.

The simplest way to generalize about the chemical reactions involved in consumption of normal fuels (such as food and wood) is to consider the basic formulae involved in *photosynthesis*. When we eat or burn something, we cause it to be *oxidized*, that is oxygen atoms from the air get attached to carbon atoms they were not attached to before. Thankfully, plants always do the exact opposite of what we humans do. The photosynthetic reaction is always presented from the plant's perspective, going from left to right. What we do when combusting things (or breathing) is simply reverse the chemistry, which is the same reaction but read from right to left. Here is how plants sequester the carbon in carbon dioxide (and, in reverse, how we release it):

$$CO_2 \quad + \quad 2 \times H_2O \quad \rightarrow \quad CH_2O \quad + \quad O_2 \qquad H_2O$$

| carbon dioxide | two bits of water | "organic" molecule | single bit of oxygen | one bit of water |

The generic term *organic*, said of molecules, includes such things as sugars and cellulose (the main constituent of wood), all compounds made entirely of combinations of carbon, hydrogen, and oxygen atoms. Looking at this reaction makes it clear why in conditions of greenhouse gas crisis, tree growth should be regarded as sacred. That what plants do in photosynthesis needs the energy of sunlight as well is a complication we cannot explain in this book, but it may be noted that when taken with the fact that carbon dioxide is a gas at STP we can grasp why plants have such intricate and extensive bodies, brimming with surface area. They need to reach out in order to get their gas and to get their light, and plants never stop trying to extend their grasp beyond their reach. This is what used to be called "vegetable dynamism," a literal and reasonable kind of "vitalism."

Trees are distinguished from "herbs," the general name for all other plants, by the fact of having woody matter, a kind of stable (chemically very inert) stuff packaged in cells that are not themselves engaged actively in metabolizing but rather serve structural and plumbing purposes. Wood taken from a tree thus does not rot readily but it also contains water, in differing amounts depending on the kind of tree, its age when cut, and the duration and conditions of storage after cutting. In order to make wood into a less variable (and lighter) fuel people millennia ago took to pre-burning wood. With the water thus distilled away, the remnant was named *charcoal*.

This is also the place to clarify just how burning can give light and not just heat. During combustion, the opposite of photosynthesis, some of the released heat gives luminosity because things are incandescing. Lumps of coal and logs glow, for awhile at least. It is interesting to read in the great technical compendium *American Gas Practice* that, in around 1930, experts were just beginning to feel sure how gas fires could either give or not give light, depending on the heat of combustion, which they had come to control quite finely. The glowing was a puzzle precisely because in a gas fire the only burning things, the gas molecules, are so very small. Taking the simplest case, the emerging idea was that when a molecule of methane is burned in the air, when its four hydrogen atoms separate from their carbon atom, some of the diatomic oxygen also present separates, and so some single oxygen atoms get to attach to single carbon atoms. Other compounds thus form and decompose rapidly in a hydrocarbon fire as the atoms that are normally not free to exist alone are frantically seeking partners and also being abruptly torn away from them, all in the heat of the moment. The key point is that some carbon atoms briefly exist unattached in the fire, and they glow for a short while. They are the wallflowers at the combustion dance that shine. The extent of the phenomenon depends on temperature, so most parts of most gas flames give little or no light. Note that the whole of this explanation is in terms of the chemical behaviour of atoms as they react with each other, combining and recombining. If we adopt another metaphor, we can say that in a fire it is as though the possible chemical reactions race to see which reaction products – intermediate or final – actually preponderate at any given moment.

What happens in nuclear power is totally different, but very confusingly the word reaction was put to use for both cases: everyday inter-chemical reactions and the intra-atomic (indeed intra-nuclear ones). It would have been nice if scientists had agreed to call the first kind "external reactions" and the second kind "internal reactions," by analogy with the combustion engines distinction. The stuff called "fuel" in the atomic sector gives off energy not in the course of a chemical reaction, rather in the course of undergoing fundamental physical change, at the level of the nucleus, which is entirely inside the atom (at least at the start). The atomic fuels are essentially some particular elements, not the molecules inside which the atoms of elements are embedded. The original metal ore that contains/becomes nuclear fuel has to be massively treated and purified to be used at all (no foraging applies here), but it is the inherent instability of the peculiar atoms in question, not the compounds they are found in, that makes them potentially useful as a source of energy. But the instability of these elements (so utterly different from the calm and collected family of paraffin molecules, for example) is why nuclear fuels also produce massively abundant and massively dangerous by-products, even before use, let alone during and after use.

This is because the atoms in question give off bits of energy more or less constantly and also inexorably. We learned to call that *radioactivity* (a word based in the Latin root of *ray*, and defined further below). So dangerously powerful are the rays that Lord Rutherford, one of their first close students, laughed off the very idea of harnessing the energy inside atoms about a century ago by declaring the only use for radioactive materials would be to make a huge uncontrollable explosion, a kind of bomb, something obviously nobody should want! Pursuing the point, we can note that the much later and rather blasé nuclear enthusiast Oppenheimer is said to have commented that the problem with hydrogen bombs is that the targets are too small.

Many naturally radioactive atoms are now lined up in a period in the table of the elements, termed the actinides (first two syllables from another ancient word for ray, Greek this time), but some heavier members of that row are so unstable that chemists are not yet absolutely certain they have perceived them. Practical work with radioactive elements mostly involves only some of the lighter ones: radium (atomic number 88), thorium (90), uranium (92), plutonium (94). The ones heavier than uranium (a.k.a. "transuranic") are all made artificially by humans bombarding their nuclei with other particles (on which strange practice see more below).

Clearly it is necessary here to start to explain some technical terms (including basic ones such as *atomic number* and *isotope*) as well as to outline the physical process that occurs in *nuclear reactions*. There are two broad ways to tame nuclear reactions to give controllable heat rather than a huge explosion. Canada chose one, more or less alone and unaided. As will be explained, that technical decision allowed Canada to play a unique and remarkably benign role in the history of nuclear power, and this book must include that episode, but the distinctive Canadian equipment for harnessing atomic power (later named "CANDU," as explained later) cannot be described simply except by reference to the more orthodox nuclear reactors developed elsewhere. First, more on nuclear physics.

The *atomic number* is a characteristic of the nucleus of an atom of an element. Inside any nucleus (of any atom of any element) there are electrons and protons, always in equal numbers, but also there may be neutrons in various quantities, including zero (of which more later). For instance, uranium is number 92 in the periodic table because in the nucleus of any atom of uranium there are 92 protons (prefix *pro-* because it is positively charged). Exactly countering the protons are 92 (negatively charged) weightless electrons, but these are not inside the nucleus, rather just fairly near it, inside the atom. The spatial zone that electrons occupy more or less defines the outer edges of an atom. Partly because they are outside the nucleus and will not stay put, it is the number of the electrons that determines the chemical (or interactive), as opposed to the physical behaviour of any element. But in addition to the 92 protons inside the nucleus of an atom of uranium there is also inside the nucleus an even larger number of neutrons (each carrying no charge, hence their "neutral" name). That is to say, an atom of an element may have a different number of neutrons compared to its fellows but the same chemical behaviour, in which case it is said to be an *isotope* of the element in question. Two atoms having a different number of protons/electrons are of different elements. Hydrogen is the smallest and also the most abundant element in the universe, and most of it is its normal isotope, which has no neutron at all, just one proton and one electron (atomic number thus deemed to be 1). All other isotopes of hydrogen and all isotopes of all other elements have at least one neutron.

Conveniently for the student, protons and neutrons all weigh the same, so the mass for the normal isotope of hydrogen is deemed 1, whereas the total mass for the isotope of uranium called 235 is due to its having not just its 92 protons but also 143 neutrons ($92 + 143 = 235$). To be candid to the reader, uranium-235 was not chosen randomly to exemplify this arithmetic. It is a peculiar isotope of a peculiar element. It responds oddly to some very general phenomena. Strong cosmic radiation, which is ubiquitous in the universe, can and occasionally does eject neutrons from atoms. With most atoms normally nobody could notice or care what the expelled "liberated" neutron then does. However, there are some exceptions. In the unusual case of atoms of uranium-235 one response to a fairly slow neutron hitting the nucleus is *fission*, a.k.a. splitting. The uranium-235 atom comes apart into two considerable chunks, each nearly half the original, plus some fast "free" neutrons on their own, but the total mass of all the fragments is lower than 235, the difference having been converted into energy, according to Einstein's terrifying equation whereby energy equals mass multiplied by the speed of light itself multiplied by the speed of light. Since that last term (*c squared*) in $E = mc^2$ is absolutely gigantic, a very small change in mass means a very, very large quantity of energy.

If some of those loose neutrons produced by that first fission hit and split other uranium-235 atoms nearby, a *chain reaction* may occur. The chain may be more or less rapid, and if fast may be unstoppable, at least until it runs out of we will call "fuel," although here the fuel is just more of itself. Controlled chain reaction is what we want happening in a "reactor." How can the reaction be controlled? The wild neutrons that cause such changes to neighbouring atoms may/can be slowed and/or absorbed by other materials (placed) in their path. Materials naturally vary in these regards and some materials do both. If one is using a material capable of both slowing and absorbing, then balancing the two becomes the issue, because absorption stops the neutron for good, when for a controlled chain reaction just some slowing is what is wanted, not a stoppage. In a detonated bomb of course the whole point is the less control the better. When nuclear material starts a chain reaction, this is also referred to as the onset of criticality. The minimum quantity of a *fissile* (splittable) material that can sustain a chain reaction is called its "critical mass." We could say that it is a certain quantity of "fuel."

In natural uranium ores, chain reactions do not happen, and that is because uranium ore (the stuff actually dug up) consists mostly of uranium isotopes that do not split when hit by a neutron. Normal concentrations in uranium ore are as follows: 0.7% is 235, almost 99.3% is 238 (so-numbered because it has 3 extra neutrons), plus a trace of 234. The atoms of the 238 isotope of uranium simply absorb the neutrons emitted by a split 235. Only if other things are encouraged to happen, some uranium-238 becomes a different element, indeed eventually the fissile isotope: plutonium-239, but that is another, military story.

The fact that nuclear reactions are unlike chemical reactions in that atoms are actually changed, not just reshuffled, makes it possible under certain conditions for a nuclear fuel to make more fuel while being used! A *breeder* reactor actually makes more fissionable material than it consumes, in the manner of a cow giving off many calves (over time). This apparent paradox, seemingly violating the law of the conservation of matter, which all non-atomic fuels (and the rest of us) absolutely obey, is explained by the fact that in nuclear reactions a change of elements happens, in a kind of cascade. In the course of being used, a nuclear fuel can make itself into another possible atomic fuel, not make more of itself. The total number of atoms does not change. They just become different elements.

Since mining and processing radioactive ores into usable states is hugely expensive, breeder reactors have some appeal to those on a tight budget, but rather unsurprisingly breeder reactors present their own special (and independently expensive) technical problems, and so they have retained a somewhat dreamy reputation. They have not (yet) been deemed really necessary and are used only in research. The non-adoption of breeders is partly because it is still open to doubt whether nuclear electricity overall is even as cheap as electricity made by burning coal or natural gas. The arcana of relative pricing of energy sources at any point in time (let alone through history) would take us far off topic. This primer opened with just one actual pre-nuclear historical example of comparing costs because variables abound, and consensus has never yet emerged on what weights to assign the various variables, let alone what time frames to use if one decides to try to include the costs of dealing with waste products! That said, it is generally agreed that the cost of the fuel for a nuclear plant is about 1% of the total costs involved. It was physicists and economists focused on energy in abstraction from apparatus and other physicalities who coined the infamous, indeed inexcusable, phrase "too cheap to meter" when first advocating atomic energy. At any rate, we are far from needing breeders (yet). Clearly nuclear fuels are not like other fuels at all! Or are they? Recall that long ago coal was seen as breedable, though different language was used. In 1816 Thomas Cooper said that burning coal to distill it gave so much valuable coke, tar, "ammoniacal liquors," and gas as to almost pay the expense of the coal used. But, he conceded, "the establishment, however, is so expensive, that it will answer only on a large scale."[14] We now know that energy abundance without massive human labour inputs comes only with integration and increased ultimate vulnerability. We must return to considering what happens on the inside of a nuclear reactor.

What is common to all peaceful uses of radioactive fuels is the need to initiate a chain reaction, but one that is completely controlled – *moderated* is actually the preferred term. The task of the moderator is to slow neutrons that have gotten away from a nucleus. If uncontrolled, nuclear fuel may either explode the containment apparatus and/or melt itself (at temperatures extremely high by terrestrial standards), at which point its behaviour becomes insanely unpredictable and extremely dangerous to life, even to rocks.

Happily, many chemical compounds can serve as moderators, and they fall into roughly three classes of effectiveness. Normal water (a.k.a. *light water*) works as a moderator and is cheap, but it is amongst the least effective compounds overall, partly because it also provides too much sheer absorption. This means it works well only with richer fuels (3% uranium-235, say, not 0.7%). Graphite and beryllium are twice as effective as moderators compared to regular water. But *heavy water* (deuterium oxide, D_2O, about which more below) is between about ten and almost a hundred times better (depending on purity). Deuterium is an isotope of hydrogen but present as only 0.0156% of the stuff that behaves chemically as hydrogen. Since deuterium is so extremely rare, heavy water has to be made in a complex plant. As an ace moderator, deuterium oxide can not only use uranium with low percentages of the critical isotope uranium-235, but it also does not need supplementing in the machine with other neutron absorbers such as boron, generally included in light water reactors as part of the emergency equipment. All this helps explain why there are two extremely different approaches to using water for the moderation problem, and they have different histories.

In this tale of two waters, one called "light," the other called "heavy," it is the reactor design using normal water that is completely tied in with the military impetus that started the entire extra-academic atomic sector going, whereas in the heavy water case, freed from military considerations at the start, a simpler engineering solution was devised. As it happens, Canada contributed the latter approach, and the distinctive Canadian reactor design was eventually adopted by many countries too poor to even consider making atomic weapons, and also unable to build huge numbers of hydroelectric dams, but nonetheless desirous of generating lots of electricity without having to import lots of fossil fuels. It might not have happened that way.

If British Prime Minister Winston Churchill had insisted on parity with the United States in the first A-bomb project, as he might have tried to (because just before the war Britain had a considerable scientific and technical lead in nuclear physics), he would have forced Canadian cooperation to take a purely military approach that might have persisted after the war. As it turned out, the Americans made their bomb alone, and after the war Britain, France, and the Soviet Union immediately went their own ways developing bombs. Canada alone seized the completely non-military possibilities of nuclear power directly. This happened as the result of political and economic deliberations beyond the scope of this book. Enmeshed in a web of alliances during the war, Canada had no need to make its own bomb after peace came, but did have a lot of uranium mining capacity. In 1953, after the Canadian work on a peaceful use of uranium was well underway, under President Truman the United States started helping many poorer countries to get nuclear power plants. Naively, it was thought these nations would not make bombs too, even though the United States supplied them directly (until 1978) with weapons-grade uranium-235! By 1965 India (mortally afraid of China) was making a bomb, and soon so was Pakistan (mortally afraid of India). The Indian bomb project, however, got underway using a Canadian reactor, exploiting the possibility inherent in using low-grade uranium ore of making the uranium-238 abundant in the low-grade fuel preparation into plutonium-239, which can be used directly for bombs (or power). So the story is not just complicated but messy. Canadian suspicions back in the mid-1950s were brushed aside by Indian officials, Canadians were glad to have the reactor sold, and the Indians went to the Americans for their supply of heavy water. Had Canada not sold them a heavy water reactor, the Indians would have still made atomic bombs some other way, maybe sooner! Exactly how much the Americans knew about the Canadian pursuit of electricity from nuclear power is another interesting topic beyond this volume.

In outlining the military version of applying nuclear physics pursued by the Americans, it is necessary first to clarify that bomb-making did not drive the American design of a controllable reactor for other than research purposes. The US government (at least in the person of Admiral Rickover) had the idea that since submarines can hide while travelling, they are ideal global-reach transports for weapons and surveillance equipment. But submarines present many technical problems, with refuelling at great distance from any naval base, exposed (however briefly) at or near the ocean surface beside a tanker ship being just two. Having a nuclear reactor on board the sub to generate electricity useful for umpteen purposes (air quality maintenance, desalinization of sea water, lighting, cooking, etc.) as well as propulsion could enable a submarine to stay underwater (fully hidden) for very long periods very far away from home. The requirement that the gear fit into an extremely manoeuvrable ship ruled out the use of deuterium oxide as moderator, because the overall apparatus for heavy water is simply too big, too "heavy." This part of the story presents as a matter of engineering only, as the design of the Canada Deuterium Uranium reactor (so named in 1958 and shortened to *CANDU*) exploited the lack of constraint due to not having to be suitable for use in a small ship moving in three dimensions. Determined to excel militarily at all costs as the overriding goal in the nuclear sector, not domestic electricity generation, the Americans involved themselves in further design and expense problems that only their military budget could possibly justify.

This unfortunate economic circumstance obtains because (as noted above) not using heavy water as moderator compels the use of a fuel richer in uranium-235. Isotope enrichment (done mostly by a kind of filtering operation after gasification of a uranium compound) is very costly. Indeed enrichment was so excessively technically demanding that for decades (until well into the 1980s) the American nuclear electricity sector generated less electricity than was consumed in enriching the fuels it was supplied with – the opposite outcome compared to that for the breeder techno-fantasy!

Compared to a light-water reactor, a CANDU can relatively easily use a greater variety of nuclear fuels, including the by-products of other types of reactors, the spent fuels. Technical reasons also make the Canadian reactor easier to use and thus easier to automate. Because there is need for speed and fine adjustments, electronic control has been particularly attractive in the nuclear power sector. Canadians pioneered in many aspects of using computers to help control reactors. Easier to repair, the Canadian reactor could reasonably be claimed to be a much safer design overall. For instance, when necessary, with a CANDU reactor it is fairly easy to add very quickly and easily (even automatically) an even more moderating material, a dose of "neutron poison" as it is fondly termed. Contrariwise, a CANDU reactor can also be operated at near breeding efficiency if modified to use thorium, a normal by-product of fission.

In normal operation with its normal fuel type, CANDU reactors are one-third more efficient in electricity generated per unit uranium ore mined. The Americans have put up with the enrichment complication because they needed to enrich nuclear materials for their bomb projects anyway. Weapons-grade uranium-235 must be enriched to around 90%! More on scaling the issues involved later. We need first to address three basic technical issues further: the nature and source of heavy water, the structural design option for a reactor that heavy water uniquely allows, and the mining and processing of uranium. As our chapter on nuclear power describes, Canada played a big role in uranium extraction during the war and afterwards, as it was the main supplier of the material for non-Soviet atomic bombs for a long time. It was only by around 1960 that the United States found inside itself good sources of uranium ores of high grade (indeed they were better than Canadian ones).

Deuterium oxide is the water that has the deuterium isotope of hydrogen instead of normal hydrogen, and it is 10% heavier than normal water as a result. Normal water, a.k.a. H_2O, could be called hydrogen oxide. The main weight in water is due to the oxygen atom in each water molecule, which is about 16 times heavier than a normal hydrogen atom, and thus 8 times heavier than the two hydrogens per molecule of water. Deuterium, the one proton and one neutron isotope of hydrogen, is technically easy to concentrate by electrolysis of regular water, though its scarcity means the scale of effort has to be huge. Because deuterium atoms have a neutron, and normal hydrogen atoms do not, deuterium oxide is not just heavier but crucially less absorbent of stray neutrons, and thus a better moderator than normal hydrogen oxide. Deuterium is so called because it has a mass double that of normal hydrogen, and *deuter* is Greek for "second."

The third isotope of hydrogen is also relevant. It has one proton and two neutrons. Even heavier than deuterium, it is called *tritium*, and all of it has been made by atomic means (none would exist naturally) and it is itself radioactive. Because hydrogen is in all of life's carbon compounds and tritium is only relatively weakly radioactive, tritium is very widely used as a tracer in biochemical research in live cells. The present author spent many years as a young man using tritium daily to study DNA repair in bacteria. The radiation from tritium does not penetrate skin, but it is not a good idea to ingest or breathe in any tritiated substances (see below why). Humans have made many other radioactive isotopes of other elements too, all by deliberate bombardment of atoms of this and that kind in controlled circumstances, in special reactors. Back to CANDU and power generation.

Of course there have been many slight modifications to the Canadian design over the decades, but the historically distinctive feature is that in the physical structure of a reactor the nuclear fuel and some coolant are placed together inside pressurized hollow rods that can be moved in and out of a relatively non-pressurized cool bath of the moderator. Both moderator and coolant are the same stuff: deuterium oxide. In the other kind of reactor, the members of the light water reactor family, the moderator and the coolant are also the same stuff – this time regular water – but the two functions of moderating and cooling happen in the same place: the fuel meets both at the same time. The meeting happens inside a container that is pressurized, so the water can get very hot without changing state to a gas (boiling away immediately). Lack of a factory able to make a strong enough container for the pressurized option is part of why

Canadians chose deuterium. In both reactor types the heat's eventual use is to make steam to spin turbines. To make electricity it is necessary only to move a magnet relative to a coil of wire or vice versa. Forcing steam confined in pipes to do work moving things like turbines is an old trick mastered more than a century ago, but it is what powers Internet searches. Steam power is more important than the Victorians could have imagined, because electricity's uses have expanded so much so recently.

From the start Canada was very important in the mining of radioactive materials. In Canada, gold and uranium mining were both done on a highly successful and very large industrial scale beginning in the interwar era very far north and west of Canada's industrial heartland. The mine sites form a kind of line along several huge northern lakes, running from Great Bear Lake (straddling the Arctic circle) southwards on through Great Slave Lake, to Lake Athabaska (just below the sixtieth parallel). These mine sites are all thousands of kilometres from the eastern Great Lakes and the St Lawrence River valley where the gold is finally spent and the electricity is mostly needed. But whereas the gold-bearing rocks were transformed into pure gold bars for the Royal Mint at the mine site up north, no such complete processing to a useable state was possible on-site for uranium. After mining, uranium ore must be concentrated locally to a considerable extent. Full refining at the mine site would mean a product too dangerous to transport through such rugged terrain and weather. In any case, the final stages are unsuitable to be performed near the mines because they are too complicated chemically and physically (simply too demanding of fossil fuel for high heat processes).

The parent mineral ore of uranium was heavy and black and called pitchblende (from a Greek word connoting deception). The uranium is chemically present in the ores as U_3O_8 (uranium oxide). This compound was often found in veins too close to be mined separately but far enough apart to mean a lot of worthless material had to be removed. Early mining for radioactive materials had focused on radium, sought mostly for medical purposes, but the *tailings* (waste stuff) from processing ores for radium turned out to be rich enough in uranium to reward processing onsite once military purposes entered the world of the atomic scientists. In all mines (as opposed to quarries) the first processing of ores is called *concentrating*. It is a matter of getting rid cheaply of as much of the unwanted matter as possible on-site. The resulting concentrate is thus low enough in mass per unit value to reward transporting it elsewhere, perhaps first somewhere more convenient for some further refinement or purification before use.

For a long period the uranium concentrate from northern Canada was shipped straight to Port Hope for final preparation. Now concentrated uranium ore first goes to a Canadian facility at Blind River, a port on the north shore of Lake Huron just south of Elliott Lake, a later (closer) uranium mine site. At Blind River that oxide (U_3O_8) is further refined to another oxide UO_3. This compound is then trucked to Port Hope midway along the north shore of Lake Ontario, where it is further modified. Two compounds are sought, in a ratio of about 4:1. Most becomes UF_6 (uranium hexafluoride), which is the gaseous form entirely exported from Canada for enrichment elsewhere, for use in light water reactors, and/or, if enriched even further, for use in atomic weapons. The smaller fraction of stuff produced at Port Hope is UO_2 (uranium dioxide) in ceramic pellet form, which is the stuff inside the fuel bundles used in CANDU reactors. That the isotopes of the uranium atoms in that fuel are still present in the same proportions as in the ground is why some Canadians call their fuel "natural uranium," a rather mystifying phrase. Enriching uranium to make weapons requires vastly larger quantities of the starting material than are needed for peaceful purposes. The scale involved can be indicated using the contemporary controversy over Iran's stockpiles of slightly enriched uranium and enriching centrifuges. Uranium enriched to 5% is ample for power generation. It is estimated that Iran has 10,000 kilograms of uranium near that percentage. It also has 200 kilograms enriched to 20%. All that enriched further to 90% would still not be quite enough to make one testable weapon, but the contention against Iran (population 80 million) is that the stock at 5% is already far bigger than could possibly be needed for peaceful purposes. The solution would be to convert it to some oxide form that is unsuitable for enrichment.[15]

Working with radioactive elements can be dangerous, because what radiates from them can damage biological molecules, and if these are not synthesized again soon in a process involving DNA repair (which occurs more or less adequately and automatically in normal cell replication), serious negative biochemical consequences may result (cancerous growth, even death). Radiation is everywhere, so the kind and quantity, especially per unit time, is what matters. In the case of biological molecules that are not replicated after their first making, most notably those in the egg cells made in a female human foetus, the radiation damage to DNA (a damage also called "mutation") is extremely important. Strictly speaking, *mutation* is just change to DNA, so not all can be called damage. Many mutations have no consequences. Ones having positive effects on the sizes of the populations of the organisms in question are in fact what drives evolution, in the power-steering sense of drive.

Radioactivity is the general name for what leaves an atom spontaneously. This diminution is also called *decay*. From a mass containing much of a radioactive atom, the leaving occurs in ratios of any of three types of streaming rays, named after the first three letters of the Greek alphabet. Alpha rays are streams of alpha particles, which are essentially nuclei of the element helium (atomic number 2), as they contain two neutrons and two protons. Beta particles are simply electrons. Gamma rays consist of photons. The last are beyond definition in this text, but we can hint at the difficulties by pointing out that "photon" names the phenomenon hovering above the unbridgeable gap between wave and particle theories of light.[16] The key point to remember is that photons can travel through matter, even several centimetres of lead, and so are very dangerous.

The element radon (86) is a radioactive gas that results from the decay of radium, and it counts as a significant mining hazard, especially as it itself decays to a fine toxic dust. All elements beyond bismuth (83) in atomic number are radioactive, but some are very rare because they are very unstable. For example, on our planet only 17 atoms of francium (87) ever exist at the same time. So radioactivity is a dauntingly complex phenomenon as well as an alarmingly dangerous one.

NOTES

1 An exception is Daintith, *Dictionary of Physical Sciences.*
2 Fouquet, *Heat, Power and Light: Revolutions in Energy Services*, 216.
3 *Canadian Geographic*, "Canada's Energy Production and Transmission." The interactive map and poster can be viewed online at the *Canadian Geographic* website, https://energyiq.canadiangeographic.ca/main/energy_map#3&-203+91&94+53&1&0&1.
4 A similarly inventive, poetic, and excellent introduction to chemistry is Atkins, *Periodic Kingdom: A Journey into the Land of the Chemical Elements.*
5 Drummond, *Progress without Planning: The Economic History of Ontario fom Confederation to the Second World War*, 93.
6 Morrison and Boyd, *Organic Chemistry.*
7 *Oxford English Dictionary*, 2nd ed.
8 *Encyclopedia Britannica*, 11th ed., 1911.
9 Cooper, *Some Information concerning Gas Lights*, 26.

10 See my essay "Adam Smith's Green Vision and the Future of Global Socialism."

11 See Mostert, *Supership*.

12 Lovelock, *The Revenge of Gaia*.

13 See, for example, Pyne, *Awful Splendour: A Fire History of Canada*.

14 Cooper, *Some Information concerning Gas Lights*, 171.

15 Mathews, "Iran: A Good Deal Now in Danger." On Iran and the centrifuge method for enriching nuclear fuels, see the very short explanation in Bernstein, "Swoo."

16 Pretor-Pinney included a brilliant introduction to the issues in his *Wave-Watcher's Companion*, 269–85.

Selected Sources of Energy Statistics: General and by Energy Carrier

GENERAL

Canada. Census of Canada, 1941, vol. 9, *Housing*; and Census of Canada, 1951, vol. 3, *Housing and Families*, have many schedules of data for energy use in Canadian homes (kinds of lighting, heating and cooking fuels, kinds of appliances, and extent of plumbing) at a variety of aggregate levels, including national, provincial, rural, and urban, and by census district across the country.

Canada, Dominion Bureau of Statistics, Decennial Census of Manufactures.

Canada, Dominion Bureau of Statistics, Department of Trade and Commerce.

Canada, Dominion Bureau of Statistics, Transportation, Communication and Other Utilities.

– *Households and the Environment: Energy Use 2011.* Catalogue no. 11-526-S. Ottawa: Statistics Canada, 2013.

– *Households and the Environment: Energy Use 2013.* Catalogue no. 16-201-X. Ottawa: Statistics Canada,

– *Households and the Environment Survey Public Use Micro Data.* Catalogue no. 16M0001XCB. Ottawa: Statistics Canada, 2010.

– *Human Activity and the Environment: Annual Statistics 2004.* Catalogue no. 16-201-XIE. Ottawa: Statistics Canada, 2004.

Canada, Natural Resources. *Energy Use Data Handbook, 1990–2008.* Catalogue no. M141-11/2008E. Ottawa: Statistics Canada, 2011.

Canada Year Books, 1867– .

Unger, Richard, and John Thistle. *Energy Consumption in Canada in the 19th and 20th Centuries*. Naples, Consiglio Nazionale delle Ricerche – Instituto di Studi sulle Societa del Mediterraneo, 2013.

United States Energy Information Administration. "Canada: International Energy Data and Analysis." http://www.eia.gov/beta/international/country.cfm?iso=CAN.

Urquhart, M.C., and K.A.H. Buckley, eds. *Historical Statistics of Canada*. Toronto: Macmillan, 1965.

Steward, F.R. "Energy Consumption in Canada since Confederation." *Energy Policy* 6, no. 3 (September 1978): 239–45.

WOOD

Canada. Census, 1870–71. "Agricultural and Industrial Schedules for Northumberland County, Ontario." Reel C-9984, CA I AK21 051 1871, Library and Archives Canada.

– Census of Canada, 1931, vol. 1.

– Census of Canada, 1941. Vol. 1, *General Review and Summary Tables*, 563–5.

– Census of Canada, 1951. Vol. 3, *Housing and Families*, "Table 25: Occupied Dwellings by Tenure Showing Principal Heating Fuel, for Counties and Census."

– First Census of Canada, 1871.

– *Sessional Papers of the Dominion of Canada*. Vol. 5, Second Session of the Seventh Parliament, Session 1892 (Ottawa: S.E. Dawson, 1892). Early Canadiana Online. http://eco.canadiana.ca/view/oocihm.9_08052_25_5.

Canadian Century Research Infrastructure / Infrastructure de recherche sur le Canada au 20ᵉ siècle (CCRI IRCS). "Selected Published Tables Data Files: Digitized Published Tables." http://ccri.library.ualberta.ca/endatabase/geography/digitizedpublictables/index.html.

Ontario. Ontario Ministry of Agriculture, Food, and Rural Affairs (OMAFRA). Land Use GIS data.

Prince Edward Island. Census of Prince Edward Island, 1861.

Western Ontario Gazetteer and Directory, 1898–99. Ingersoll, ON: Ontario Publishing & Advertising, 1899.

WIND

Assiniboia. Census of Assiniboia, 1849, table 5.

– Census of Assiniboia, 1856, table 5.

Canada. Census of Canada, 1871, vol. 4.
– Census of Canada, 1891, vol. 4.
– Census of Canada, 1911, vol. 3.
– Census of Industrial Establishments, 1871.
Canadian Industry in 1871 Project (CANIND71). Guelph, ON:
    University of Guelph. http://www.canind71.uoguelph.ca/.
CanWEA. "Vision/Mission." http://canwea.ca/about-canwea/
    visionmission/.

WATER

Bloomfield, G.T., and Elizabeth Bloomfield. *Water Wheels and Steam*
    Engines: Powered Establishments of Ontario. Guelph: University of
    Guelph, 1989.
Canada. Environment and Climate Change Canada. "Frequently Asked
    Questions." http://www.ec.gc.ca/eau-water/default.
    asp?lang=En&n=1C100657-1.
Canadian Census of Industrial Establishments, 1871. http://www.
    canind71.uoguelph.ca/index.shtml.

COAL

Canada. Census of Canada, 1941. Vol. 9, *Housing.*
– *Report of the Royal Commission on Coal.* Ottawa: Edmond Cloutier,
    1947.
– *Statistical Review of Coal in Canada.* Ottawa: Energy, Mines and
    Resources, 1984.
Canada. Tables of Trade and Navigation. Published every year in the
    Sessional Papers of the Province of Canada for the years 1858–67 and
    Sessional Papers of the Dominion of Canada 1867 into the 1920s.
Canada, Statistics Canada. *Electric Power Generation, Transmission*
    and Distribution. Ottawa: Ministry of Industry, 2009.
– Historical Statistics of Canada: Main Index. http://www.statcan.
    gc.ca/pub/11-516-x/index-eng.htm.
    The following can be found using the URL index above on the
    Statistics Canada website:
    Series Q1–5, *Canadian Production of Coal, 1867 to 1976.*
Series Q6–12, *Utilization of Coal, 1867 to 1976.*
Coal Association of Canada. "Coal Trade." http://www.coal.ca/wp-
    content/uploads/2012/04/2010-NRCan-Coal-Report.pdf.

Index Mundi. "Canada Coal Consumption by Year." http://www.
  indexmundi.com/energy.aspx?country=ca&product=coal&graph=
  consumption.
McLeish, John. *The Production of Coal and Coke in Canada during the
  Calendar Year 1909*. Ottawa: Government Printing Bureau, 1910. Also
  data in this series for 1910, 1913, 1915, 1916, 1918, 1919, and 1920.
Milner, W.C. *Coal: Analysis of the Trade between Canada and United
  States*. Ottawa: Mortimer, 1904.
Wynne-Roberts, R.O. *Report on Coal and Power Investigation*. Regina:
  J.W. Cram, 1913.

ELECTRICITY

Canada, Dominion Bureau of Statistics. "Central Electric Stations,"
  Census of Industry, 1919–
Canada, Statistics Canada. *Electric Power Generation, Transmission
  and Distribution*. 2007. Catalogue no. 57-202-X. Ottawa: Statistics
  Canada, 2009.
– Historical Statistics of Canada: Main Index. http://www.statcan.
  gc.ca/pub/11-516-x/index-eng.htm.
  The following can be found using the main index:
  Series Q75–80, *Electric Generating Capacity and Output by Type of
  Ownership, 1919 to 1976*.
  Series Q81–84, *Electric Generating Capacity by Type of Prime Mover,
  1917 to 1976*.
  Series Q85–91, *Electrical Generation by Utilities and Industrial
  Establishments, by Type of Prime Mover, 1919 to 1976*.
  Series Q92–96, *Production and Trade in Electrical Energy, 1919 to
  1975*.
  Series Q97–101, *Electrical Utilities: Number of Customers by Class,
  1920 to 1975*.
  Series Q102–106, *Electrical Utility Sales by Class of Customer,
  Selected Years, 1930 to 1975*.
  Series Q107–113, *Electric Utilities Revenues by Class of Customer*.
Manitoba. *A Farm Electrification Programme: Report of Manitoba
  Electrification Enquiry Commission, 1942*. Winnipeg: King's Printer,
  1943.
Manitoba Power Commission. *22nd Annual Report of the Manitoba
  Power Commission, Year Ended 30 November 1941*. Winnipeg:
  Manitoba Power Commission, 1945.

LIQUID PETROLEUM

British Columbia. *Report of the Commissioner, Royal Commission on Gasoline Pricing Structure*. Victoria, Queen's Printer, 1966.

Canada, Dominion Bureau of Statistics, Special Surveys Division. *Household Facilities and Equipment*. 1953–1980.

Canada, National Energy Board. *Annual Reports*.

– *Energy Briefing Note: Canadian Energy Overview 2012: Energy July 2013*. Ottawa: National Energy Board, 2013.

Canada, Statistics Canada. Historical Statistics of Canada: Main Index. http://www.statcan.gc.ca/pub/11-516-x/index-eng.htm.
The following can be found using the main index:
Series Q13–18, *Production of Crude Petroleum by Province, 1943 to 1975*.
Series Q19–25, *Production and Trade of Crude Petroleum, 1868 to 1976*.
Series Q59–63, *Proven Crude Oil Reserves, 1950 to 1975*.
Series Q131–136, *Principal Statistics of the Petroleum and Natural Gas Industry 1929–1976*.
Series Q149–159, *Canadian Refinery Shipments of Petroleum Products, 1949–1976*.
Series Q55–58, *Exploratory and Development Drilling in Western Canada, 1947–1976*.

Canada, Statistics Canada, Environment Accounts and Statistics Division. Households and the Environment Survey. "Table 1: Lawn Mower Use in Canada and Provinces, 2007." http://www.statcan.gc.ca/pub/16-002-x/2010001/article/lawnmowers-tondeuses/tbl/tbl001-eng.htm.

Ontario. *Annual Report of the Department of Public Highways, 1918*. Toronto: King's Printer, 1919.

Ontario, Department of Transport. *Annual Report 1970*. Toronto: Queen's Printer.

NATURAL AND MANUFACTURED GAS

Canada. Census of Canada, 1901. Vol. 3, *Manufacturing*, "Table 15: Manufactures of Provinces, Compared by Establishments and Products for 1881, 1891 and 1901."

Canada. Dominion Bureau of Statistics, *Coke and Gas Industry*, 1925–59.

– *Coke and Gas Industry Annual Reports*, 1929 and 1930.
– *Manufacture of the Non-Metallic Minerals in Canada* (semi-annual statistics) 1919–41.
– Department of Trade and Commerce. *Household Equipment*. Ottawa: King's Printer, November 1948.
Canada, Statistics Canada. Historical Statistics of Canada: Main Index. http://www.statcan.gc.ca/pub/11-516-x/index-eng.htm.
  The following can be found using main index:
  Series Q26–30, *Production of Natural Gas by Province, 1943 to 1975*.
  Series Q31–37, *Production and Trade in Natural Gas by Province, 1892 to 1976*.
  Series Q64–69, *Proven Natural Gas Reserves, 1955 to 1975*.
  Series Q114–117, *Gas Utilities, Number of Customers by Class, 1958–1976*.
  Series Q118–125, *Gas Utilities Sales (Volume and Amount) by Class of Customer, 1958–1976*.
– *Human Activity and the Environment: Annual Statistics 2004*. Catalogue no. 16-201-XIE. Ottawa: Statistics Canada, 2004.
United States Energy Information Administration. "Canada: International Energy Data and Analysis." http://www.eia.gov/beta/international/country.cfm?iso=CAN.

NUCLEAR POWER

Canada, Statistics Canada. *Electric Power Generation, Transmission and Distribution*. 2007. Catalogue no. 57-202-X. Ottawa: Statistics Canada, 2009.
– Historical Statistics of Canada: Main Index. http://www.statcan.gc.ca/pub/11-516-x/index-eng.htm.
  The following can be found using the main index:
  Series Q81–84, *Electric Generating Capacity by Type of Prime Mover, 1917 to 1976*.
  Series Q85–91, *Electrical Generation by Utilities and Industrial Establishments, by Type of Prime Mover, 1919 to 1976*.
Canadian Environmental Assessment Agency. *Decommissioning of Uranium Mine Tailings Management Areas in the Elliot Lake Area*. Report of the Environmental Assessment Panel. Ottawa: Minister of Supply and Services Canada, 1996.
Denison Environmental Services. "Closed Mines." http://www.denisonenvironmental.com/Closed%20Mines.html.

Ontario Water Resource Commission. *Summary Report on Water Pollution from the Uranium Mining Industry in Ontario.* Toronto: Ontario Ministry of the Environment, 1970. https://archive.org/stream/summaryreportonwoosnsn18803/summaryreportonwoosnsn18803_djvu.txt.

Ontario Water Resources Commission (OWRC). *Water Pollution from the Uranium Mining Industry in the Elliot Lake and Bancroft Areas.* Vol. 1, *Summary.* Toronto: Ontario Ministry of the Environment, October 1971. https://archive.org/details/waterpollutionfro1onta.

Selected Energy Consumption Statistics

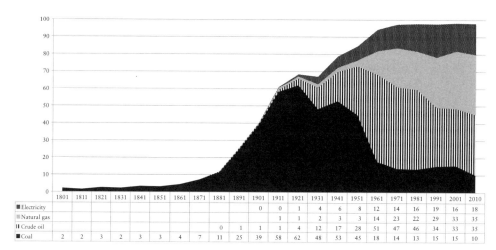

	1801	1811	1821	1831	1841	1851	1861	1871	1881	1891	1901	1911	1921	1931	1941	1951	1961	1971	1981	1991	2001	2010
■ Electricity											0	0	1	4	6	8	12	14	16	19	16	18
▓ Natural gas												1	1	2	3	3	14	23	22	29	33	35
▌▌ Crude oil									0	1	1	1	4	12	17	28	51	47	46	34	33	35
■ Coal	2	2	3	2	3	3	4	7	11	25	39	58	62	48	53	45	18	14	13	15	15	10

Figure A3.1   Mineral energy consumption in Canada, 1800–2010 (% of all energy consumed – coal, oil, electricity, natural gas)

Source: Unger and Thistle, *Energy Consumption in Canada in the Nineteenth and Twentieth Centuries*, 127–32.

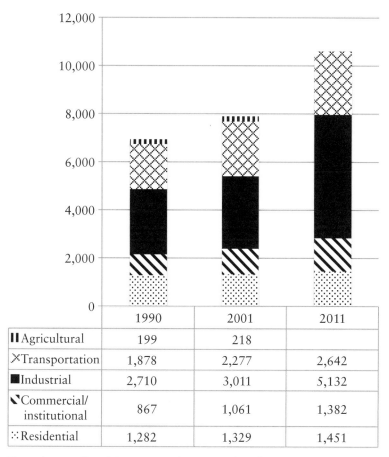

	1990	2001	2011
|| Agricultural	199	218	
✗ Transportation	1,878	2,277	2,642
■ Industrial	2,710	3,011	5,132
◥ Commercial/ institutional	867	1,061	1,382
∴ Residential	1,282	1,329	1,451

Figure A3.2a   Canada's energy use by sector, selected years, 1990–2011 (petajoules)

Sources: 1990, 2001, 2008 data from Canada, Natural Resources, *Energy Use Data Handbook, 1990–2008*; 2011 data from Canada, National Energy Board, *Energy Briefing Note: Canadian Energy Overview 2012*, 6.

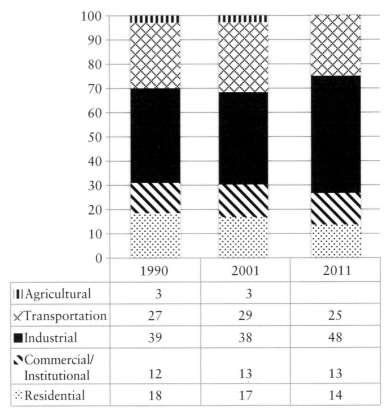

	1990	2001	2011
‖Agricultural	3	3	
×Transportation	27	29	25
■Industrial	39	38	48
◥Commercial/ Institutional	12	13	13
∴Residential	18	17	14

Figure A3.2b   Canada's energy use by sector, selected years, 1990–2011 (%)

Sources: 1990, 2001, 2008 data from Canada, Natural Resources, *Energy Use Data Handbook, 1990–2008*; 2011 data from Canada, National Energy Board, *Energy Briefing Note: Canadian Energy Overview 2012, 6*.

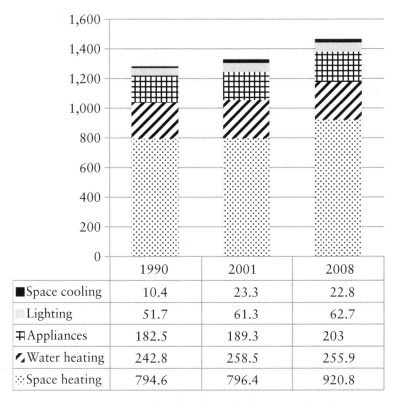

	1990	2001	2008
■ Space cooling	10.4	23.3	22.8
Lighting	51.7	61.3	62.7
⊞ Appliances	182.5	189.3	203
◢ Water heating	242.8	258.5	255.9
⋰ Space heating	794.6	796.4	920.8

Figure A3.3a   Energy use in Canadian homes by function, selected years, 1990–2008 (petajoules)

Sources: 1990–2001 data from "Table 1: Canada's Secondary Energy Use by Sector, End-Use and Sub-Sector (Petajoules)," Canada, Natural Resources, *Energy Use Data Handbook, 1990–2008*; 2008 data from Canada, National Energy Board, *Energy Briefing Note: Canadian Energy Overview 2012*, 6.

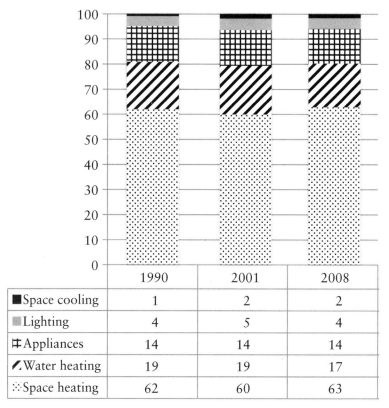

	1990	2001	2008
■Space cooling	1	2	2
▧Lighting	4	5	4
⊞Appliances	14	14	14
◪Water heating	19	19	17
⋮Space heating	62	60	63

Figure A3.3b Energy use in Canadian homes by function, selected years, 1990–2008 (%)

Sources: 1990–2001 data from "Table 1: Canada's Secondary Energy Use by Sector, End-Use and Sub-Sector (Petajoules)," Canada, Natural Resources, *Energy Use Data Handbook, 1990–2008*; 2008 data from Canada, National Energy Board, *Energy Briefing Note: Canadian Energy Overview 2012*, 6.

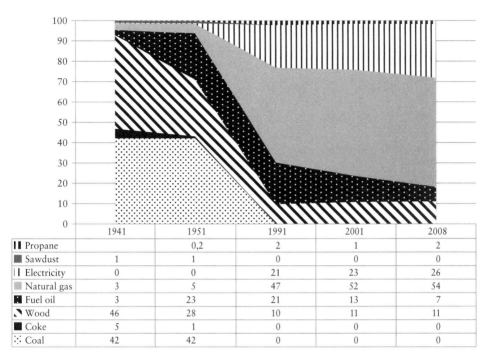

	1941	1951	1991	2001	2008
‖ Propane		0,2	2	1	2
▦ Sawdust	1	1	0	0	0
‖ Electricity	0	0	21	23	26
▦ Natural gas	3	5	47	52	54
▦ Fuel oil	3	23	21	13	7
◣ Wood	46	28	10	11	11
■ Coke	5	1	0	0	0
∴ Coal	42	42	0	0	0

Figure A3.4   Percentage of home heating provided by energy carriers, selected years, 1941–2008

Sources: 1941 Census, vol. 9, *Housing*, table 11; 1951 Census of Canada, vol. 8, *Housing and Families*, table 24; 1990, 2001, 2008 data from Canada, Natural Resources, *Energy Use Data Handbook, 1990–2008*, table 2, 31. This is the total amount used by households for heating, based on a sample.

Table A3.1 Oats (bu.) per farm horse, by province, 1851–1921

	1851	1861	1871	1881	1891	1901	1911	1921
Prince Edward Island	–	118.2	–	112.9	78.2	135.2	145.1	115.1
Nova Scotia	48.0	47.2	42.2	32.8	24.0	37.6	48.4	50.2
New Brunswick	64.0	75.1	68.0	62.2	50.6	77.9	84.7	87.0
Quebec	48.6	70.3	59.7	73.0	49.1	104.6	91.0	110.8
Ontario	56.5	56.2	45.3	68.1	61.1	122.2	110.7	145.7
Manitoba	–	–	–	75.9	96.5	64.6	108.2	111.4
Saskatchewan	–	–	–	5.5[a]	26.7[a]	34.4[a]	116.1	87.2
Alberta	–	–	–	–	–	–	41.5	103.4
British Columbia	–	–	–	9.7	21.1	38.6	30.7	30.4
Canada	–	–	50.8	66.5	56.1	96.0	94.4	105.7

Note: The data for Prince Edward Island 1861 and 1871 are in Canada Census Report, 1871, and the number for 1871 is missing – it should be 134.1. Note also that the data for New Brunswick and Nova Scotia 1851 and 1861 are in Canada Census Report, 1871.
[a] Includes Alberta

Sources: *Province of Canada Census Reports*, 1851, 1861; *Canada Census Reports*, 1871–1921.

Preliminary List of Communities with Manufactured Gas Plants in Canada

Colin A.M. Duncan and R.W. Sandwell

Place name	Company name	Start date	End date	Population near start date	Notes
Anyox, BC	Granby Consolidated Mining, Smelting & Power Company Ltd, 1919–35	1919	1935	~3,000 in 1914	Coking focus. Coastal, far north, near Alaska panhandle.
Barrie, ON	Barrie Gas Company Ltd	1878	1923	4,854 in 1881	
	Barrie Gas Department	1923	1935		
	Barrie Water Light & Gas Commission	1935	1939		
	Barrie Public Utilities Commission	1939	1939		Renamed Barrie Public Utilities Commission when plant closed in 1939.
Belleville, ON	Belleville Gas Company	1854	1923	6,277 in 1861	Belleville Light Department listed in *International Gas Journal*, 1914.
	Corporation of the City of Belleville Gas Department	1923	1932		
	Ontario Shore Gas Company Ltd	1932	1937		In 1932 Ontario Shore Gas Company, Ltd, took over. By 1937 it sent gas to the Belleville Public Utilities Commission, which ran the plant until 1947.

Place name	Company name	Start date	End date	Population near start date	Notes
	Belleville Public Utilities Commission	1937	1947		
Berlin (Kitchener), ON	Berlin Gas Company	1882	1903	4,054 in 1881	Purchased by the Berlin Public Utilities Commission in 1903. Called the Kitchener Light Commission 1919–23, renamed Kitchener Public Utilities Commission in 1924. Plant operated until 1958, also serving Waterloo in its final three decades (see below).
	Kitchener Light Commission, Kitchener Public Utilities Commission	1903	1958		Berlin Public Utilities Commission called the Kitchener Light Commission 1919–23, renamed Kitchener Public Utilities Commission 1924.
Brampton, ON	Brampton Gas Company	1888	1902	3,252 in 1891	
	Equitable Gas	1902	1917		Equitable Gas Company Ltd took over. The plant closed in 1917, apparently.
Brandon, MB	Canada Gas & Electric Corporation	1914	1948	13,839 in 1911	Listed in *International Gas Journal*, 1914, still operating in 1919, renamed the Manitoba Power Commission by 1932, delisted in 1948.
Brantford, ON	Brantford Gas Company	1860	1911?	6,251 in 1861	Ran a plant until 1911, then perhaps changed to natural gas.
Brockville, ON	Brockville Gas Company	1853	1948	4,112 in 1861	Mentioned in the *Journal of the Legislative Assembly*, 1853; in 1921 became Brockville Public Utilities Commission; switched by the city to propane in 1948, and to natural gas in 1958.
Calgary, AB	Calgary Gas Co.	1904?	?	43,704 in 1911	Eugene Cost converted manufactured gas plant to natural gas when he discovered the latter around 1909.
Carberry, MB	Carberry Gas Company	1919	1924	794 in 1921	Acetylene only. On railway line.
Charlottetown, PEI	Charlottetown Gas Light Company	1853	?	7,872 in 1871	Mentioned in the *Journal of the House of Assembly of Prince Edward Island*, 1853.

Place name	Company name	Start date	End date	Population near start date	Notes
Chatham, ON	Chatham Gas Company	1873	1923	5,873 in 1871	Never included in government listings of gas manufacturing plants, but otherwise identified as operating until 1929, so may have been a natural gas treatment facility in the era before the transcontinental pipeline, or so its location suggests.
Chatham, NB	Chatham Gas Light Company	?	?	5,762 in 1881	Listed in 1878 *American Gas Companies Directory*.
Cobourg, ON	Cobourg Gasworks	1857	1937	5,000 in 1861	Renamed the Cobourg Gas, Light and Water Company and then the Cobourg Utilities Corporation. At some point supply was taken over by Hydro Electric Power Commission of Ontario, but gas-making stopped in 1937.
Coleman, AB	International Coal and Coke	1919	1951		In 1932 had 90 beehives, in 1937 there were 100 in use. Near Crow's Nest Pass.
Cornwall, ON	Cornwall Gas & Light Company	1882	1929?	5,436 in 1881	By 1923 the supplier was named the Stormont Electric and Power Company, but it was gone by 1929.
Deseronto, ON		1887	1920	3,388 in 1891	Deseronto got gas by-products from a chemical factory until 1920.
Deloraine, MB	Deloraine Gas Plant	1919	1927	685 in 1921	Acetylene only. On railway line. In 1925 said to be run by Canadian Carbide Company.
Dundas, ON	Dundas Gas Light Company	1863	1909	3,135 by 1871	In 1909 the plant was shut by Dominion Gas Company, which brought gas in from Brantford. In the early 1920s Dundas was supplied by the Wentworth County Gas Company, itself not a gas manufacturer, as government listings did not include it.
Fernie, BC	Crow's Nest Pass Coal Company	1923	1935	4,500 in 1918	486 beehive ovens in 1928. Later joined with Michel. Together, they had 942 beehives in 1932.
Fredericton, NB	Fredericton Gas Light Company	?	?	6,218 in 1881	Listed in 1878 *American Gas Companies Directory*.

Place name	Company name	Start date	End date	Population near start date	Notes
Galt (now Cambridge), ON	Galt Gaslight Company	1887	1911	7,535 in 1891	
Guelph, ON	Guelph Gas Company	1871	1923	6,878 in 1871	The plant was taken over by the Guelph Board of Light and Heat Commission by 1923.
	Guelph Board of Light and Heat Commission	1923	1957		Changed to "... Commissioners" in 1932. Active until 1957.
Halifax, NS	Halifax Gas and Light	1843	1902	20,749 in 1851	Bought out by People's Heat and Light in 1896, closed in 1902, sold assets to Electric Tramway Company, which was listed in *International Gas Journal*, 1914.
	Nova Scotia Light & Power Company Ltd	1919	1952		Nova Scotia Tramways and Power Company started in 1916, and then changed name.
Hamilton, ON	Hamilton Gas Light Company	1853	1919	19,096 in 1861	Listed in *International Gas Journal*, 1914.
	United Gas & Fuel Company of Hamilton Ltd	1919	1923		1919–23 Hamilton was supplied by the United Gas & Fuel Company of Hamilton Ltd, company later engulfed by the new steel industry, and the dedicated gas-plant shut ca 1925.
	Hamilton ByProduct Coke Oven Ltd	1924	1959		United Gas & Fuel Company distributed gas from this plant.
	Hamilton Steel Company of Canada	1924	1959		No local coal for this huge industry.
	Ontario Coke Oven Division	1944	1947		Created in response to munitions pressure.
	Dominion Foundries & Steel Ltd	1951	1959?		Still there in 1959.
Hamiota, MB	Hamiota Gas Plant	1919	1929	609 in 1921	Acetylene only. On railway line. From 1928–29 Shawinigan Chemicals ran the plant.
Harbor Grace, NL	Gas Company of Harbour Grace	1852	?	3,610 in 1836	Mentioned in the *Journal of the House of Assembly of Newfoundland*, fourth session of the fourth General Assembly, 1852.
Ingersoll, ON	Ingersoll Gas Light Company	1876	1915	4,022 in 1871	By 1932 gas storage and distribution done by Dominion Natural Gas Company

Place name	Company name	Start date	End date	Population near start date	Notes
Kingston, ON	Kingston Gas Department	1850		~12,000 ca 1850	Mentioned in the *Journals of the Legislative Assembly*, 1848. Listed in *International Gas Journal*, 1914.
Ladysmith, Cumberland, and Union Bay, BC	Canadian Collieries Ltd	1919	1924		Coking focus. All on Vancouver Island.
Lindsay, ON	Consumers Gas Company	1881	?	5,080 in 1881	
	Lindsay Electric Light Company	?	1890		Lindsay Electric Light Company gained control and shut the gas-plant in 1890.
Listowel, ON	Listowel Gas & Electric Light Company	1891	1915	2,587 in 1891	
London, ON	London's City Gas Company	1853	1935	11,555 by 1961	Made gas, switched to local natural gas ca 1935.
Michel, BC	Crow's Nest Pass Coal Company	1928	1959		100 beehives in 1939, but 1,072 in reserve. Last beehive operation in Canada. 20 miles north of Fernie.
Miniota, MB	Miniota Gas Plant	1919	1927		Acetylene only. On railway line. By 1925 run by Canadian Carbide Company.
Moncton, NB	Moncton Gas Light Company	?	?	4,569 in 1881	Listed in 1878 *American Gas Companies Directory*.
Montreal, QC	Montreal Gas Company	1841	1935?	44,591 in 1844	Montreal Light, Heat and Power Consolidated listed in *International Gas Journal*, 1914. By 1935 listed as buying from Montreal Coke and then no longer listed.
	Montreal Coke & Manufacturing Company	1928	1959		No local coal but much industry. Renamed Quebec Natural Gas Corporation in 1957.
Moosomin, SK	Moosomin Gas Company	(1920s?)	?	1,099 in 1921	Acetylene only. On railway line.
Morris, MB	Acetylene Construction	1919	1932	796 in 1921	Plant run by Acetylene Construction out of Montreal renamed Canadian Carbide. After 1927 Shawinigan Chemicals took over. This town had acetylene only and was on the railway line.

Place name	Company name	Start date	End date	Population near start date	Notes
Nanaimo, BC	Nanaimo Gas Company	1914	?	8,306 in 1911	Listed in *International Gas Journal*, 1914.
Napanee, ON	Napanee Gas Company	1876	1878	3,680 in 1881	Listed in *International Gas Journal*, 1914. After two years' operation, gas-plant taken over by Napanee Water and Electric Light Company.
	Napanee Water and Electric Light Company	1878	1911		
	Seymour Power and Electric Company	1911	1916		
	Hydro Electric Power Commission of Ontario	1916	1921		The Hydro Electric Power Commission of Ontario took over, only to shut it in 1921.
Nelson, BC	Nelson Gas Company	1919	1957	5,230 in 1921	Switched to propane in 1950.
New Westminster, BC	Cunningham Hardware Company Gas Manufacturers	1919	1926	14,495 in 1921	The source was called New Westminster Gas Company Ltd in 1926, then ceased operation.
Oshawa, ON	Oshawa Gas Company	1903	1916	4,394 in 1901	Oshawa's original gasworks were purchased by Oshawa Gas Company in 1903. In 1916 the Hydro Electric Power Commission of Ontario bought the works.
	Hydro Electric Power Commission of Ontario	1916	1928		Public Utilities started only in 1928 when the gas-plant was shut and gas acquired from a Consumers Gas storage facility from Ontario Shore Gas, which switched Oshawa to propane in 1951, was renamed Shorgas Ltd Whitby in 1956, but gone by 1958.
Ottawa, ON	Bytown Consumers Gas Company	1854	1937	14,669 by 1861	The Bytown Consumers Gas Company ran a plant that was relocated in 1920 by the Ottawa Gas Company, by then so named, which was listed in *International Gas Journal*, 1914. This plant was briefly reclassified as a coke plant 1937–39 for reasons unknown. By 1940 reverted status.

Place name	Company name	Start date	End date	Population near start date	Notes
	Ottawa Light, Heat and Power	1944	1950		Overseeing firm renamed Ottawa Light, Heat and Power in 1944.
	Inter-Provincial Utilities	1950	1957		Another name change, to Inter-Provincial Utilities Ltd, in 1950, still active in 1957.
Owen Sound, ON	Owen Sound Public Utilities Commission	1888	1948	7,497 in 1891	Owen Sound plant taken over by Owen Sound Public Utilities Commission in 1888, but by 1940 reclassified as a coke plant, gone by 1948.
Peterborough, ON	Unknown	1869	1906	4,611 in 1871	
	Hydro Electric Power Corporation of Ontario	1906	1927		
	Peterborough Public Utilities Commission	1928	1948		Peterborough Public Utilities Commission took on gas-making in 1928, switched to propane in 1948, shut in 1950 (but on natural gas by 1958).
Picton, NS	Picton Gas Light Company			3,403 in 1881	Listed in 1878 *American Gas Companies Directory*.
Port Hope, ON	Port Hope Gas Light Company	1859	1938	4,162 in 1861	Renamed the Port Hope Gas & Light Company and then Port Hope Gas Company. Taken over by Hydro Electric Power Commission of Ontario. Eventually shut in 1938.
Port Stanley, ON	Southern Ontario Gas Company; Dominion Gas Company	1920?	1950?		Port Stanley had a refinery engaged in gasification 1945–58. During the 1920s and 1930s the Southern Ontario Gas Company had been active there, as was the Dominion Gas Company from 1930s until the 1950s.
Quebec City, QC	Quebec Power Company (Gas Department)	1914	1959?	68,840 in 1901	Listed as Quebec Railway, Light, Heat & Power Company Ltd in 1919, but in 1928–29 as at St Malo, under its new name. Still listed in 1959, but from 1956 with a Verdun address, with the same equipment!

Place name	Company name	Start date	End date	Population near start date	Notes
Sarnia, ON	Sarnia Consumers Gas Company	1884	1893	3,874 in 1881	In 1893 taken over by the Sarnia Gas and Electric Light Company, which switched the town to natural gas in 1909.
Sault Ste Marie, ON	Western Gas and Light Company			21,092 in 1921	Got coal by ship, presumably.
	Algoma Steel Corporation	1923	1959		Coking focus.
	Great Northern Gas	1925	?		This company distributed to the city and from 1937 to other places too.
Shawinigan, QC	Shawinigan Chemicals			10,625 in 1921	History is unclear, but perhaps consumers got acetylene gas.
Sherbrooke, QC	Sherbrooke Light, Heat and Power Company	1914?	?	~10,000 in 1891	Listed in *International Gas Journal*, 1914.
	City of Sherbrooke Gas Department				City of Sherbrooke Gas Department, renamed Corporation of Sherbrooke in 1925, by 1929 renamed City of Sherbrooke Electric Department, switched to propane in 1948. By 1959 Sherbrooke was one of only six non-coke plants listed for all Canada.
Simcoe, ON	Simcoe Gas and Water	1891	1910	2,674 in 1891	Developed an agreement with Dominion Natural Gas Company in 1906, plant shut around 1910.
Souris, MB	Consumers Gas Company Ltd	1919	1929	1,710 in 1921	Acetylene only. On railway line.
Sorel, QC	Sorel Gas Light Company	?	?	5,791 in 1881	Listed in 1878 *American Gas Companies Directory*.
St Catharines, ON	St Catharines & Welland Canal Gas Light Company	1853	1903	6,284 in 1861	Listed in the *Journal of Legislative Assembly*, 1853. In 1903, the St Catharines Gas Company took over. In 1912 the city took over, and the plant closed in 1928.
St Hyacinthe, QC	St Hyacinthe Gas and Electric Light Company	1914	?	9,200 in 1901	Listed in *International Gas Journal*, 1914.
St John, NB	St John Gas Light Company	?	?	4,133 in 1881	

Place name	Company name	Start date	End date	Population near start date	Notes
	St John Railway Company	1914?	?		Listed in 1878 *American Gas Companies Directory* and in *International Gas Journal*, 1914.
	New Brunswick Power Company	1919?	1948		
St John's, NL	St John's Gas Light Company	1848	1951	14,945 in 1836	Mentioned in the *Journal of the House of Assembly of the Province of New Brunswick*, 1848. Listed in *Journal of the House of Assembly of Newfoundland*.
St Thomas, ON	St Thomas Gas Company	1877	1923?	8,367 in 1881	
	St Thomas Gas Department	1923	1936		The St Thomas Gas Department was making the gas by 1923, but after a name change to Gas Commissioners of St Thomas, the plant closed in 1936.
Stratford, ON	Stratford Gas Company	1875	1925	4,313 in 1871	
	Stratford Public Utility Commission	1925	1953		Gas making stopped in 1953.
Sydney, NS	British Empire Steel Corporation	1919	1924	22,545 in 1921	Renamed Dominion Iron & Steel Company in 1925.
	Dominion Iron and Steel Company	1925	1930		Called Dominion Steel and Coals Corporation Ltd by 1930, still there 1959.
Toronto, ON	Consumers Gas Company	1848	1955	~30,000 in 1851	Added two plants in 1909.
Trois-Rivières, QC	City Gas & Electric Corporation	?	?		The government's separate propane listing started with it in 1943 and it stayed on until 1959. A wood-burning gas-manufacturing plant was run by Riche Gas long before, but in 1909 it was bought and taken over by a natural gas start-up, but fields were too small. City had had natural gas street lighting in 1856, but no gas delivery from about 1910 to 1930, when a spherical storage tank for gas was built.

Place name	Company name	Start date	End date	Population near start date	Notes
Vancouver, BC	Vancouver Gas Company	1914	1935	100,401 in 1911	Listed in *International Gas Journal*, 1914, joined in 1926 by BC Electric Power and Gas Company, and shifted to coke status in 1935. In 1951 renamed BC Electric Company, shut in 1951.
Victoria, BC	BC Electric Railway Company Ltd (Victoria Gas Company)	1919	1959		Operating in 1919, switched to propane (brought by rails and ship?) in 1954, still listed 1959.
Waterloo, ON	Waterloo Gas Company	1889	1894	2,941 in 1891	Changed its name to Waterloo Consumers Gas Company
	Waterloo Consumers Gas Company; Waterloo Water and Light Commission Gas Department	1894	1927?		1914–20 the plant was run by the Waterloo Gas Department, called the Waterloo Water and Light Commission Gas Department 1924–27, but thereafter (or soon) Waterloo customers came to be served from Kitchener's plant, although some think the Waterloo site was still active, perhaps as a storage facility.
Windsor, ON	Windsor Gas Company	1871	1930	4,253 in 1871	It presumably competed with a local natural gas supply, at least for a while, as the information about Sarnia is suggestive.
Winnipeg, MB	Winnipeg Electric Railway	1914	1957	136,035 in 1911	Listed in *International Gas Journal*, 1914. No local coal but much industry. Dropped the word *Railway* from its title in 1932. Renamed Winnipeg & Central Gas Company in 1952, gone by 1958.
Woodstock, ON	Woodstock Gas Light Company	1876	1919	3,982 in 1871	
Yarmouth, NS	Yarmouth Gas Light Company	?	?	3,485 in 1881	Listed in 1878 *American Gas Company Directory*.

Bibliography

NOTE: *Appendix 2 contains a list of useful published statistical sources pertaining to energy, historical and contemporary, including a listing by kind of energy carrier.*

Adams, Frank D. "Committee on Minerals: Our Mineral Resources and the Problem of the Proper Conservation." In *Sixth Annual Report of the Commission of Conservation*. Ottawa, 1915.

Ahuvia, Aaron, and Elif Izberk-Bilgin. "Limits of the McDonaldization Thesis: eBayization and Ascendant Trends in Post-Industrial Consumer Culture." *Consumption, Markets & Culture* 14, no. 4 (December 2011): 361–84.

Allardyce, Gilbert. "'The Vexed Question of Sawdust': River Pollution in Nineteenth-Century New Brunswick." In *Consuming Canada: Readings in Environmental History*, ed. Chad Gaffield and Pam Gaffield, 119–30. Toronto: Copp Clark, 1995.

American Electrical Directory, 1886, The. Fort Wayne, IN: Page Taylor and Company, 1886.

American Hoist and Derrick. *American Hoist and Derrick Catalogue, 1907–08.* St Paul, MN: American Hoist and Derrick, 1908.

Anastakis, Dimitry. *Autonomous State: The Struggle for a Canadian Auto Industry from OPEC to Free Trade.* Toronto: University of Toronto Press, 2013.

– "A 'War on Pollution'? Canadian Responses to the Automotive Emissions Problem, 1970–1980." *Canadian Historical Review* 90, no. 1 (March 2009): 99–136.

Anders, Howard S. "The Dust Menace and Municipal Diseases." Transactions of the American Climatological Association 27 (1911): 276–88.

Anderson, David. "Levittown Is Burning! The 1979 Levittown Pennsylvania Gas Line Riot and the Decline of the Blue-Collar American Dream." *Labor: Studies in Working Class History in the Americas* 2, no. 3 (2005): 47–66.

Anderson, Letty. "Water-Supply." In *Building Canada: A History of Public Works*, ed. Norman R. Ball, 195–200. Toronto: University of Toronto Press, 1988.

Ankli, Robert E., H. Dan Helsberg, and John Herd Thompson. "The Adoption of the Gasoline Tractor in Western Canada." *Canadian Papers in Rural History* 2 (1980): 9–39.

Apps, Jerry. *Horse-Drawn Days: A Century of Farming with Horses*. Madison, WI: Wisconsin Historical Society, 2010.

Archibald, E.S. *Preparing Farm Horses for Summer Work*. Ottawa: Department of Agriculture, [1917].

Archibald, E.S., and G.B. Rothwell. *The Feeding of Horses*. Ottawa: Department of Agriculture, 1916.

Armstrong, Christopher, Matthew Evenden, and H.V. Nelles. *The River Returns: An Environmental History of the Bow*. Montreal and Kingston: McGill-Queen's University Press, 2009.

Armstrong, Christopher, and H.V. Nelles. *Monopoly's Moment: The Organization and Regulation of Canadian Utilities, 1830–1930*. Toronto: University of Toronto Press, 1986.

– *Wilderness and Waterpower: How Banff National Park Became a Hydroelectric Storage Reservoir*. Calgary: University of Calgary Press, 2013.

Armstrong, Robert. *Structure and Change: An Economic History of Quebec*. N.p.: Gage, 1984.

Atkins, P.W. *Periodic Kingdom: A Journey into the Land of the Chemical Elements*. New York: Basic, 1995.

Atkinson, Miriam, and Monica E. Mulrennan. "Local Protest and Resistance to the Rupert Diversion Project, Northern Quebec." *Arctic* 62, no. 4 (2009): 468–80.

Auditor General of Canada. *Report of the Auditor General of Canada*. Ottawa: Queen's Printer, 1983, chap. 9, s. 9.11.

Avallone, Eugene A., Theodore Baumeister, and Ali Sadegh, eds. *Marks' Standard Handbook for Mechanical Engineers*. 11th ed. New York: McGraw Hill, 2007.

B9 Energy Group. "Flagships of the Future." 2015. http://www.b9energy.com/B9Shipping/FlagshipsoftheFuture/tabid/5069/language/en-US/Default.aspx.

Babin, Ronald. *The Nuclear Power Game*. Montreal, Black Rose Books, 1985.

Baird, H.C., Son, and Company. *Clay-Working Machinery and Supplies Catalogue No. 12, 1910–1919*. Parkhill, ON: Baird.

Bakker, Karen. "Introduction." In *Eau Canada: The Future of Canada's Water*, ed. Karen Bakker, 1–16. Vancouver: UBC Press, 2007.

Baldwin, Peter C. *Domesticating the Street: Reform of Public Space, 1850–1930.* Columbus: Ohio State University Press, 1999.

Bamford, Donald A. *Freshwater Heritage: A History of Sail on the Great Lakes 1670–1918.* Toronto: Dundurn, 2007.

Barclay, Harold B. *The Role of the Horse in Human Culture.* London: J.A. Allan, 1980.

Barker, T.C. "The Delayed Decline of the Horse in the Twentieth Century." In *Horses in European Economic History*, ed. F.M.L. Thompson, 101–12. Reading: British Agricultural History Society, 1983.

Barnard, Ed. A. *Manuel d'agriculture: le livre des Cercles Agricoles.* Montreal: Eusèbe Senécal, 1895.

Barnes, A.S.L. *The Heating of Houses, Coal and Electricity Compared.* Ottawa: Privy Council for Scientific and Industrial Research, 1918.

Baron, Martin. "L'éloge de *la Grise:* Le cheval et la culture populaire au Québec (1850–1960)." MA thesis, Université de Sherbrooke, 1997.

Baskerville, Peter, and Eric Sager. *Unwilling Idlers: The Urban Unemployed and Their Families in Late Victorian Canada.* Toronto: University of Toronto Press, 1998.

Basov, Vladimir. "Top 10 Oil and Gas Producing Countries in 2012." 14 June 2013. *Mining.com.* http://www.mining.com/top-10-oil-and-gas-producing-countries-in-2012-24585/.

Battagello, Dave. "Wind Turbine Plant Shuts Down Windsor Operations." *Windsor Star,* 29 March 2012.

BC Hydro. "Right of Way Management." 2013. https://www.bchydro.com/energy-in-bc/our_system/right_of_way_management.html.

– "Site C Clean Energy Project." 2013. http://www.bchydro.com/energy-in-bc/projects/site_c.html?WT.mc_id=rd_sitec.

BC Hydro Power Pioneers. *Gaslights to Gigawatts: A Human History of BC Hydro and Its Predecessors by the BC Hydro Power Pioneers.* Vancouver: BC Hydro Power Pioneers, 1998.

Bédard, Roger. "Le Frontière du Labrador et le Contrat de Churchill Falls Corporation." *Action Nationale* 58, no. 1 (1968): 63–78.

Bélanger, Andrée. "Évolution du cheptel équin et de la culture équestre dans la vallée du Saint-Laurent sous l'influence britannique, 1760–1850." MA thesis, Université Laval, 2010.

Belich, James. *Replenishing the Earth: The Settler Revolution and the Rise of the Anglo-World, 1783–1939.* Oxford: Oxford University Press, 2009.

Bell, Becki. *The Little Book of Horse Poop.* Rough and Ready, CA: Palfrey Media Publishing, 2005.

Bellamy, Matthew. *Profiting the Crown: Canada's Polymer Corporation, 1942–1990*. Montreal and Kingston: McGill-Queen's University Press, 2005.

Bellavance, Claude. *Shawinigan Water and Power, 1898–1963: Formation et décline d'un groupe industriel au Québec*. Montreal: Boréal, 1994.

Bellavance, Claude, Roger Levasseur, and Yvon Rousseau. "De la lute antimonopoliste a la promotion de la grande entreprise l'essor de deux institutions: Hydro-Québec et Desjardins, 1920–1965." *Recherches Sociographiques* 40, no. 3 (1999): 551–78.

Bellis, Mary. "Machines to Cut Grains." aboutmoney. inventors.about.com/od/rstartinventions/a/reaper_2.htm.

Belyea, Barbara, ed. *Columbia Journals / David Thompson*. Montreal and Kingston: McGill-Queen's University Press, 2007.

Benidickson, Jamie, "John Rudolphus Booth." In *Canada's Entrepreneurs: From the Fur Trade to the 1929 Stock Market Crash*, ed. J. Andrew Ross and Andrew D. Smith, 328–36. Toronto: University of Toronto Press, 2011.

Bercuson, David J., ed. *Alberta's Coal Industry, 1919*. Calgary: Historical Society of Alberta, 1978.

Bernier, Paul. *Le cheval canadien*. Montreal: Septentrion, 1992.

Bernstein, Jeremy. "Swoo." *London Review of Books*, 31 July 2014, 12.

Beutler, Corinne. "L'outillage agricole dans les inventaires paysans de la région de Montréal reflète-t-il une transformation de l'agriculture entre 1792 et 1835?" In *Sociétés villageoises et rapports villes-campagnes au Québec et dans la France de l'ouest, XVIIᵉ–XXᵉ siècles*, ed. François Lebrun and Normand Séguin, 121–30. Trois-Rivières: Centre de Recherche en Études Québécoises, Université du Québec à Trois-Rivières, 1987.

Binnema, Theodore. *Common and Contested Ground: A Human and Environmental History of the Northwestern Plains*. Norman, OK: University of Oklahoma Press, 2001.

Black, Brian. "Oil for Living: Petroleum and American Mass Consumption." *Journal of American History* (Spring 2012): 40–50.

Black, Henry. *Canadian Scientists and Inventors: Biographies of People Who Shaped Our World*. Markham, ON: Pembroke Publishers, 2008.

Blanchard, Jim. *Winnipeg: Diary of a City*. Winnipeg: University of Manitoba Press, 2005.

Bleakney, J. Sherman. *Sods, Soil, and Spades: The Acadians at Grand Pré and Their Dykeland Legacy*. Montreal and Kingston: McGill-Queen's University Press, 2004.

Bloomfield, Elizabeth, and G.T. Bloomfield. *Patterns of Canadian Industry in 1871: An Overview Based on the First Census of Canada*. Guelph, ON: Department of Geography, University of Guelph, 1990.

Bloomfield, G.T. "I Can See a Car in That Crop: Motorization in Saskatchewan." *Saskatchewan History* 37, no. 1 (1984): 25–31.

Bloomfield, G.T., and Elizabeth Bloomfield. "'Our Prosperity Rests upon Manufactures': Industry in the Central Canadian Urban System, 1871." *Urban History Review* 22, no. 2 (June 1994): 75–96.

– *Water Wheels and Steam Engines: Powered Establishments of Ontario.* Guelph, ON: University of Guelph, 1989.

Blouin, Claude. "La mécanisation de l'agriculture entre 1830 et 1890." In *Agriculture et colonisation au Québec*, ed. Normand Séguin, 93–111. Montreal: Boréal Express, 1980.

Board of Health. "Report on the Origin and Progress of the Epizootic among Horses in 1872, with a Table of Mortality in New York. Illustrated with Maps." In *Third Annual Report of the Board of Health of the Health Department of the City of New York April 11, 1872, to April 30, 1873*, 250–91. New York: D. Appleton, 1873.

Bodansky, Daniel. "The History of the Global Climate Change Regime." In *International Relations and Global Climate Change*, ed. Urs Luterbacher and Detlef F. Sprinz, 23–40, Cambridge, MA: MIT Press, 2001.

Bomberg, Mark, and Donald Onysko. "Heat, Air and Moisture Control in Walls of Canadian Houses: A Review of the Historic Basis for Current Practices." *Journal of Building Physics* 26, no. 1 (July 2002): 3–31:

Bonnell, Jennifer L. *Reclaiming the Don: An Environmental History of Toronto's Don River Valley.* Toronto: University of Toronto Press, 2014.

Bonnett, John. *Emergence and Empire: Innis, Complexity, and the Trajectory of History.* Montreal and Kingston: McGill-Queen's University Press, 2013.

Boon, Christopher J. *A Preliminary Analysis of Factors Affecting Modal Selection for the Movement of Western Canadian Coal to Ontario.* Kingston, ON: Canadian Institute of Guided Ground Transport, Queen's University, 1978.

Bothwell, Robert. *Eldorado: Canada's National Uranium Company.* Toronto: University of Toronto Press, 1984.

– *Nucleus: A History of Atomic Energy of Canada.* Toronto: University of Toronto Press, 1988.

Bott, Robert D. *Evolution of Canada's Oil and Gas Industry.* Calgary: Canadian Centre for Energy Information, 2012.

Bouchard, Gérard. "L'agriculture saguenayenne entre 1840 et 1950: l'évolution de la technologie." *Revue d'Histoire de l'Amérique Française* 43, no. 3 (1990): 353–80.

Bowen, Lynne. *Whoever Gives Us Bread: The Story of Italians in British Columbia.* Vancouver: Douglas and McIntyre, 2011.

Bradley, Ben. "By the Road: Fordism, Automobility, and Landscape Experience in the BC Interior." PhD diss., Queen's University, 2012.

Brandon, Robert. *A History of Dresden*. N.p., 1954.

Bratt, Duane. *The Politics of CANDU Exports*. Toronto: University of Toronto Press, 2006.

Breen, David H. "1947: The Making of Modern Alberta." In *Alberta Formed, Alberta Transformed*, ed. Michael Payne, Donald Wetherell, and Catherine Cavanaugh, 538–63. Calgary: University of Alberta Press, 2005.

– *Alberta's Petroleum Industry and the Conservation Board*. Edmonton: University of Alberta Press, 1992.

– *The Canadian Prairie West and the Ranching Frontier, 1874–1924*. Toronto: University of Toronto Press, 1983.

Brewer, Priscilla J. *From Fireplace to Cookstove: Technology and the Domestic Ideal in America*. Syracuse, NY: Syracuse University Press, 2000.

British Columbia. *Report of the Commissioner, Royal Commission on Gasoline Pricing Structure*. Victoria: Queen's Printer, 1966.

Brown, Ernest C. *American Gas Company Directory*. New York: Progressive Age, 1890, 1904, 1918.

Brown, Richard. *The Coal Fields and Coal Trade of the Island of Cape Breton*. London: Sampson Low, Marston, Low and Searle, 1871.

Brown, Stanley P., Wayne C. Miller, and Jane M. Eason. *Exercise Physiology: Basis of Human Movement in Health and Disease*. Baltimore, MD: Lippincott Williams & Wilkins, 2006.

Burnett, John. *Plenty and Want: A Social History of Diet in England from 1815 to the Present Day*. London: Thomas Nelson, 1966.

Burton, Tony, Nick Jenkins, David Sharpe, and Ervin Bossanyi. *Wind Energy Handbook*. New York: Wiley, 2011.

Buxton, William F., ed. *Harold Innis and the North: Appraisals and Contestations*. Montreal and Kingston: McGill-Queen's University Press, 2013.

Cadham, John. "The Canadian Nuclear Industry: Status and Prospects." Nuclear Energy Futures Paper No. 8, 9 November 2009. https://www.cigionline.org/publications/2009/11/canadian-nuclear-industry-status-and-prospects.

Cairns, George Alexander. "Memories of My Life and Times." Unpublished, 1999.

Campbell, Marjorie Wilkins. *The North West Company*. Toronto: Macmillan, 1957.

Canada. *Census of Canada, 1931*. Vol. 1, *Population*, table 5, "Population, Rural and Urban Counties or Census Divisions, 1851–1931."

– *Census of Canada, 1941*. Vol. 1, *Population*, table 10, "Population by Census Sub-division, 1871–1941."

– *Census of Canada, 1941*, Vol. 9, *Housing*, table 18, "Occupied Dwellings with Specified Conveniences, 1941."
– *Census of Canada, 1951*. Vol. 1, *Population: General Characteristics*, table 14, "Population for Counties and Subdivisions, Rural and Urban, 1951 and 1941" (1941 definitions).
– *Census of Canada, 1951*. Vol. 3, *Housing and Families*, table 25, "Occupied Dwellings by Tenure Showing Principal Heating Fuel, for Counties and Census."
– *Sessional Papers of the Dominion of Canada*. Vol. 5, *Second Session of the Seventh Parliament, Session 1892*. Ottawa: S.E. Dawson, 1892. http://eco.canadiana.ca/view/oocihm.9_08052_25_5.
– *Sessional Papers of the Province of Canada*, for the years ending 30 June 1867 and 30 June 1868. Tables of the Trade and Navigation of the Province of Canada.
– *Sessional Papers of the Dominion of Canada*, for the years 1867–1890. Tables of Trade and Navigation of the Province of Canada.
Canada, Census and Statistics Office. *Census of the Manufactures of Canada, Bulletin 2, Canada 1906*. Ottawa: King's Printer, 1907.
Canada, Department of Agriculture. *Census of Canada, 1901*. Vol. 3, *Manufactures*. Ottawa: Department of Agriculture, 1905.
Canada, Department of Trade and Commerce. *Report, Sessional Papers of the Dominion of Canada for the Years 1914, 1915, 1916*.
Canada, Dominion Bureau of Statistics. "Central Electric Stations." In *Census of Industry, 1939*. Ottawa: King's Printer, 1936.
– *Eighth Census of Canada, 1941*. Vol. 9, *Housing*. Ottawa: Edmond Cloutier, 1949.
– *Household Equipment November 1948: Cooking Facilities, Washing Machines, Refrigerators, Vacuum Cleaners, and Radios in Canadian Homes*. Ottawa: Department of Trade and Commerce, 1948.
– *Household Facilities and Equipment*. Cat. no. 64-202. Ottawa: Queen's Printer, 1953–80.
– *Ninth Census of Canada, 1951*. Vol. 3, *Housing and Families*. Ottawa: Edmond Cloutier, 1953.
– *Seventh Census of Canada, 1931*. Vol. 1, *Populations*. Ottawa: J.O. Patenaude, 1949.
Canada, Energy Mines and Resources. *Energy in Canada: A Background Paper*. Ottawa: Energy, Mines and Resources Canada, 1987.
– "Statistical Review of Coal in Canada." Ottawa: Energy, Mines and Resources, 1984.
Canada, National Energy Board. *Annual Report*. Ottawa: National Energy Board, 1967.

– *Energy Briefing Note: Canadian Energy Overview 2012.* Ottawa: National
    Energy Board, 2013.
Canada, Natural Resources. *Energy Use Data Handbook, 1990–2008.* Cat.
    no. M141-11/2008E PDF. Ottawa: Statistics Canada, 2011.
– "Government of Canada Invests in Wind Energy in Southern Ontario."
    News release, 28 June 2011. http://news.gc.ca/web/article-eng.
    do?nid=608529.
Canada, Royal Commission on Canada's Economic Prospects. *Hearings Held
    at St John's, 18 October 1955.* Ottawa: Queen's Printer.
– *Hearings Held at Vancouver, 30 November 1955.* Ottawa: Queen's Printer.
Canada, Royal Commission on Coal. *Report of the Canada Royal
    Commission on Coal, 1946.* Ottawa: Edward Cloutier, 1947.
Canada, Statistics Canada. *Electric Power Generation, Transmission and
    Distribution.* Cat. no. 57-202-X. Ottawa: Statistics Canada, 2007.
– "Energy Supply and Demand, by Fuel Type," 2009. http://www.statcan.
    gc.ca/tables-tableaux/sum-som/l01/cst01/prim72-eng.htm.
– *Human Activity and the Environment: Annual Statistics 2004.* Cat.
    no. 16-201-XIE. Ottawa: Statistics Canada, 2004. http://publications.gc.ca/
    Collection-R/Statcan/16-201-XIE/0000416-201-XIE.pdf.
– "Study: Lawns and Gardens and the Environment." 2007. http://www.
    statcan.gc.ca/daily-quotidien/070926/dq070926b-eng.htm.
– "Table 1: Lawn Mower Use, Canada and Provinces 2007." http://www.
    statcan.gc.ca/pub/16-002-x/2010001/article/lawnmowers-tondeuses/tbl/
    tbl001-eng.htm.
– Series Q85–91, *Electrical Generation by Utilities and Industrial
    Establishments, by Type of Prime Mover, 1919 to 1976.* http://www.statcan.
    gc.ca/access_acces/archive.action?l=eng&loc=Q85_91-eng.csv.
– Series Q149–159, *Canadian Refinery Shipments of Petroleum Products,
    1949 to 1976 (Thousands of Barrels).* http://www.statcan.gc.ca/access_
    acces/archive.action?l=eng&loc=Q149_159-eng.csv.
Canadian Century Research Infrastructure / Infrastructure de recherche sur
    le Canada au 20ᵉ siècle (CCRI/IRCS). "Census of Canada, Contextual Data,
    Geography: Digitized Published Tables." https://ccri.library.ualberta.ca/
    endatabase/geography/digitizedpublictables/index.html.
Canadian Coalition for Nuclear Responsibility (CCNR). "Uranium: A
    Discussion Guide." March 2007. http://www.ccnr.org/nfb_uranium_0.html.
Canadian Environmental Assessment Agency. *Decommissioning of Uranium
    Mine Tailings Management Areas in the Elliot Lake Area.* Report of the
    Environmental Assessment Panel. Ottawa: Minister of Supply and Services
    Canada, 1996.

Canadian Federation of Boards of Trade and Municipalities. *The Georgian Bay Canal and Nova Scotia Coal.* Canada: s.n., 1909.

Canadian Geographic. "Canada's Energy Production and Transmission." https://energyiq.canadiangeographic.ca/main/ energy_map#3&-203+91&94+53&1&0&1.

CanWEA. "Big Growth in Small Wind: CanWEA." *APPrO,* 1 August 2013: http://magazine.appro.org/index.php?option=com_content&task=view&id= 1231&Itemid=44.

– "Small Wind Energy." http://canwea.ca/wind-facts/small-wind-energy/.

– "Wind Vision 2025: Powering Canada's Future." 2013. http://canwea.ca/ pdf/windvision/Windvision_summary_e.pdf.

Carlson, G.L. *Studies in Horse Breeding.* Winnipeg: s.n., 1920.

Carlson, Hans. *Home Is the Hunter: The James Bay Cree and Their Land.* Vancouver: UBC Press, 2004.

Caron, Carl. "De Manic-Outardes à la Baie James: la gestion des choix techniques à Hydro-Québec." In *Grands Projets et Innovations Technologiques au Canada,* ed. Philippe Faucher, 95–124. Montreal: Presses de l'Université de Montréal, 2000.

Castaneda, Christopher. *Invisible Fuel: Manufactured and Natural Gas in America, 1800–2000.* New York: Twayne Publishers, 1999.

– "Natural Gas." In *Concise Encyclopedia of History of Energy,* ed. Cutler J. Cleveland, 163–74. San Diego, CA: Elsevier, 2009.

CBC News Toronto. "Agent Orange 'Widely Used' in Ontario." 28 February 2011.

– "Darlington Nuclear Plant Gets Renewed Licence." 23 December 2015. http://www.cbc.ca/news/canada/toronto/ darlington-nuclear-plant-renewed-licence-1.3379378.

Centre for Media and Democracy. "Patrick Moore: Media Coverage That Doesn't Disclose Moore's Nuclear Consultancy Work." 27 February 2012. http://www.sourcewatch.org/index.php/ Patrick_Moore:_Media_coverage_that_doesn%27t_disclose_Moore%27s_ nuclear_consultancy_work.

Chamberlin, J. Edward. *Horse: How the Horse Has Shaped Civilizations.* New York: Blue Bridge, 2006.

Chandler, Alfred, "Anthracite Coal and the Beginnings of the Industrial Revolution in the United States." *Business History Review* 46, no. 2 (1972): 141–81.

Chatsko, Paul. *Developing Alberta's Oil Sands: From Karl Clark to Kyoto.* Calgary: University of Calgary Press, 2004.

Christian, David. *Maps of Time: An Introduction to Big History.* Berkeley, CA: University of California Press, 2004.

Christopherson, Robert W. *Geosystems: An Introduction to Physical Geography,* 5th ed. Upper Saddle River, NJ: Pearson Education, 2005.

Churchill, Jason. "Pragmatic Federalism: The Politics behind the 1969 Churchill Falls Contract." *Newfoundland Studies* 15, no. 2 (1999): 215–46.

Clark, Andrew Hill. *Three Centuries and the Island: A Historical Geography of Settlement and Agriculture in Prince Edward Island, Canada.* Toronto: University of Toronto Press, 1959.

Clark, Claudia. *Radium Girls: Women and Industrial Health Reform, 1910–1935.* Chapel Hill: University of North Carolina Press, 1997.

Clark, D.C. Letter to the editor. *Gas Industry: Heat, Light, Power* 13, no. 1 (1913): 77.

Clark, Nancy, Cato Coleman, Kerri Figure, Tom Mailhot, and John Zeigler. "Food for Trans-Atlantic Rowers: A Menu Planning Model and Case Study." *International Journal of Sport Nutrition and Exercise Metabolism* 13 (2003): 227–42.

Clarke, John. *The Ordinary People of Essex: Environment, Culture, and Economy on the Frontier of Upper Canada.* Montreal and Kingston: McGill-Queen's University Press, 2010.

Clendinning, Anne. *Demons of Domesticity: Women and the English Gas Industry, 1889–1939.* Hampshire: Ashgate Publishing, 2004.

Coates, Ken. "The Power to Transform: The Kemano Power Project and the Debate about the Future of Northern British Columbia." *Journal of Northern Studies* 1, nos 1–2 (2007): 31–50.

Cobban, Timothy W. *Cities of Oil: Municipalities and Petroleum Manufacturing in Southern Ontario, 1860–1960.* Toronto: University of Toronto Press, 2013.

Cole, Arthur H. "The Mystery of Fuel Wood Marketing in the United States." *Business History Review* 44, no. 3 (1970): 339–59.

Collins, E.J.T. "Dietary Change and Cereal Consumption in Britain in the Nineteenth Century." *Agricultural History Review* 23, no. 2 (1975): 97–115.

Colpitts, George. "Environment, Pemmican, and Trade in a Northern Great Plains Bioregion." *Prairie Forum* 37 (Fall 2012): 83–102.

– *Game in the Garden: A Human History of Wildlife in Western Canada.* Vancouver: UBC Press, 2000.

– "Moose-Nose and Buffalo Hump: The Amerindian-European Food Exchange in the British North American Fur Trade to 1840." In *Dining on Turtles: Food Feasts and Drinking in History,* ed. Diane Kirkby and Tanja Luckins, 64–81. Houndmills, UK: Palgrave Macmillan, 2007.

– "Provisioning the HBC: Market Economies in the British Buffalo Commons in the Early Nineteenth Century." *Western Historical Quarterly* 43, no. 2 (Summer 2012): 179–203.

Combe, F.A., and J.T. Farmer. "The Relation of Electricity to Coal." In *Symposium on Fuel and Coal, McGill University, 1931, Proceedings*, 288–329. Montreal: s.n., 1932.

Cooper, Laurie Armstrong, and Douglas Clay. *History of Logging and River Driving in Fundy National Park: Implications for Ecological Integrity of Aquatic Ecosystems*. Alma, NB: [Parks Canada Atlantic Region], 1997.

Cooper, Thos. *Some Information concerning Gas Lights*. Philadelphia: John Conrad, 1816.

Coronil, Fernando. *The Magical State: Nature, Money and Modernity in Venezuela*. Chicago: University of Chicago Press, 1997.

Courier Mail. "Cold Wave in America," 12 February 1934.

Courteau, Dick. "Horse Power: A Practical Suggestion That Would Transform the Way We Live." *Orion Magazine* (September/October 2007). https://orionmagazine.org/article/horse-power/.

Courville, Serge, Jean-Claude Robert, and Normand Séguin. *Atlas historique du Québec: le pays laurentien au XIXᵉ siècle. Les morphologies de base*. Sainte-Foy: Les Presses de l'Université Laval, 1995.

Craig, Beatrice. *Backwoods Consumers and Homespun Capitalists: The Rise of Market Culture in Eastern Canada*. Toronto: University of Toronto Press, 2009.

Creighton, Donald. *Empire of the St Lawrence*, Toronto: Macmillan, 1956.

Crisman, Kevin J., and Arthur B. Cohn. *When Horses Walked on Water: Horse-Powered Ferries in Nineteenth-Century America*. Washington, DC: Smithsonian Institution Press, 1998.

Cronon, William. *Changes in the Land: Indians, Colonists and the Ecology of New England*. New York: Hill and Wang, 1983.

Crosby, Alfred W. *Children of the Sun: A History of Humanity's Unappeasable Appetite for Energy*. New York: W.W. Norton, 2006.

– *Ecological Imperialism: The Biological Expansion of Europe, 900–1900*. *Cambridge*: Cambridge University Press, 2004.

Cruikshank, Ken, and Nancy Bouchier. "Blighted Areas and Obnoxious Industries: Constructing Environmental Inequality on an Industrial Waterfront, Hamilton, Ontario, 1890–1960." *Environmental History* 9, no. 3 (2004): n.p.

Daintith, John, ed. *Dictionary of Physical Sciences*. London: Pan, 1976.

Darmstadter, Joel, Joy Dunkerley, and Jack Alterman. *How Industrial Societies Use Energy: A Comparative Analysis*. Baltimore, MD: Johns Hopkins University Press, 1977.

David Suzuki Foundation, Pembina Institute. *Is Natural Gas the Climate Change Solution for Canada?* 2011. http://www.davidsuzuki.org/publications/reports/2011/is-natural-gas-a-climate-change-solution-for-canada/.

Davies, K.G., ed. *Peter Skene Ogden's Snake Country Journal, 1826–27.* London: Hudson's Bay Record Society, 1961.

Davis, Donald. "Dependent Motorization: Canada and the Automobile to the 1930s." *Journal of Canadian Studies* 21 (1986): 106–32.

De Blois, Solange. "Les Moulins de Terrebonne (1720–1755) ou les hauts et les bas d'une entreprise seigneuriale." *Revue d'histoire de l'Amérique Française* 51, no. 1 (1997): 39–70.

Dean, Joanna. "Species at Risk: C. tetani, the Horse and the Human." In *Animal Metropolis: Histories of Human-Animal Relations in Urban Canada,* ed. Christabelle Sethna, Darcy Ingram, and Joanna Dean. Calgary: University of Calgary Press, in press.

Dechêne, Louise. *Habitants and Merchants in Seventeenth-Century Montreal.* Montreal and Kingston: McGill-Queen's University Press, 1992.

– *Habitants et marchands de Montréal au XVIIᵉ Siècle.* Paris: Plon, 1974.

– *Le Partage des subsistances au Canada sous le Régime français.* Montréal: Boréal, 1994.

– *Le peuple, l'État et la guerre au Canada sous le Régime français.* Montreal: Boréal, 2008.

– "Observations sur l'agriculture du Bas-Canada au début du XIXᵉ siècle." In *Évolution et éclatement du monde rural: France-Québec XVIIᵉ-XVIIIᵉ Siècles,* ed. Jean-Pierre Wallot and Joseph Goy, 189–202. Montreal: Presses de l'Université de Montréal, 1986.

Deloges, Yvon, and Alain Gelly. *The Lachine Canal: Riding the Waves of Industrial and Urban Development, 1860–1950.* Sillery, QC: Septentrion, 2002.

de Mille, George. *Oil in Canada West: The Early Years.* Calgary: Northwest Printing, 1970.

de Molina, Manuel González, and Victor M. Toledo. *The Social Metabolism: A Socio-Ecological Theory of Historical Change.* London: Springer, 2014.

Denison Environmental Services. "Closed Mines." http://www.denisonenvironmental.com/Closed%20Mines.html.

Derry, Margaret E. *Horses in Society: A Story of Animal Breeding and Marketing Culture, 1800–1920.* Toronto: University of Toronto Press, 2006.

Desbiens, Caroline. *Power from the North: Territory, Identity, and the Culture of Hydroelectricity in Quebec.* Vancouver: UBC Press, 2013.

Desjardins, Pauline. "Navigation and Waterpower: Adaptation and Technology on Canadian Canals." In "Waterpower, the Lachine Canal, and the Industrial

Development of Montreal." Special issue, *IA: The Journal of the Society for Industrial Archaeology* 29, no. 1 (2003): 21–47.

Dessureault, Christian. "L'inventaire après décès et l'agriculture bas-canadienne." *Material History Bulletin* 17 (1983): 127–38.

Dessureault, Christian, with John A. Dickinson. "Farm Implements and Husbandry in Colonial Quebec, 1740–1840." In *New England/New France, 1600–1850*, ed. Peter Benes, 100–21. Boston: Boston University Press, 1992.

Dewar, Elaine. "Nuclear Reaction." *Canadian Geographic* (May/June 2005): 69–84.

Dewhurst, J. Frederick and Associates. *America's Needs and Resources: A New Survey*. New York: Twentieth Century Fund, 1955.

Directory of Gas Light Companies and Their Officers in the United States, Canada, South America and Cuba. N.p.: W. Goodwin, 1878.

Dobak, William A. "Killing the Canadian Buffalo, 1821–1881." *Western Historical Quarterly* 27 (Spring 1996): 39–48.

Dodd, Susan. *The Ocean Ranger: Remaking the Promise of Oil*. Halifax: Fernwood, 2012.

Doern, G.B., A. Dorman, and R.W. Morrison. "Canadian Nuclear Energy Policy." *Energy Studies Review* 11, no. 2 (2003). https://energystudiesreview.ca/esr/article/view/448.

– *Canadian Nuclear Energy Policy: Changing Ideas, Institutions, and Interests*. Toronto: University of Toronto Press, 2001.

Dolphin, Frank, and John Dolphin. *Country Power: The Electrical Revolution in Rural Alberta*. Edmonton: Plains Publishing, 1993.

Dominion Coal Company. *Memorandum Respecting Reciprocity in Coal with the United States of America*. Halifax: s.n., 1910.

Dorion, Marie-Josée. "L'électrification du monde rural québécois." *Revue d'histoire de l'Amérique Française* 54 (2000): 3–37.

Drouard, Alain. "Reforming Diet at the End of the Nineteenth Century in Europe." In *Food and the City in Europe Since 1800*, ed. Peter Lumnel Atkins and Derek J. Peter Oddy, 215–25. Abingdon, UK: Ashgate Publishing Group, 2008.

Drummond, Ian M. *Progress without Planning: The Economic History of Ontario from Confederation to the Second World War*. Toronto: University of Toronto Press, 1987.

Duff, J. Clarence. *Toronto Then and Now*. Toronto: Fitzhenry and Whiteside, 1984.

Dufous, Jules. "Le Projet Grande-Baleine et L'Avenir des Peuples Autochtones au Québec." *Cahiers de Géographie du Québec* 40, no. 110 (1996): 233–52.

Duncan, Colin A.M. "Adam Smith's Green Vision and the Future of Global Socialism." In *New Socialisms: Futures beyond Globalization*, ed. Robert Albritton, Shannon Bell, John R. Bell, and Richard Westra, 90–104. London: Routledge, 2004.

Durant, Darrin. "Burying Globally, Acting Locally: Control and Co-option in Nuclear Waste Management." *Science and Public Policy,* August 2007, 515–28.

Dusyk, Nichole. "The Transformative Potential of Participatory Politics: Energy Planning and Emergent Sustainability in British Columbia, Canada." PhD diss., University of British Columbia, 2013.

Dwyer, Alan D. "Wood and Coal: A Change of Fuel," *History Today* 26 (September 1976): 598–607.

Easterbrook, W.T., and M.H. Watkins, eds. *Approaches to Canadian Economic History*. Ottawa: Carleton University Press, 1986.

Edgerton, David. *The Shock of the Old: Technology and Global History since 1900*. London: Profile Books, 2008.

Edwards, Gordon. "Uranium: The Deadliest Metal." Canadian Coalition for Nuclear Responsibility, March 2007. http://www.ccnr.org/uranium_deadliest.html.

Eggleston, Wilfrid. *National Research in Canada: The NRC 1916–66*. Toronto: Clarke Irwin, 1978.

Ellis Keystone Industrial Works. *The Ellis Keystone Industrial Works Catalogue*. Pottstown, PA: Ellis Keystone, ca 1886.

Elofson, W.M. *Cowboys, Gentlemen and Cattle Thieves: Ranching on the Western Frontier*. Montreal and Kingston: McGill-Queen's University Press, 2000.

EnergyRealities.org. "Per Capita Energy Consumption." http://www.energy realities.org/chapter/meeting-our-needs/item/per-capita-energy-consumption/erp327B7C729A3B31D2B.

Engels, Mary-Louise. *Rosalie Bertell: Scientist, Eco-Feminist, Visionary*. Toronto: Women's, 2005.

Ens, Gerhard J. "Fatal Quarrels and Fur Trade Rivalries: A Year of Living Dangerously on the North Saskatchewan, 1806–07." In *Alberta Formed: Alberta Transformed*, ed. Michael Payne, Donald Wetherell, and Catherine Cavanaugh, 1:133–159. Edmonton: University of Alberta Press, 2005.

Environment and Climate Change Canada. "Frequently Asked Questions." http://www.ec.gc.ca/eau-water/default.asp?lang=En&n=1C100657-1.

Environment Canada. "Dams and Diversions." 2010. https://www.ec.gc.ca/eau-water/default.asp?lang=En&n=9D404A01-1.

Envirotech. "Are Animals Technology?" 2001. http://www.envirotechweb.org/wp-content/uploads/2007/05/animaltech.pdf.

Epperson, Bruce. "Fighting Traffic." *Technology and Culture* 50, no. 4 (2009): 982–6.

Epstein, Eric. "Top Hydroelectric Producing Nations." *Rock the Capital*, 30 January 2012. http://www.rockthecapital.com/01/30/ top-hydroelectric-producing-nations/.

Estrin, David, and John Swaigen. *Environment on Trial: A Guide to Ontario Environmental Law and Policy,* 3rd ed. Toronto: Emond Montgomery Publications, 1993.

European Wind Energy Association. *Wind Energy in Europe.* Brussels: European Wind Energy Association, 1991.

Evans, Francis T. "Roads, Railways and Canals: Technical Choices in 19th-Century Britain." *Technology and Culture* 22, no. 1 (1981): 1–34.

Evans, William. *A Treatise on the Theory and Practice of Agriculture.* Montreal, 1835.

Evenden, Matthew. *Allied Power: Mobilizing Hydroelectricity during Canada's Second World War.* Toronto: University of Toronto Press, 2015.

– "Aluminum, Commodity Chains, and the Environmental History of the Second World War." *Environmental History* 16, no. 1 (January 2011): 69–93.

– *Fish versus Power: An Environmental History of the Fraser River.* New York: Cambridge University Press, 2004.

– "Mobilizing Rivers: Hydro-electricity, the State and the Second World War in Canada." *Annals of the Association of American Geographers* 99, no. 5 (2009): 845–55.

Fay, C.R., and H.A. Innis. "The Maritime Provinces." In *The Cambridge History of the British Empire*, ed. J. Holland Rose, A.P. Newton, and E.A. Benian, 6:657–71. New York: Macmillan, 1930.

Feehan, James. "Smallwood, Churchill Falls, and the Power Corridor through Quebec." *Acadiensis* 40, no. 2 (2011): 112–27.

Fiege, Mark. *The Republic of Nature: An Environmental History of the United States.* Seattle: University of Washington Press, 2012.

Filey, Mike. *A Toronto Album: Glimpses of a City That Was.* Hamilton, ON, Dundurn, 2001.

Fleming, Keith. "Hydroelectricity." In *Oxford Companion to Canadian History*, ed. Gerald Hallowell, 300–1. Oxford: Oxford University Press, 2004.

– *Power at Cost: Ontario Hydro and Rural Electrification, 1911–58.* Montreal and Kingston: McGill-Queen's University Press, 1992.

Fleming, R.B. "The Trolley Takes Command: 1892–1894." *Urban History Review* 19, no. 3 (1991): 218–25.

Flink, James. *The Automobile Age.* Cambridge, MA: MIT Press, 1988.

Foley, Gerald. *The Energy Question*. 2nd ed. Harmondsworth, UK: Penguin Books, 1981.

"Forestry Commission: Summary of Preliminary Report of Mr J.H. Morgan." *Sessional Papers*, 48 Vic (1885), vol. 7, no. 13, p. 6.

Forkey, Neil S. "Damning the Dam: Ecology and Community in Ops Township, Upper Canada." *Canadian Historical Review* 79, no. 1 (March 1998): 68–99.

Forsythe, Mark, and Greg Dickson. *The Trail of 1858: British Columbia's Gold Rush Past*. Madeira Park, BC: Harbour Publishing, 2007.

Fossum, John Erik. *Oil, the State and Federalism: The Rise and Demise of Petro-Canada as a Statist Impulse*. Toronto: University of Toronto Press, 1997.

Fouquet, Roger. *Heat, Power and Light: Revolutions in Energy Services*. Cheltenham, UK: Edward Elgar Publishing, 2008.

– "The Slow Search for Solutions: Lessons from Historical Energy Transitions by Sector and Service." *Energy Policy* 38 (2010): 6586–96.

Fox, William, Bill Brooks, and Janice Tyrwhitt. *The Mill*. Toronto: McClelland and Stewart, 1976.

Frank, David. "Class Conflict in the Coal Industry Cape Breton 1922." In *Essays in Canadian Working Class History*, ed. Gregory S. Kealey and Peter Warrian, 173–177. Toronto: McClelland and Stewart, 1976.

Fraser, Derek. *A History of Modern Leeds*. Manchester: Manchester University Press, 1980.

Freeman, Neil. *The Politics of Power: Ontario Hydro and Its Government, 1906–1995*. Toronto: University of Toronto Press, 1996.

Froschauer, Karl. *White Gold: Hydroelectric Power in Canada*. Vancouver: UBC Press, 1999.

Frost, Robert L. Review of *L'Électricité et ses consommateurs. Actes du Quatrième colloque de l'Association pour l'histoire de l'électricité en France. Journal of Economic History* 50, no. 1 (March 1990): 200–2.

Furness, Zack. *One Less Car: Bicycling and the Politics of Automobility*. Philadelphia: Temple University Press, 2010.

Gad, Gunter. "Location Patterns of Manufacturing: Toronto in the Early 1880s, Urban History." *Urban History Review / Revue d'histoire urbaine* 22, no. 2 (June 1994): 111–38.

Gazette (Montreal). "St Lawrence Clean Up Starts after Barge Spills Bunker Oil," 25 June 1976.

Geoghegan, John J. "Designers Set Sail, Turning to Wind Power to Help Power Cargo Ships." *New York Times*, 27 August 2012.

Gérin-Lajoie, Antoine. *Jean Rivard*. Translated by Vida Bruce. Toronto: McClelland and Stewart, 1977.

Gerriets, Marilyn. "The Impact of the General Mining Association on the Nova Scotia Coal Industry, 1826–1850." *Acadiensis* 21, no. 1 (Autumn 1991): 54–84.

Giangrande, Carol. *The Nuclear North: The People, the Regions and the Arms Race.* Toronto: House of Anansi, 1983.

Gibson, Robert B. "From Wreck Cove to Voisey Bay: The Evolution of Federal Environmental Assessment in Canada." *Impact Assessment and Project Appraisal* 20, no. 3 (2002): 151–9.

Gidney, Norman. "From Coal to Forest Products: The Changing Resource Base of Nanaimo, B.C." *Urban History Review* 7, no. 1 (June 1978): 18–47.

Gillis, R. Peter. "Rivers of Sawdust: The Battle over Industrial Pollution in Canada, 1865–1903." In *Canadian Environmental History: Essential Readings,* ed. David Freeland Duke, 265–83. Toronto: Canadian Scholars' Press, 2006. Originally published in the *Journal of Canadian Studies* 21, no. 1 (1986).

Global Wind Energy Council. "Global Wind Statistics 2012." http://www.gwec.net/wp-content/uploads/2013/02/GWEC-PRstats-2012_english.pdf.

Globe. "Heat with an Oil Right," 26 July 1926.

– "Opening of the Street Railway: The Dejeuner, Concert and Ball," 11 September 1861.

– "The Street Railway," 15 April 1880.

Globe and Mail. "How to Care for Your Oil-Burning Furnace," 7 October 1954.

– "Open-Air Horse Parade," 2 July 1904.

Glynn, Joseph. *Rudimentary Treatise on the Construction of Cranes and Machinery for Raising Heavy Bodies, for the Erection of Buildings, and for Hoisting Goods.* London, 1854.

Goad, Charles E. *Fire Insurance Plan of Ailsa Craig, Ont.* Toronto: Chas. E. Goad, 1890 (1903).

– *Fire Insurance Plan of Crediton, Ont.* Toronto: Chas. E. Goad, 1896.

– *Fire Insurance Plan of London, Ont.* 1881. Toronto: Chas. E. Goad, 1888.

– *Fire Insurance Plan of Stratford, Ont.* 1908. Toronto: Chas. E. Goad, 1925.

Goktepe, A. Salim. "Energy Systems in Sport." In *Amputee Sports for Victims of Terrorism,* ed. Centre of Excellence Defense against Terrorism, 24–31. Amsterdam: IOS, 2007.

Goldring, Philip. "MacKintosh, William." In *Dictionary of Canadian Biography,* 7:567. Toronto: University of Toronto Press, 1988.

Gooday, Graeme. *Domesticating Electricity: Technology, Uncertainty and Gender, 1880–1914.* London: Pickering and Chatto, 2008.

Gough, Barry M., ed. *The Journal of Alexander Henry the Younger, 1799–1814*. 2 vols. Toronto: Champlain Society, 1988.

Gow, Sandy. *Roughnecks, Rock Bits and Rigs: The Evolution of Oil Well Drilling Technology in Alberta, 1883–1970*. Calgary: University of Calgary Press, 2005.

Graham, Fiona. "Sailing into the Future of Global Trade." BBC News, 28 December 2012. http://www.bbc.co.uk/news/business-20792058.

Grahame, Gordon Hill. *Short Days Ago*. Toronto: MacMillan, 1972.

Grandy, Leah. "The Era of the Urban Horse: Saint John, New Brunswick, 1871–1901." MA thesis, University of New Brunswick, 2004.

Grant, Hugh. "Canada's Petroleum Industry: An Economic History, 1900–1960." PhD diss., University of Toronto, 1986.

Grant, Shelagh. *Polar Imperative: A History of Arctic Sovereignty in North America*. Vancouver: Douglas and McIntyre, 2010.

Greene, Ann Norton. *Horses at Work: Harnessing Power in Industrial America*. Cambridge, MA: Harvard University Press, 2008.

Greer, Allan. *Peasant, Lord and Merchant: Rural Society in Three Quebec Parishes 1740–1840*. Toronto: University of Toronto Press, 1985.

Greig, A.R. *Wall Insulation*. Bulletin 1. Saskatoon: College of Engineering, University of Saskatchewan, 1922.

Griffith, J.W. *The Uranium Industry: Its History, Technology and Prospects*. Ottawa: Mineral Resources Division, Department of Energy, Mines and Resources, 1967.

Griffiths, Naomi E.S. *From Migrant to Acadian: A North American Border People, 1604–1756*. Montreal and Kingston: McGill-Queen's University Press, 2004.

Grigg, David. "The Nutritional Transition in Western Europe." *Journal of Historical Geography* 21, no. 3 (1995): 247–61.

Guillet, Edwin. *The Story of Canadian Roads*. Toronto: University of Toronto Press, 1966.

Hall, P.M. "Disposal of Manure." Paper read before the Section of Public Officials, American Public Health Association, September 1913.

Hamelin, Jean, and Yves Roby. *Histoire Économique du Québec*. Montreal: Fides, 1966.

Hamilton, Henry E. *Incidents and Events in the Life of Gurdon Saltonstall Hubbard 1802–1886*. Chicago: Graff, Newberry Library, 1997.

Hamilton, Shane. *Trucking Country: The Road to America's Wal-Mart Economy*. Princeton, NJ: Princeton University Press, 2008.

Hammond, Lorne. "Marketing Wildlife: The Hudson's Bay Company and the Pacific Northwest, 1821–1849." In *Canadian Environmental History:*

Essential Readings, ed. David Freland Duke, 203–22. Toronto: Canadian Scholars', 2006.

Hanlan, Jas P. "Special Campaigns." *Gas Industry: Heat, Light, Power* 13, no. 1 (1913): 68–79.

Hardy, René, and Normand Séguin. *Forêt et société en Mauricie*. Montreal: Boréal Express, 1984.

Hargreaves, Mark, and Lawrence Spriet, eds. *Exercise Metabolism*, 2nd ed. Champaign, IL: Human Kinetics, 2006.

Harland, John. *Seamanship in the Age of Sail*. London: Conway Maritime, 1984.

Harley, C. Knick. "Aspects of the Economics of Shipping, 1850–1913." In *Change and Adaptation in Maritime History: The North Atlantic Fleets in the Nineteenth Century*, ed. Lewis R. Fischer and Gerald E. Panting, 169–86. St John's: Maritime History Group, 1985.

Harper, Douglas. *Changing Works: Visions of a Lost Agriculture*. Chicago: University of Chicago Press, 2001.

Harries, Kate. "Nuclear Reaction: Accusations of Nuclear Fallout Divide a Small Ontario Town." *Walrus*, September 2012.

Harris, Richard. *Creeping Conformity: How Canada Became Suburban, 1900–1960*. Toronto: University of Toronto Press, 2004.

Harris, Richard Colebrook. *The Reluctant Land: Society, Space and Environment in Canada before Confederation*. Vancouver: UBC Press, 2008.

– *The Seigneurial System in Early Canada: A Geographical Study*. Montreal and Kingston: McGill-Queen's University Press, 1984.

Harvey, Mark. "Sound Politics: Wilderness, Recreation, and Motors in the Boundary Waters, 1945–64." *Minnesota History* 58, no. 3 (Fall 2002): 130–45.

Hatheway, Allen W. *Remediation of Former Manufactured Gas Plants and Other Coal-Tar Sites*. Boca Raton, FL: CRC, ca 2012.

Hausman, William, J. Peter Hertner, and Mira Wilkins. *Global Electrification: Multinational Enterprise and International Finance in the History of Light and Power, 1878–2007*. Cambridge: Cambridge University Press, 2008.

Hawley, John A., and Will G. Hopkins. "Aerobic Glycolytic and Aerobic Lipolytic Power Systems." *Sports Medicine* 19, no. 4 (1995): 240–50.

Hayes, Derek. *Historical Atlas of Toronto*. Vancouver: Douglas and McIntyre, 2008.

Hearn, Chester G. *Tracks in the Sea: Matthew Fontaine Maury and the Mapping of the Oceans*. Camden, ME: International Marine, 2002.

Hecht, Gabrielle. *The Radiance of France: Nuclear Power and National Identity after World War II*. Cambridge, MA: MIT Press, 1998.

Henry, Alexander the Elder. *Travels and Adventures: In Canada and the Indian Territories*. Edmonton: M.G. Hurtig, 1969.

Henry, W.A. *Feeds and Feeding: A Hand-Book for the Student and Stockman*, 11th ed. Madison, WI: printed by author, 1911.

Heriot, George. *Travels through the Canadas*. London: Richard Phillips, 1805.

Heron, Craig. *Working in Steel: The Early Years in Canada, 1883–1935*. Toronto: McClelland and Stewart, 1988.

Hersey, John. *Hiroshima*. New York: Alfred Knopf, 1946.

Hills, R.L., and A.J. Pacey. "The Measurement of Power in Early Steam-Driven Textile Mills." *Technology and Culture* 13, no. 1 (1972): 25–43.

Hilton, George. "Transport Technology and the Urban Pattern." *Journal of Contemporary History* 4, no. 3 (July 1969): 123–35.

Hirt, Paul W. *The Wired Northwest: The History of Electric Power, 1870s–1970s*. Lawrence: University of Kansas Press, 2012.

Hoberg, George. "Sleeping with the Elephant: The American Influence on Canadian Environmental Assessment." *Journal of Public Policy* 11, no. 1 (1991): 107–31

Holley, I.B., Jr. "Blacktop: How Asphalt Paving Came to Urban United States." *Technology and Culture* 44, no. 4 (2003): 703–33.

Holman, Eugene. "Freedom and Energy Go Together," 15 April 1951. *Empire Club of Canada Addresses*. Toronto: Empire Club of Canada. http:// speeches.empireclub.org/62016/data?n=1.

Hornsby, Stephen J. *Surveyors of Empire: Samuel Holland. K/W/F/ Des Barres, and the Making of the Atlantic Neptune*. Montreal and Kingston: McGill-Queen's University Press, 2011.

"Horse and the Bicycle, The." *Scientific American* 73, no. 3 (1895): 43.

Howison, John. *Sketches of Upper Canada, Domestic, Local and Characteristic*. Edinburgh: Oliver and Boyd, 1821.

Hribal, Jason. "Animals Are Part of the Working Class Reviewed." *Borderlands E-Journal* 11, no. 2 (2012). http://www.borderlands.net.au/vol11no2_2012/ hribal_animals.htm.

Hughes, Thomas P. *Networks of Power: Electrification in Western Society, 1880–1930*. Baltimore, MD: Johns Hopkins University Press, 1983.

– "Technological Momentum." In *Does Technology Drive History? The Dilemma of Technological Determinism*, ed. Merritt Roe Smith and Leo Marx, 101–13. Cambridge, MA: MIT Press, 1994.

Hunter, Louis C. *A History of Industrial Power in the United States, 1780–1930*. Vol. 1, *Waterpower*. Charlottesville: University Press of Virginia, 1979.

– *A History of Industrial Power in the United States, 1780–1930*. Vol. 2, *Steam Power*. Charlottesville: University Press of Virginia, 1985.

Hurley, Andrew. "Creating Ecological Wastelands: Oil Pollution in New York City, 1870–1900." *Journal of Urban History* 20, no. 3 (May 1994): 340–64.

Hutcheon, N.B., and G.O. Handegord. "Evolution of the Insulated Wood-Frame Wall in Canada." Division of Building Research, paper no. 946. Ottawa: National Research Council Canada, 1980.

Iacovetta, Franca, Valerie J. Korinek, and Marlene Epp, eds. *Edible Histories, Cultural Politics: Towards a Canadian Food History*. Toronto: University of Toronto Press, 2012.

Independent Electricity System Operator (IESO). "Ontario Demand Peaks," n.d. http://www.ieso.ca/Pages/Participate/Settlements/Ontario-Demand-Peaks-Archive.aspx.

IndexMundi. "Canada Coal Consumption by Year." http://www.indexmundi.com/energy.aspx?country=ca&product=coal&graph=consumption.

– "Canadian Consumption of Natural Gas, 1980–2012." http://www.indexmundi.com/energy.aspx?country=ca&product=gas&graph=consumption.

– "Canadian Production of Natural Gas, 1980–2012." http://www.indexmundi.com/energy.aspx?country=ca&product=gas&graph=production.

Ingram, Darcy. "'We Are No Longer Freaks': The Cyclists' Rights Movement in Montreal." *Sport History Review 43 (2012): 18–42*.

Innis, Harold Adams. *The Fur Trade in Canada*. 1930. Toronto: University of Toronto Press, 1999.

– *Peter Pond: Fur Trade and Adventurer*. Toronto: Irwin and Gordon, 1930.

Integrated Power Technology Corporation. "Sail Manufacturers." http://www.intpowertechcorp.com/pub_sail.htm.

Intera Technologies. *Inventory of Coal Gasification Plants in Ontario*. Vol. 1. Prepared for the Ontario Ministry of the Environment, Waste Management Branch, April 1987. http://guelph.ca_wp_uploads_part-a-inventory-of-Coal-Gasification.pdf.

Jackson, John N. *The Welland Canals and Their Communities: Engineering, Industrial, and Urban Transformation* Toronto: University of Toronto Press, 1997.

Jackson, John N., and Fred A. Addis. *The Welland Canals: A Comprehensive Guide*. St Catharines, ON: Lincoln Graphics, 1982.

Jehlen, Myra. *American Incarnation: The Individual, the Nation, and the Continent*. Cambridge, MA: Harvard University Press, 1986.

Jellison, Katherine. *Entitled to Power: Farm Women and Technology, 1913–63*. Chapel Hill: University of North Carolina Press, 1993.

Jennings, John. *The Canoe: A Living Tradition*. Toronto: Firefly Books, 2002.

Johnston, James F.W. *Notes on North America, Agricultural, Economical, and Social*. Vol. 2. Edinburgh: William Blackwood and Sons, 1851.

Jones, Christopher F. "A Landscape of Energy Abundance: Anthracite Coal Canals and the Roots of American Fossil Fuel Dependence, 1820–1860." *Environmental History* 15 (July 2010): 449–84.

– "The Carbon-Consuming Home: Residential Markets and Energy Transitions." *Enterprise & Society* 12, no. 4 (December 2011): 790–823.

– *Routes of Power: Energy and Modern America.* Cambridge, MA: Harvard University Press, 2014.

Jones, Robert Leslie. "French-Canadian Agriculture in the St Lawrence Valley, 1817–1850." In *Approaches to Canadian Economic History*, ed. W.T. Easterbrook and M.H. Watkins, 110–26. Toronto: McClelland and Stewart, 1967.

– *History of Agriculture in Ontario, 1613–1880.* Toronto: University of Toronto Press, 1946.

– "The Old French-Canadian Horse: Its History in Canada and the United States." *Canadian Historical Review* 28, no. 2 (June 1947): 125–54.

Jones, Van. *The Green Collar Economy.* New York: Harper Collins, 2008.

Jørgensen, Dolly, Finn Arne Jørgensen, and Sara B. Pritchard, eds. *New Natures: Joining Environmental History with Science and Technology Studies.* Pittsburgh: University of Pittsburgh Press, 2013.

Judson, Adoniram B. "History and Course of the Epizootic among Horses upon the North American Continent in 1872–1873." *Public Health Papers and Reports* 3 (1875): 88–109.

Kanarek, Harold K. "The Pennsylvania Anthracite Strike of 1922." *Pennsylvania Magazine of History and Biography* 99, no. 2 (April 1975): 207–25.

Kander, Astrid, Paolo Malanima, and Paul Warde. *Power to the People: Energy in Europe over the Last Five Centuries.* Princeton, NJ: Princeton University Press, 2014.

Karamanski, Theodore J. *Schooner Passage: Sailing Ships and the Lake Michigan Frontier.* Detroit: Wayne State University Press, 2000.

Kaye, Barry. "Flour Milling at Red River: Wind, Water and Steam." *Manitoba History* 2 (1981): 12–20.

Keefer, Thomas C. *The Canals of Canada: Prospects and Influence.* Toronto: Andrew H. Armour, 1850.

Keeling, Arn. "'Born in an Atomic Test Tube': Landscapes of Cyclonic Development at Uranium City, Saskatchewan." *Canadian Geographer* 54, no. 2 (2010): 228–52.

Keeling, Arn, and John Sandlos. "Claiming the New North: Mining and Colonialism at the Pine Point Mine, Northwest Territories, Canada." *Environment and History* 18, no. 1 (February 2012): 5–34.

– "Environmental Justice Goes Underground? Historical Notes from Canada's Northern Mining Frontier." *Environmental Justice* 2, no. 3 (2009): 117–25.

Kelly, K. "The Changing Attitude of Farmers to Forest in 19th Century Ontario." *Ontario Geography* 8 (1974): 64–77.

Kelly-Gagnon, Michel, Germain Belzile, and Youri Chassin. "A Plea for a Quebec-Alberta Dialogue." Montreal Economic Institute Research Paper, May 2011.

Kendell, Chet. "Economics of Farming with Horses – Assumptions." Rural Heritage. www.ruralheritage.com/back_forty/economics_assumptions.htm.

Kennedy, Michael. "Fraser River Placer Mining Landscapes." *BC Studies* 160 (Winter 2008/9): 35–66.

Kenny, James L., and Andrew Secord. "Engineering Modernity: Hydro-electric Development in New Brunswick, 1945–1970." *Acadiensis* 39, no. 1 (2010): 3–26.

– "Public Power for Industry: A Re-examination of the New Brunswick Case, 1940–1960." *Acadiensis* 30, no. 2 (2001): 84–108.

Kenny, Stephen. "'Cahots' and Catcalls: An Episode of Popular Resistance in Lower Canada at the Outset of the Union." *Canadian Historical Review* 65, no. 2 (1984): 184–208.

Kheraj, Sean. "Living and Working with Animals in Nineteenth Century Toronto." *Urban Explorations: Environmental Histories of the Toronto Region*, ed. L. Anders Sandberg, Stephen Bocking, Colin Coates, and Ken Cruikshank, 120–40. Hamilton: L.R. Wilson Institute for Canadian History, 2013.

King, Lionel Bradley. "The Electrification of Nova Scotia, 1884–1973: Technological Modernization as a Response to Regional Disparity." PhD diss., University of Toronto, 1999.

Kiple, Kenneth F. *A Movable Feast: Ten Millennia of Food Globalization.* Cambridge: Cambridge University Press, 2007.

Kirkby, Diann, and Tanja Luckins, eds. *Dining on Turtles: Food Feasts and Drinking in History.* Houndmills, UK: Palgrave Macmillan, 2007.

Kirkpatrick, Robert B. *Their Last Alarm: Honouring Ontario's Firefighters.* Burnstown, ON: General Store Publishing House, 2002.

Klein, Naomi. *This Changes Everything: Capitalism vs the Climate.* Toronto: Knopf Canada, 2014.

Krech, Shepard, III. "On the Aboriginal Population of the Kutchin." *Arctic Anthropology* 15, no. 1 (1978): 89–104.

Krensky, Stephen. *Four against the Odds: The Struggle to Save Our Environment.* New York: Scholastic, 1992.

Kruse, Timothy Messer. "The Best Dressed Workers in New York City: Liveried Coachmen of the Gilded Age." *Labor History* 37, no. 1 (1995): 5–27.

Krywulak, Tim. *Fuelling Progress: One Hundred Years of the Canadian Gas Association, 1907–2007*. Ottawa: Canadian Gas Association, 2007.

Kuletz, Valerie L. *The Tainted Desert: Environmental and Social Ruin in the American West*. New York: Routledge, 1998.

Kunnas, Jan. "Fire and Fuels: CO_2 and SO_2 Emissions in the Finnish Economy, 1800–2005." PhD diss., Florence, European University Institute, 2009.

Kunstler, James Howard. *The Geography of Nowhere: The Rise and Decline of America's Man-Made Landscape*. New York: Touchstone, 1993.

Lai, Ping, and Nancy C. Lovell. "Skeletal Markers of Occupational Stress in the Fur Trade: A Case Study from a Hudson's Bay Company Fur Trade Post." *International Journal of Osteoarchaeology* 2 (1992): 221–34.

Laidlaw, George. *Reports & Letters on Light Narrow Gauge Railways by Sir Charles Fox and Son*. Toronto, 1867.

Lajeunesse, Adam. "The New Economics of North American Arctic Oil." *American Review of Canadian Studies* 43, no. 1 (2013): 107–22.

Lambert, Estelle V., David P. Speechly, Steven C. Dennis, and Timothy D. Noakes. "Enhanced Endurance in Trained Cyclists during Moderate Intensity Exercise following 2 Weeks Adaptations to a High Fat Diet." *European Journal of Applied Physiology* 69 (1994): 287–93.

Langdon, John. *Horses, Oxen and Technological Innovation: The Use of Draught Animals in English Farming from 1066–1500*. Cambridge: Cambridge University Press, 1986.

Laperle, Dominique. *Le grain, le meule et les vents: Le Métier de meunier en Nouvelle-France*. Sainte Foy, QC: Les Editions GID, 2003.

Larkin, F. Daniel. "Essay about the Erie Canal." New York State Archives. N.d. http://btceriecanalcourse.weebly.com/uploads/1/4/8/4/14847764/dewitt_clinton_and_the_erie_canal_larkin.pdf.

Law, James. "Influenza in Horses." In *Report of the Commissioner of Agriculture for the Year of 1872*, 203–48. Washington, DC, 1874.

Lawr, D.A. "The Development of Ontario Farming, 1870–1914: Patterns of Growth and Change." *Ontario History* 64, no. 4 (1972): 239–51.

Laxer, James. *Canada's Energy Crisis*. Toronto: James Lorimer, 1975.

– *Oil and Gas*. Toronto: James Lorimer, 1983.

Leavitt, Clyde. "Wood Fuel to Relieve the Coal Shortage in Eastern Canada." Ottawa Commission of Conservation Canada, 1918.

Lecain, Timothy. *Mass Destruction: The Men and Giant Mines That Wired America and Scarred the Planet*. New Brunswick, NJ: Rutgers University Press, 2009.

Leddy, Lianne. "Poisoning the Serpent: Uranium Exploitation and the Serpent River First Nation, 1953–1988." In *The Natures of Empire and the Empires of Nature*, ed. Karl Hele, 125–48. Waterloo, ON: Wilfrid Laurier University Press, 2013.

Lee, David. *Lumber Kings and Shantymen: Logging, Lumber and Timber in the Ottawa Valley.* Toronto: James Lorimer, 2006.

Legget, Robert F. *Canals of Canada.* Vancouver: Douglas, David, and Charles, 1976.

Lepperd, L.B. "MPC (Maximum Permissible Concentration) Objectives for Drinking Water Contaminated by Certain Uranium-Thorium Daughter Mixtures." Reproduced in full in Ontario Water Resources Commission, *Industrial Wastes Survey of Stanrock Mines Limited, Elliot Lake Ontario.* Vol. 1, 1969. Toronto, October 1971. Archives of Ontario, TD227.06 A575 1971 VNI I HDL.

Lessing, Rudolf. "Coal Ash and Clean Coal. Lecture I, Delivered November 23, 1925." *Journal of the Royal Society of Arts* 74, no. 3817 (15 January 1926).

Leung, Felicity. *Direct Drive Waterpower in Canada, 1607–1910.* Ottawa: Environment Canada, Parks, 1980s[?].

– *Grist and Flour Mills in Ontario: From Millstones to Rollers, 1780s–1880s.* Ottawa: National Historic Parks and Sites Branch, Parks Canada, Environment Canada, 1981.

Lew, Byron. "The Diffusion of Tractors on the Canadian Prairies: The Threshold Model and the Problem of Uncertainty." *Explorations in Economic History* 37 (2000): 189–216.

Lewis, E. "St. Elmo, Efficiency in the Advertising and New Business Dept." *Gas Industry: Heat, Light, Power,* 85–6.

Lezius, Walter G. "Geographic Aspects of Coal Cargoes from Toledo." *Economic Geography* 10, no. 4 (October 1934): 374–81.

Liddicoat, Wallace L. Waterwheels in the Service of British Columbia's Pioneers. Altona, MB: Friesens, 1996.

Lindmark, Magnus, and Lars Fredrik Andersson. "Household Firewood Consumption in Sweden during the Nineteenth Century." *Journal of Northern Studies* 2 (2010): 55–78.

Lizée, Erik. "Betrayed: Leduc, Manning and Surface Rights in Alberta, 1947–55." *Prairie Forum* 35, no. 1 (Spring 2010): 77–100.

– "Rhetoric and Reality: Albertans and Their Oil Industry under Peter Lougheed." MA thesis, University of Alberta, 2010.

Loney, Martin. "The Construction of Dependency: The Case of the Grand Rapids Hydro Project." *Canadian Journal of Native Studies* 7, no. 1 (1987): 57–78.

Long, Alan. "Emory Creek: The Environmental Legacy of Gold Mining on the Fraser River." *British Columbia History* 39, no. 3 (Spring 2007): 8–10.

Loo, Tina. "Disturbing the Peace: Environmental Change and the Scales of
    Justice on a Northern River." *Environmental History* 12 (2007): 895–919.
– "People in the Way: Modernity, Environment and Society on the Arrow
    Lakes." *BC Studies* 142/3 (2004): 161–96.
Lorain, John. *Nature and Reason Harmonized in the Practice of Husbandry.*
    Philadelphia: Carey and Lea, 1825.
Lorinc, John. "On with the Wind." *Canadian Geographic Magazine* 129, no. 3
    (June 2009). http://www.canadiangeographic.ca/magazine/jun09/wind_
    power.asp.
Lovell, Nancy C., and Aaron A. Dublenko. "Further Aspects of Fur Trade Life
    Depicted in the Skeleton." *International Journal of Osteoarchaeology* 9
    (1999): 248–56.
Lovelock, James. *The Revenge of Gaia.* London: Allen Lane, 2006.
Lowe, Peter. *Animal Powered Systems.* New York: John Wiley and Sons, 1986.
Lower, A.R.M. *Settlement and the Forest Frontier in Eastern Canada*; and
    Harold A. Innis, *Settlement on the Mining Frontier.* Toronto: Macmillan,
    1936.
Lutz, John. *Makuk: New History of Aboriginal-White Relations.* Vancouver:
    UBC Press, 2008.
Macdonald, Catherine. "Water Power and the Transformation of Canada
    1600–1960: A Synthesis of National Museum of Science and Technology
    Research Reports, 1989–1991." Ottawa: NMST Contract 2104-1-044, 1992.
Macdonald, Colin. "Rocks to Reactors: Uranium Exploration and the
    Market." Paper presented at the World Nuclear Association Annual
    Symposium, London, 5–7 September 2001.
Macdonald, Edith. "The Scribe." In *Golden Jubilee 1869–1919: A Book to
    Commemorate the Fiftieth Anniversary of the T. Eaton Company*, 169–70.
    Toronto: T. Eaton, 1919.
MacDonald, R. Bruce. *North Star of Herschel Island: The Last Canadian
    Arctic Fur Trading Ship.* Victoria, BC: Friesen, 2012.
MacDowell, L.S. "The Elliot Lake Uranium Miners' Battle to Gain
    Occupational Health and Safety Improvements in the Post-War Period."
    Labour / Le Travail 69 (Spring 2012): 91–118.
MacDowell, Laurel. *An Environmental History of Canada.* Vancouver: UBC
    Press, 2012.
MacEachern, Alan. *The Institute of Man and Resources: An Environmental
    Fable.* Charlottetown: Island Studies, 2003.
– "No Island Is an Island: A History of Tourism on Prince Edward Island,
    1870–1939." MA thesis, Queen's University, 1991.
MacEwan, Grant. *Heavy Horses: Highlights of Their History.* Whitewater, WI:
    Heart Prairie, 1991.

MacEwan, Paul. *Miners and Steelworkers: Labour in Cape Breton*. Toronto: Hekkert, 1976.

MacFadyen, Joshua. "Fuel Wood." In *Encyclopedia of American Environmental History*, ed. Kathleen A. Brosnan, 1:344–6. New York: Facts on File, 2011.

MacFadyen, Joshua, and William Glen. "Top-Down History: Delimiting Forests, Farms, and the Agricultural Census on Prince Edward Island Using Aerial Photography, c. 1900–2000." In *Historical Geographic Information Systems in Canada*, ed. Jennifer Bonnell and Marcel Fortin, 197–224. Canadian History & Environment Series. Calgary: University of Calgary Press, 2014.

Macfarlane, Daniel. "'Caught between Two Fires': St Lawrence Seaway and Power Project, Canadian-American Relations, and Linkage." *International Journal* 67, no. 2 (2012): 465–82.

– *Negotiating a River: Canada, the United States and the Creation of the St Lawrence Seaway*. Vancouver: UBC Press, 2014.

Mackenzie, Alexander. *Voyages from Montreal: On the River St Laurence through the Continent of North America, to the Frozen and Pacific Oceans, in the Years 1789 and 1793: With a Preliminary Account of the Rise, Progress and Present State of the Fur Trade of That Country*. New York: Evert Duyckinck, 1803.

Mackenzie, Fred T. *Our Changing Planet: An Introduction to Earth System Science and Global Environmental Change,* 3rd ed. Upper Saddle River, NJ: Pearson Education, 2003.

Mackenzie, W.D.C. "Oil for Western Canada." *Western Business and Industry* (May 1953): 50–2.

MacKinnon, Robert. "Agriculture and Rural Change in Nova Scotia, 1851–1951." *Canadian Papers in Rural History* 10 (1996): 231–73.

Macphail, Andrew. *The Master's Wife*. Toronto: McClelland and Stewart, 1977.

Major, J. Kenneth. *Animal Powered Machines*. Aylesbury, UK: Shire Publications, 1985.

– "Animal-Powered Machinery in the Medieval Period." *Society for the History MTS Proceedings*, 5 November 1987. Oxford: University College.

– "The Pre-Industrial Sources of Power: Muscle Power." *History Today* 30, no. 3 (1980). http://www.historytoday.com/j-kenneth-major/ pre-industrial-sources-power-muscle-power.

Malone, Patrick M. *Water Power in Lowell: Engineering and Industry in Nineteenth Century America*. Baltimore, MD: Johns Hopkins University Press, 2009.

Manitoba Electrification Enquiry Commission. *A Farm Electrification Programme: Report of Manitoba Electrification Enquiry Commission, 1942*. Winnipeg: King's Printer, 1943.

Manitoba Power Commission. *22nd Annual Report of the Manitoba Power Commission, Year Ended 30 November 1941*. Winnipeg: Manitoba Power Commission, 1945.

Manore, Jean L. *Cross-Currents: Hydroelectricity and the Engineering of Northern Ontario*. Waterloo, ON: Wilfrid Laurier University Press, 1999.

Manwell, J.F., J.G. McGowan, and A.L. Rogers. *Wind Energy Explained*. New York: Wiley, 2002.

Marchetti, Cesare. "The Future of Natural Gas: A Darwinian Analysis." *Technological Forecasting and Social Change* 31 (1987): 155–71.

"Market for Small Wind Turbines in Canada: Forecast, Trends and Opportunities, 2006–2021." TaiYou Research, 2012. http://www.market-research.com/Taiyou-Research-v3862/Small-Wind-Turbines-Canada-7194806/.

Marland, G., T.A. Boden, and R.J. Andres. "Global, Regional, and National CO_2 Emissions." In T.A. Borden, D.P. Kaiser, R.J. Sepanski, and F.W. Stoss, eds., *Trends 93: A Compendium of Data on Global Change*, 505–84. Oak Ridge, TN: Carbon Dioxide Information Analysis Centre, Oak Ridge National Laboratory, 1994.

– *Global, Regional, and National Fossil-Fuel CO_2 Emissions*. Oak Ridge, TN: Carbon Dioxide Information Analysis Centre, Environmental Sciences Division, Oak Ridge National Laboratory. http://cdiac.ornl.gov/trends/emis/overview.html.

Martin, Thibault, and Steven Hoffman, eds. *Power Struggles: Hydroelectric Development and First Nations in Manitoba and Quebec*. Winnipeg: University of Manitoba Press, 2008.

Martin-Neilsen, Janet. "South over the Wires: Hydro-electricity Exports from Canada." *Water History* 1 (2009): 109–29.

Massell, David. *Amassing Power: J.B. Duke and the Saguenay River, 1897–1927*. Montreal and Kingston: McGill-Queen's University Press, 2000.

– "'As Though There Was No Boundary': The Shipshaw Project and Continental Integration." *American Review of Canadian Studies* (Summer 2004): 187–222.

– *Québec Hydropolitics: The Peribonka Concessions of the Second World War*. Montreal and Kingston: McGill-Queen's University Press, 2011.

– "A Question of Power: A Brief History of Hydroelectricity in Quebec." In *Quebec Questions: Quebec Studies for the Twenty-First Century*, ed. Stéphan Gervais, Christopher Kirkey, and Jarrett Rudy, 338–56. Oxford: Oxford University Press, 2011.

Massey, Samuel. *Dumb Animals: A Plea for Man's Dumb Friends, Being the Substance of an Address Delivered in Salem Church by Rev. Samuel Massey*. Montreal: 1888.

Mastert, Noel. *Supership*. New York: Knopf, 1974.

Mathews, Jessica T. "Iran: A Good Deal Now in Danger." *New York Review of Books*, 20 February 2014.

McArdle, William D., Frank I. Katch, and Victor L. Katch. *Essentials of Exercise Physiology*, 3rd ed. Vol. 1. Baltimore, MD: Lippincott, Williams and Wilkins, 2006.

McCalla, Douglas. *Consumers in the Bush: Shopping in Rural Upper Canada, 1808–1861*. Montreal and Kingston: McGill-Queen's University Press, 2015.

– *Planting the Province: The Economic History of Upper Canada*. Toronto: University of Toronto Press, 1993.

McClelland, Peter D. "The New Brunswick Economy in the Nineteenth Century." PhD diss., Harvard University, 1966.

McCusker, John J., and Russell R. Menard. *The Economy of British North America, 1607–1789*. Chapel Hill: University of North Carolina Press, 1985.

McCutcheon, Sean. *Electric Rivers: The James Bay Project*. Montreal: Black Rose, 1991.

McDougall, John N. *Fuels and the National Policy*. Toronto: Butterworths, 1982.

McEwan, Andrew. "'Our Bugbear of War': The Development of Canadian Army Veterinary Practices in the Great War." Paper presented to the History of Medicine Days Conference, University of Calgary, 9 March 2013.

McFarlin, Michael J. "Top 10 Natural Gas Producers by Country." *Futures: Stock, Commodity, Options and Forex Strategies for the Modern Trader*, 16 February 2012. http://www.futuresmag.com/2012/02/15/top-10-natural-gas-producers-country-slideshow.

McIlwraith, Thomas Forsyth, and Edward K. Muller, eds. *North America: The Historical Geography of a Changing Continent*. Lanham, MD: Rowman & Littlefield, 2001.

McInnis, Marvin. "A Contrarian's View of Canadian Economic History." Manuscript, 2013.

McKee, Oliver. "The Horse and the Motor." *Lippincott's Monthly Magazine* 7 (January–June 1896): 379–84.

McKenzie, Roderic. "Réminiscences." In *Les bourgeois de la Compagnie du Nord-Ouest*, ed. L.R. Masson, 1:7–66. New York: Antiquarian, 1960.

McLeish, John. *The Production of Coal and Coke in Canada during the Calendar Year 1909*. Ottawa: Government Printing Bureau, 1910.

– *The Production of Coal and Coke in Canada during the Calendar Year 1910*. Ottawa: Government Printing Bureau, 1911.

– *The Production of Coal and Coke in Canada during the Calendar Year 1916*. Ottawa: Government Printing Bureau, 1917.

– *The Production of Coal and Coke in Canada during the Calendar Year 1918*. Ottawa: J de LaBroquerie Taché, 1919.

– *The Production of Coal and Coke in Canada during the Calendar Year 1920*. Ottawa: F.A. Acland, 1921.

– *The Production of Coal and Coke in Canada during the Calendar Year 1919*. Ottawa: Thomas Mulvet, 1921.

McNally, Larry. "The Relationship between Transportation and Water Power on the Lachine Canal in the Nineteenth Century." In *Critical Issues in the History of Canadian Science, Technology and Medicine*, ed. Richard A. Jarrell and Arnold E. Roos, 76–88. Thornhill, ON: HSTC Publications, 1983.

– "Roads, Streets, and Highways." In *Building Canada: A History of Public Works*, ed. Norman Ball, 31–58. Toronto: University of Toronto Press, 1988.

McNeill, J.R. *Something New under the Sun: An Environmental History of the Twentieth Century World*. New York: W.W. Norton, 2000.

McPhee, John. *The Curve of Binding Energy: A Journey into the Awesome and Alarming World of Theodore B. Taylor*. New York: Farrar, Strauss & Giroux, 1974.

McShane, Clay, and Joel Tarr. "The Horse as Technology: The City Animal as Cyborg." In *Horses and Humans: The Evolution of the Equine/Human Relationship*, ed. Sandra L. Olsen, 365–75. Oxford: British Archeological Reports, 2006.

– *The Horse in the City: Living Machines in the Nineteenth Century*. Baltimore, MD: Johns Hopkins University Press, 2007.

Mellin, Robert. *Tilting: House Launching, Slide Hauling, Potato Trenching, and Other Tales from a Newfoundland Fishing Village*. New York: Princeton Architectural Press, 2003.

Mellor, Robynne. "A Comparative Case Study of Uranium Mine and Mill Tailings Regulation in Canada and the United States." Paper presented in the History Department, Georgetown University, 18 December 2014.

Ménard, André, and Danielle J. Marceau. "Simulating the Impact of Forest Management Scenarios in an Agricultural Landscape of Southern Quebec, Canada, Using a Geographic Cellular Automata." *Landscape and Urban Planning* 79, nos 3–4 (March 2007): 253–65.

Mendell, Brooks C., and Amanda H. Lang. *Wood for Bioenergy: Forests as a Resource for Biomass and Biofuels*. Durham, NC: Forest History Society, 2012.

Merchant, Carolyn. *American Environmental History: An Introduction*. New York: Columbia University Press, 2007.

Meyer, Carrie E. "The Farm Debut of the Gasoline Engine." *Agricultural History* 87, no. 3 (Summer 2013): 287–313.

Milner, W.C. *Coal: Analysis of the Trade between Canada and United States.* Ottawa: Mortimer, 1904.

Miquelon, Dale, "Les Forges Saint-Maurice." In *Canadian Encyclopedia.* http://www.thecanadianencyclopedia.ca/en/article/les-forges-saint-maurice/.

Mitchell, David. *WAC Bennett and the Rise of British Columbia.* Vancouver: Douglas & McIntyre, 1983.

Mitchell, Timothy. "Carbon Democracy." *Economy and Society* 38, no. 3 (August 2009): 399–432.

– *Carbon Democracy: Political Power in the Age of Oil.* London: Verso, 2011.

Mochuruk, Jim. *Formidable Heritage: Manitoba's North and the Cost of Development, 1870 to 1930.* Winnipeg: University of Manitoba Press, 2004.

Moloney, Paul. "Electricity Demand Tops Record." *Toronto Star,* 7 January 1988.

Mom, Gijs. *The Electric Vehicle: Technology and Expectations in the Automobile Age.* Baltimore, MD: Johns Hopkins University Press, 2004.

Mom, Gijs, and David A. Kirsch. "Technologies in Tension: Horses, Electric Trucks and the Motorization of American Cities, 1900–1925." *Technology and Culture* 42, no. 3 (2001): 489–518.

Monaghan, David W. *Canada's New Main Street: The TransCanada Highway as Idea and Reality, 1912–1956.* Ottawa: Canada Science and Technology Museum, 2002.

Moodie, Susanna. *Roughing It in the Bush.* Toronto: Coles Publishing, 1974.

Moody Labour Saving Agricultural Implements. *Autumn Catalogue.* Terrebonne, QC: Moody, n.d.

Morgan, Jerome J. *A Textbook of American Gas Practice.* 2 vols. Maplewood, NJ: Jerome J. Morgan, 1931–5.

Morgan, Nigel. "Infant Mortality, Flies and Horses in Later-Nineteenth-Century Towns: A Case Study of Preston." *Continuity and Change* 17, no. 1 (May 2002): 97–132.

Morris, Eric. "From Horse Power to Horsepower." *Access* 30 (Spring 2007). http://www.accessmagazine.org/articles/spring-2007/horse-power-horsepower/.

Morrison, Robert Thornton, and Robert Neilson Boyd. *Organic Chemistry.* 3rd ed. Boston: Allyn & Bacon, 1973.

Morritt, Hope. *Rivers of Oil: The Founding of North America's Petroleum Industry.* Kingston: Quarry, 1993.

Morse, Eric W. *Fur Trade Canoe Routes of Canada, Then and Now.* Toronto: University of Toronto Press, 1989.

Morse, Kathryn. *The Nature of Gold: An Environmental History of the Klondike Gold Rush.* Seattle: University of Washington Press, 2003.

– "There Will Be Birds: Images of Oil Disasters in the Nineteenth and Twentieth Centuries." *Journal of American History* 99, no. 1 (June 2012): 124–34.

Morton, Arthur S., ed. *The Journal of Duncan M'Gillivray of the North West Company at Fort George on the Saskatchewan, 1794–5.* Toronto: Macmillan Canada, 1929.

Morton, A.S. *A History of the Canadian West to 1870–71.* Toronto: University of Toronto Press, 1973.

Morton, Peter D. "Street Rivals: Jaywalking and the Invention of the Motor Age Street." *Technology and Culture* 48, no. 2 (2007): 331–59.

Mostert, Noel. *Supership.* New York: Warner, 1975.

Mouat, Jeremy. *The Business of Power: Hydro-electricity in Southeastern British Columbia, 1897–1997.* Victoria: Sono Nis, 1997.

– "The Politics of Coal: A Study of the Wellington Miners' Strike of 1890–91." *BC Studies* 77 (Spring 1988): 3–29.

Muise, Del, Rosemary Langhout, and Ronald H. Walder. "From Firewood to Coal: Fuelling the Nation to 1891." In *Historical Atlas of Canada*, vol. 2, plate 49. Toronto: University of Toronto Press, 1993.

Muldrew, Craig. *Food, Energy and the Creation of Industriousness: Work and Material Culture in Agrarian England, 1550–1780.* Cambridge University Press, 2011.

Murton, James, Dean Bavington, and Carly Dokis, eds. *Subsistence under Capitalism: Historical and Contemporary Perspectives.* Montreal and Kingston: McGill-Queen's University Press, 2016.

Musgrove, Peter. *Wind Power.* Cambridge: Cambridge University Press, 2010.

Natural Resources Canada. "Government of Canada Invests in Wind Energy in Southern Ontario." News release, 28 June 2011.

– *Stand-Alone Wind Energy Systems: A Buyer's Guide.* Ottawa: Natural Resources, 2003.

Nelles, H.V. *The Politics of Development: Forests, Mines and Hydro-electric Power in Ontario, 1849–1941.* Toronto: Macmillan Canada, 1974.

Nelles, H.V., and Christopher Armstrong. *Monopoly's Moment: The Organization and Regulation of Canadian Utilities, 1830–1930.* Philadelphia: Temple University Press, 1986.

Nemeth, Tammy. "Canada Oil and Gas Relations, 1958 to 1974." PhD diss., University of British Columbia, 2007.

Netherton, Alexander. "From Rentiership to Continental Modernization: Shifting Policy Paradigms of State Intervention in Hydro in Manitoba, 1922–1977." PhD diss., Carleton University, 1993.

– "The Political Economy of Canadian Hydro-electricity: Between Old 'Provincial Hydros' and Neoliberal Regional Energy Regimes." *Canadian Political Science Review* 1, no. 1 (June 2007): 107–24.

1927 Edition of the T. Eaton Co. Catalogue for Spring and Summer, The. Toronto: Musson Book, 1971.

Noakes, T.D. "Physiological Models to Understand Exercise Fatigue and the Adaptations That Predict or Enhance Athletic Performance." *Scandinavian Journal of Medicine & Science in Sports* 10 (2000): 123–45.

Nordhaus, William D. "Do Real-Output and Real-Wage Measures Capture Reality? The History of Lighting Suggests Not." In *The Economics of New Goods,* ed. Timothy F. Bresnahan and Robert J. Gordon, 27–70. Chicago: University of Chicago Press, 1996.

Norrie, Kenneth, Douglas Owram, and J.C. Herbert Emery. *A History of the Canadian Economy.* 4th ed. Toronto: Thomson Nelson, 2008.

Norris, John. "The Vancouver Island Coal Miners, 1912–1914: A Study of an Organizational Strike." *BC Studies* 45 (Spring 1980): 56–72.

North, Douglass C. North. "Ocean Freight Rates and Economic Development, 1750–1913." *Journal of Economic History* 18, no. 4 (1958): 537–55.

Nute, Grace Lee. *Voyageur.* St Paul: Minnesota Historical Society, 1987.

Nye, David E. *Consuming Power: A Social History of American Energies.* Cambridge, MA: MIT Press, 1998.

– *Electrifying America: Social Meanings of a New Technology.* Cambridge, MA: MIT Press, 1992.

O'Brien, P.K. "Agriculture and the Industrial Revolution." *Economic History Review,* n.s., 30, no. 1 (February 1977): 166–81.

O'Dea, William T. *The Social History of Lighting.* London: Routledge and Kegan Paul, 1958.

Olmstead, Alan L., and Paul W. Rhode. *Creating Abundance: Biological Innovation and American Agricultural Development.* New York: Cambridge University Press, 2008.

Olsen, Sandra L. "Introduction." In *Horses and Humans: The Evolution of the Equine/Human Relationship,* ed. Sandra L. Olsen, 1–10. Oxford: Archeopress, 2006.

Olson, Sherry H. *The Depletion Myth: A History of Railroad Use of Timber.* Cambridge, MA: Harvard University Press, 1971.

Ommer, Rosemary E. "The Decline of the Eastern Canadian Shipping Industry, 1880–95." *Journal of Transport History* 5, no. 1 (March 1984): 25–44.

Ontario Ministry of Agriculture, Food and Rural Affairs (OMAFRA). "Land Use Geographical Information Systems." http://www.omafra.gov.on.ca/english/landuse/gis/portal.htm.

Ontario, Department of Mines. *Bulletin 155, Being the Report of the Special Committee on Mining Practices at Elliot Lake.* Part 2, *Accident Review, Ventilation, Ground Control and Related Subjects.* Toronto: Frank Fogg, 1961.

Ontario, Department of Public Highways. *Annual Report 1918*. Toronto: King's Printer, 1919.

Ontario Water Resources Commission. "Water Pollution from the Uranium Mining Industry in the Elliot Lake and Bancroft Areas." Toronto: Ontario Water Resources Commission, October 1971.

Ormsby, Margaret A. *British Columbia: A History*. Vancouver: MacMillan Canada, 1958.

Osborn, P.G. *A Concise Law Dictionary for Students and Practitioners*. London: Sweet & Maxwell, 1927.

Ottawa Citizen. "A Walker," 20 February 1867.

Otter, Chris. "The British Nutrition Transition and Its Histories." *History Compass* 10/11 (2012): 812–25.

"Outsider." Letter to editor, *Ottawa Citizen*, 25 March 1867.

Paquet, Gilles, and Jean-Pierre Wallot. "Structures sociales et niveaux de richesse dans les campagnes du Québec: 1792–1812." *Material History Bulletin* 17 (1983): 25–44.

Parks, Noreen. "The Ecological Costs of Coal Ash Waste." *Frontiers in Ecology and the Environment* 10, no. 8 (October 2012): 400.

Parr, Joy. *Domestic Goods: The Material, the Moral and the Economic in the Postwar Years*. Toronto: University of Toronto Press, 1999.

– "'Lostscapes': Found Sources in Search of a Fitting Representation." *Journal of the Association for History and Computing* 7, no. 2 (August 2004): n.p.

– "Modern Kitchen, Good Home, Strong Nation." *Technology and Culture* 43, no. 4 (October 2002): 657–67.

– *Sensing Changes: Technologies, Environments, and the Everyday, 1953–2003*. Vancouver: UBC Press, 2010.

Passfield, Robert, "Waterways." In *Building Canada: A History of Public Works*, ed. Norman R. Ball, 113–42. Toronto: University of Toronto Press, 1988.

Patton, M.J. "The Coal Resources of Canada." *Economic Geography* 1, no. 1 (March 1925): 73–88.

Patzek, Tad W. "Ethanol from Corn: Clean Renewable Fuel for the Future, or Drain on Our Resources and Pockets." AA Dordrecht: Kluwer Academic Publishers, 2003.

Peers, Laura. *The Ojibwa of Western Canada, 1780–1870*. Winnipeg: University of Manitoba Press, 1994.

Perren, Richard. *Taste, Trade and Technology: The Development of the International Meat Industry since 1840*. Surrey, UK: Ashgate Publishing, 2006.

Peyton, Jonathan. "Corporate Ecology: B.C. Hydro's Stikine-Iskut Project the Unbuilt Environment." *Journal of Historical Geography* 37, no. 3 (July 2011): 358–69.

Pierce, Josephine H. "Dog and Horse Power." *Chronicle of the Early American Industries Association* 2 (March 1938): 25–8.

R.F. Pierce. "The Sale of Gas for Illumination." *Gas Industry: Heat, Light, Power* 13, no. 1 (1913): 64–5.

Pimentel, David, and Tad W. Patzek. "Biofuel: Ethanol Production Using Corn, Switchgrass, and Wood: Biodiesel Production Using Soybean and Sunflower." *Natural Resource Research* 14, no. 1 (March 2005): 65–7.

Piper, Liza. *The Industrial Transformation of Subarctic Canada.* Vancouver: UBC Press, 2009.

Podruchny, Carolyn. *Making the Voyageur World: Travelers and Traders in the North American Fur Trade.* Lincoln: University of Nebraska Press, 2006.

Pomeranz, Kenneth. *The Great Divergence: China, Europe and the Making of the Modern World Economy.* Princeton, NJ: Princeton University Press, 2000.

Pomfret, Richard. "The Mechanization of Reaping Nineteenth-Century Ontario: A Case Study of the Pace and Causes of the Diffusion of Embodied Technical Change." In *Perspectives on Canadian Economic History*, ed. Douglas McCalla, 81–95. Toronto: Copp Clark Pitman, 1987.

Porter, Marilyn. "'She Was Skipper of the Shore Crew': Notes on the History of the Sexual Division of Labour in Newfoundland." *Labour / Le Travail* 15 (Spring 1985): 105–23.

Power. VHS recording. Directed by Magnus Isaacson. Produced by Glen Salzman and Mark Zannis. Cineflix Productions and National Film Board of Canada, 1996.

Pratt, Terry. *Dictionary of Prince Edward Island English.* Toronto: University of Toronto Press, 1996.

Pretor-Pinney, Gavin. *The Wave-Watcher's Companion.* New York: Perigee, 2010.

Priamo, Carol. *Mills of Canada.* Toronto: McGraw-Hill Ryerson, 1976.

Pritchard, Sara B. "Japan Forum: An Envirotechnical Disaster: Nature, Technology and Politics at Fukushima." *Environmental History* 17, no. 2 (2012): 219–243.

Prudham, Scott, Gunter Gad, and Richard Anderson. "Networks of Power: Toronto's Waterfront Energy Systems from 1840 to 1970." In *Reshaping Toronto's Waterfront*, ed. Gene Desfor and Jennefer Laidley, 175–202. Toronto: University of Toronto Press, 2011.

Purdy, G.A. *Petroleum: Prehistoric to Petrochemicals.* Vancouver: Copp Clark, 1957.

Pursell, Carroll. "Parallelograms of Perfect Form: Some Early Brick-Making Machines." *Smithsonian Journal of History* 3 (Spring 1968): 19–27.

Pyne, Stephen J. *Awful Splendor: A Fire History of Canada.* Vancouver: UBC Press, 2011.

Radforth, Ian. *Bushworkers and Bosses: Logging in Northern Ontario, 1900–1980.* Toronto: University of Toronto Press, 1987.

Rajala, Richard A. *Clearcutting the Pacific Rain Forest: Production, Science, and Regulation.* Vancouver: UBC Press, 1998.

– "The Forest Industry in Eastern Canada: An Overview." In *Broadaxe to Flying Shear: The Mechanization of Forest Harvesting East of the Rockies,* ed. C. Ross Silversides, 123–30. Ottawa: National Museum of Science and Technology, 1997.

– "'This Wasteful Use of a River': Log Driving, Conservation, and British Columbia's Stellako River Controversy, 1965–1972." *BC Studies* 165 (Spring 2010): 31–74.

Rasid, Harun. "The Effects of Regime Regulation by the Gardiner Dam on Downstream Geomorphic Processes in the South Saskatchewan River." *Canadian Geographer* 23, no. 2 (June 1979): 140–58.

Ray, Arthur J. *Indians in the Fur Trade: Their Role as Trappers, Hunters, and Middlemen in the Lands Southwest of Hudson Bay, 1660–1870.* Toronto: University of Toronto Press, 1998.

– "The Northern Great Plains: Pantry of the Northwestern Fur Trade, 1774–1885." *Prairie Forum* 9 (1984): 263–80.

– "Periodic Shortages, Native Welfare, and the Hudson's Bay Company, 1670–1930." In *Interpreting Canada's North: Selected Readings,* ed. Kenneth S. Coates and William R. Morrison, 5–7. Toronto: Copp Clark Pitman, 1989.

Rekmans, Loraine, Keith Lewis, and Anabel Dwyer, eds. *This Is My Homeland: Stories of the Effects of Nuclear Industries by People of the Serpent River First Nation.* Cutler, ON: Serpent River First Nation, 2003.

Reuss, Martin, and Stephen Cutcliffe, eds. *The Illusory Boundary: Environment and Technology in History.* Charlottesville: University of Virginia Press, 2010.

Reynolds, R.V., and Albert H. Pierson. "Fuel Wood Used in the United States, 1630–1930." *USDA Circular* 641 (February 1942): 1–20.

Reynolds, Terry S. *Stronger Than a Hundred Men: A History of the Vertical Water Wheel.* Baltimore, MD: Johns Hopkins University Press, 1983.

Richardson, Ronald E., George H. McNevin, and Walter G. Rooke. *Building for People: Freeway and Downtown: New Frameworks for Modern Needs.* Toronto: Ryerson, 1970.

Righter, Robert W. *Wind Energy in America: A History.* Norman: University of Oklahoma Press, 1996.

Roberts, David. *In the Shadow of Detroit: Gordon M. McGregor, Ford of Canada, and Motoropolis*. Detroit: Wayne State University Press, 2006.

Roberts, Kenneth G., and Philip Shackleton. *The Canoe: A History of the Craft from Panama to the Arctic*. Toronto: Macmillan, 1983.

Roberts, Paul. *The End of Oil: On the Edge of a Perilous New World*. New York: Mariner Books, Houghton Mifflin, 2005.

Robertson, Barbara R. *Sawpower: Making Lumber in the Sawmills of Nova Scotia*. Halifax: Nimbus Publications and the Museum of Nova Scotia, 1986.

Robertson, J. Ross. *Landmarks of Toronto*. Vol. 2. Toronto: J.R. Robertson, 1894.

– *Robertson's Landmarks of Toronto* (1896), 2:763. Wikimedia Commons, https://commons.wikimedia.org/wiki/Category:Robertson%27s_Landmarks_of_Toronto#/media/File:Toronto%27s_first_Ferry_Horse_Boat.jpg.

Rome, Adam. *The Bulldozer in the Countryside: Suburban Sprawl and the Rise of American Environmentalism*. Cambridge: Cambridge University Press, 2001.

Roosenberg, Richard. "Animal-Driven Shaft Power Revisited." Tillers International, 1992. http://www.tillersinternational.org/farming/resources_techguides/Animal-DrivenShaftPowerRevisitedTechGuide.pdf.

Rosano, Michela. "On with the Wind: Wind Energy in Canada Timeline." *Canadian Geographic Magazine*, June 2009. http://www.canadiangeographic.ca/magazine/jun09/wind_power_timeline.asp.

Rosenberg, David M., F. Berkes, R.A. Bodaly, R.E. Hecky, C.A. Kelly, and J. Rudd. "Large-Scale Impacts of Hydroelectric Development." *Environmental Reviews* 5, no. 1 (1997): 27–54.

Ross, Kristin. *Fast Car, Clean Bodies: Decolonization and the Reordering of French Culture*. Cambridge, MA: MIT Press, 1996.

Ross, Susan M. "Steam or Water Power? Thomas C. Keefer and the Engineers Discuss the Montreal Waterworks in 1852." In "Waterpower, the Lachine Canal, and the Industrial Development of Montreal." Special issue, *IA: The Journal of the Society for Industrial Archaeology* 29, no. 1 (2003): 49–64.

Ross, Victor. *Petroleum in Canada*. Toronto: Southam, 1917.

Ross Company, E.W. *Catalogue*. Springfield, OH: E.W. Ross, 1896.

Rothwell Machine Company. *Catalogue of the Latest Improved Brick Machinery and Brick Makers Supplies*. Hamilton, ON: Rothwell Machine, n.d.

Roy, Patricia. "The British Columbia Electric Railway Company, 1897–1928: A British Company in British Columbia." PhD diss., University of British Columbia, 1970.

- "Direct Management from Abroad: The Formative Years of the British Columbia Electric Railway." *Business History Review* 47, no. 2 (1973): 239–59.

Russell, Edmund, James Allison, Thomas Finger, and John K. Brown. "The Nature of Power: Synthesizing the History of Technology and Environmental History." *Technology and Culture* 52, no. 2 (April 2011): 246–59.

Russell, Loris. *A Heritage of Light: Lamps and Lighting in the Early Canadian Home.* Toronto: University of Toronto Press, 1968.

Sager, Eric W. *Seafaring Labour: The Merchant Marine of Atlantic Canada 1820–1914.* Montreal and Kingston: McGill-Queen's University Press, 1989.

Sager, Eric W., with Gerald E. Panting. *Maritime Capital: The Shipping Industry in Atlantic Canada, 1820–1914.* Montreal and Kingston: McGill-Queen's University Press, 1990.

Sahlins, Marshall. *Stone Age Economics.* Chicago: Aldine-Atherton, 1972.

Sandlos, John, and Arn Keeling. "Zombie Mines and the (Over)burden of History." *Solutions* 4, no. 3 (June 2013): 80–3. http://thesolutionsjournal. com/node/23361.

Sandwell, R.W. "An Introduction to Lighting in Canada, 1840–1920." *The Extractive Industries and Society: An International Journal.* Forthcoming.

- "Canada's First Oil Boom: Kerosene Lighting 1846–1900." *The Extractive Industries and Society: An International Journal* (forthcoming).

- *Canada's Rural Majority, 1870–1940.* Toronto, University of Toronto Press, 2016.

- "Mapping Fuel Use in Canada: Exploring the Social History of Canadians' Great Fuel Transformation." In *Historical GIS in Canada*, ed. Jennifer Bonnell and Marcel Fortin, 239–68. Calgary: University of Calgary Press, 2014.

- "Notes towards a History of Rural Canada, 1870–1940." In *Social Transformation in Rural Canada: Community, Cultures, and Collective Action,* ed. John R. Parkins and Maureen G. Reed, 21–42. Vancouver: UBC Press, 2013.

- "Pedagogies of the Unimpressed: Re-Educating Ontario Women for the Mineral Economy, 1900–1940." *Ontario History* 107, no. 1 (Spring 2015): 36–59.

- "Rural Households, Subsistence and Environment on the Canadian Shield 1901–1941." In *Subsistence under Capitalism: Historical and Contemporary Perspectives,* ed. James Murton, Dean Bavington, and Carly Dokis, 121–53. Montreal and Kingston: McGill-Queen's University Press, 2016.

Sasges, Michael. "Colliers and Cowboys: Imagining the Industrialization of British Columbia's Nicola Valley." MA research project, Simon Fraser University Liberal Studies Programme, 2013.

Saskatchewan. Inquiry into the Practicability of Producing Power at Coal Centres and Distributing It throughout the Province. R.O. Wynne-Roberts, *Report on Coal and Power Investigation, 1912.*

Saskatchewan Research Council. *Annual Report, 2009.* http://www.src.sk.ca/ resource%20files/annual%20report%202009-10.pdf.

Sathyajith, Matthew. *Wind Energy: Fundamentals, Resources Analysis, and Economics.* New York: Springer, 2006.

Sawyer, Peter, ed. *The Oxford Illustrated History of the Vikings.* New York: Oxford University Press, 1997.

Scanlan, T.F., C.K. Bayne, and D.R. Johnson. "Investigation of Attic-Insulation Effectiveness by Using Actual Energy-Consumption Data." In *Thermal Insulation, Materials, and Systems for Energy Conservation in the '80s,* ed. Francis A. Govan, 19–34. Philadelphia: American Society for Testing and Materials International, 1983. http://www.astm.org/DIGITAL_LIBRARY/ STP/SOURCE_PAGES/STP789.htm

Schivelbusch, Wolfgang. *Disenchanted Night: The Industrialization of Light in the Nineteenth Century.* Translated from the German by Angela Davies. Berkeley: University of California Press, 1988.

Schrepfer, Susan R., and Philip Scranton. *Industrialising Organisms: Introducing Evolutionary History.* New York: Routledge, 2004.

Schurr, Sam. *Electricity in the American Economy: Agent of Technological Progress.* New York: Greenwood, 1990.

Schurr, Sam H., and Bruce C. Netschert, with Vera F. Eliasberg, Joseph Lerner, and Hans. H. Lansberg. *Energy in the American Economy, 1850–1975.* Baltimore, MD: Johns Hopkins University Press, 1960.

Séguin, Robert-Lionel. *La civilisation traditionelle de l'"habitant" aux 17e et 18e Siècles.* Ottawa: Fides, 1967.

Shayt, David H. "Stairway to Redemption: America's Encounter with the British Prison Treadmill." *Technology and Culture* 30, no. 4 (October 1989): 908–38.

Shell Oil Co., Chemical Division. *The Canadian Petrochemical Industry.* Toronto: Ryerson, 1956.

Sheppard, Mary, ed. *Oil Sands Scientists: The Letters of Karl A. Clark, 1920–1949.* Edmonton: University of Alberta Press, 1989.

Shortt, Adam, and Arthur G. Doughty, eds. *Documents Relating to the Constitutional History of Canada, 1759–1791.* Ottawa: J. de L. Taché, 1918.

Sieferle, R.P. *The Subterranean Forest: Energy Systems and the Industrial Revolution.* Cambridge: White Horse, 1982.

Silversides, C.S. *Broadaxe to Flying Shear: The Mechanization of Forest Harvesting East of the Rockies.* Ottawa: National Museum of Science and Technology, 1997.

Simmons, Ian Gordon. *Environmental History: A Concise Introduction.* Oxford: Blackwell, 1993.

Sims, Gordon H.E. *A History of the Atomic Energy Control Board.* Ottawa: Atomic Energy Control Board, 1981.

Sinclair, Peter. *Energy in Canada.* Don Mills, ON: Oxford University Press, 2011.

– "Who Founded Greenpeace? Not Patrick Moore." Climate Denial Crock of the Week. 1 July 2014. http://climatecrocks.com/2014/07/01/who-founded-greenpeace-not-patrick-moore/.

Skeoch, Alan E. "Developments in Plowing Technology in Nineteenth-Century Canada." *Canadian Papers in Rural History* 3 (1982): 156–77.

Skinner, Claiborne. *Regional Perspectives on Early America: Upper Country: French Enterprise in the Colonial Great Lakes.* Baltimore, MD: Johns Hopkins University Press, 2008.

– "The Sinews of Empire: The Voyageurs and the Carrying Trade of the *Pays d'en Haut*, 1681–1754." PhD diss., University of Illinois at Chicago, 1991.

Sloan, W.A. "The Native Response to the Extension of the European Traders into the Athabasca and Mackenzie Basin, 1770–1814." *Canadian Historical Review* 60, no. 3 (1979): 281–99.

Smil, Vaclav. "America's Oil Imports: A Self-Inflicted Burden." *Annals of the Association of American Geographers* 101, no. 4 (July 2011): 712–16.

– *Energy at the Crossroads: Global Perspectives and Uncertainties.* Cambridge, MA: MIT Press, 2003.

– *Energy, Food, Environment: Realities, Myths, Opinions.* Oxford: Clarendon, 1987.

– *Energy in Nature and Society: General Energetics of Complex Systems.* Cambridge, MA: MIT Press, 2008.

– *Energy in World History.* Boulder, CO: Westview, 1994.

– *Energy Transitions: History, Requirements, Prospects.* Santa Barbara: ABC-CLIO, 2010.

– *Made in the USA: The Rise and Retreat of American Manufacturing.* Cambridge, MA: MIT Press, 2013.

– "On Energy Transitions." Big Ideas Podcast. http://podbay.fm/show/129166905/e/1345869000?autostart=.

Smith, L.G. "Taming B.C. Hydro: Site C and the Implementation of the B.C. Utilities Commission Act." *Environmental Management* 12, no. 4 (1988): 429–43.

Sobey, Douglas. *Early Descriptions of the Forests of Prince Edward Island: II. The British and Post-Confederation Periods (1758–c.1900). Part A: The*

Analyses. Charlottetown: PEI Department of Environment, Energy and
  Forestry, 2006.
Somerville, Alexander. *Montreal Gazette,* 10 October 1860. Reprinted in the
  Gazette (Upper Canada Village, ON), 15 June 1977. Quoted in Felicity
  Leung, *Direct Drive Waterpower in Canada, 1607–1910.* Ottawa:
  Environment Canada, Parks, 1980, 59.
South, Peter, and Raj S. Rangi. *The Performance and Economics of the
  Vertical-Axis Wind Turbine Developed at the National Research Council,
  Ottawa, Canada.* St Joseph, MI: American Society of Agricultural Engineers,
  1973.
Sprague, John B. "Great Wet North? Canada's Myth of Water Abundance." In
  Eau Canada: The Future of Canada's Water, ed. Karen Bakker, 23–36.
  Vancouver: UBC Press, 2007.
Stadfeld, Bruce. "Electric Space: Social and Natural Transformations in British
  Columbia's Hydro-electricity Industry to World War II." PhD diss.,
  University of Manitoba, 2002.
Stanley, Anna. "Citizenship and the Production of Landscape and Knowledge
  in Contemporary Canadian Nuclear Fuel Waste Management." *Canadian
  Geographer* 52, no. 1 (2008): 64–82.
– "Labours of Land: Domesticity, Wilderness and Dispossession in the
  Development of Canadian Uranium Markets." *Gender, Place and Culture*
  20, no. 2 (2013): 195–217.
– "Risk, Scale and Exclusion in Canadian Nuclear Fuel Waste Management."
  ACME: An International E-Journal for Critical Geographies 4, no. 2 (2008)
  208–9.
Starr, Richard. *Power Failure?* Halifax: Formac, 2011.
Stein, Jeremy. "Time, Space and Social Discipline: Factory Life in Cornwall,
  Ontario, 1867–1893." *Journal of Historical Geography* 21, no. 3 (1995):
  278–99.
Steinberg, Ted. *Nature Incorporated: Industrialization and the Waters of New
  England.* Cambridge: Cambridge University Press, 1991.
Stevens, Peter A. "'Roughing It in Comfort': Family Cottaging and Consumer
  Culture in Postwar Ontario." *Canadian Historical Review* 94, no. 2 (June
  2013): 234–62.
Stevenson, R.D., and R.J. Wassersburg. "Horsepower from a Horse." *Nature,*
  15 July 1993, 195.
Steward, F.R. "Energy Consumption in Canada since Confederation." *Energy
  Policy* 6, no. 3 (September 1978): 239–45.
Stewart, Hilary. *Indian Fishing: Early Methods on the Northwest Coast.* 1977.
  Vancouver: Douglas & McIntyre, 1982.

Stoll, Steven. "Farm against Forest." In *American Wilderness: A New History*, ed. Michael Lewis, 55–72. Oxford: Oxford University Press, 2007.

Storey, Robert. "From the Environment to the Workplace ... and Back Again? Occupational Health and Safety Activism, 1970s–2000+." *Canadian Review of Sociology and Anthropology* 41, no. 4 (November 2004): 419–47.

Story, G.M., W.J. Kirwin, and J.D.A. Widdowson, eds. *Dictionary of Newfoundland English*. Toronto: University of Toronto Press, 1982.

Styran, Roberta M., and Robert R. Taylor. *This Great National Object: Building the Nineteenth-Century Welland Canals*. Montreal and Kingston, McGill-Queen's University Press, 2012.

Sutherland, Neil. *Growing Up: Childhood in English Canada from the Great War to the Age of Television*. Toronto: University of Toronto Press, 1997.

Svitil, Kathy. "Wind-Turbine Placement Produces Tenfold Power Increase, Researchers Say." phys.org, 13 July 2011. http://phys.org/news/2011-07-wind-turbine-placement-tenfold-power.html.

Swainson, Neil. *Conflict over the Columbia: The Canadian Background to an Historic Treaty*. Montreal and Kingston: McGill-Queen's University Press, 1979.

Talbot, Edward Allen. *Five Years' Residence in the Canadas*. Vol. 1. London: Longman, Hurst, Ree, Orme, Brown, and Green, 1824.

Tann, Jennifer. "Horse Power 1780–1880." In *Horses in European Economic History: A Preliminary Canter*, ed. F.M.L. Thompson, 26–30. Reading: British Agricultural History Association, 1983.

Tarr, Joel. "The Horse: Polluter of the City." In *The Search for the Ultimate Sink: Urban Pollution in Historical Perspective*, ed. Joel A. Tarr, 323–34. Akron, OH: University of Akron Press, 1996.

– "A Note on the Horse as an Urban Power Source." *Journal of Urban History* 25, no. 3 (1999): 434–48.

– "Toxic Legacy: The Environmental Impact of the Gas Industry in the United States." *Technology and Culture* 44, no. 1 (January 2014): 107–47.

– "Transforming an Energy System: The Evolution of the Manufactured Gas Industry and the Transition to Natural Gas in the United States, 1870–1954." In *The Governance of Large Technical Systems*, ed. O. Coutard, 19–37. London: Routledge, 1999.

Tataryn, Lloyd. *Dying for a Living*. Montreal, Deneau and Greenberg Publishers, 1979.

Telleen, Maurice. *The Draft Horse Primer: A Guide to the Care and Use of Work Horses and Mules*. Emmaus, PA: Rodale, 1977.

Thomas, T.H. "Animal Power Production, Mechanisms for Linking Animals to Machines." In *Animal Traction for Agricultural Development*, ed. P. Starkey.

Eschborn, Germany: GTZ/Vieweg, 1990. http://www.fao.org/wairdocs/ilri/
x5455b/x5455boz.htm

Thompson, F.M.L. "Horses and Hay." In *Horses in European Economic History: A Preliminary Canter,* ed. F.M.L. Thompson, 50–72. Reading: British Agricultural History Society, 1983.

– "Nineteenth Century Horse Sense." *Economic History Review* 29, no. 1 (1976): 60–81.

Thomson, Lesslie R. "Some Economic Aspects of the Canadian Coal Problem." In *Symposium on Fuel and Coal, McGill University, 1931, Proceedings,* 350–464. Montreal: s.n., 1932.

Thornton, Patricia, and Sherry Olsen. "Mortality in Late Nineteenth Century Montreal: Geographic Pathways of Contagion." *Population Studies* 65, no. 2 (2011): 157–81.

Thrift, Nigel. "Inhuman Geographies: Landscapes of Speed, Light and Power." In *Writing the Rural: Five Cultural Geographies,* ed. Paul Cloke, Marcus Doel, David Matless, Martine Phillips and Nigel Thrift, 193–209. London: Paul Chapman, 1994.

Tillman, David A. *Wood as an Energy Resource.* New York: Academic, 1978.

Tomory, Leslie. "Building the First Gas Network: 1812–1820." *Technology and Culture* 52, no. 1 (January 2011): 75–102.

– "The Environmental History of the Early British Gas Industry, 1812–1830." *Environmental History* 17 (January 2012): 29–54.

– *Progressive Enlightenment: The Origins of the Gaslight Industry, 1780–1820.* Cambridge, MA: MIT Press, 2011.

Toronto Grey and Bruce Railway. *Hand-book of Useful Information, Respecting the Line, and the Country Tributary to It.* Toronto: Globe Printing, 1872.

Toronto Telegram. "Four Horses Die in Fire at Stables," 13 December 1933.

Traill, Catharine Parr. *The Backwoods of Canada.* London: Charles Knight, 1839.

– *Lost in the Backwoods: A Tale of the Canadian Forest.* London: T. Nelson and Sons, 1882.

Tucker E.J., ed. *75 Years 1848–1923: The Consumers' Gas Company of Toronto.* Toronto: Consumers' Gas, 1923.

Tudor, Sean. *A Brief History of Wind Power Development in Canada, 1960s–1990s.* Ottawa: Canadian Science and Technology Museum, 2010.

Tulchinsky, Gerald J.J. *The River Barons: Montreal Businessmen and the Growth of Industry and Transportation, 1837–1853.* Toronto: University of Toronto Press, 1977.

Tyrell, J.B., ed. *Journals of Samuel Hearne and Philip Turnor between the Years 1774 and 1792.* Toronto: Champlain Society, 1934.

Uekoetter, Frank. "Fukushima, Europe, and the Authoritarian Nature of Nuclear Technology." *Environmental History* 17, no. 2 (2012): 277–84.

Unger, Richard, and John Thistle. *Energy Consumption in Canada in the 19th and 20th Centuries.* Naples: Consiglio Nazionale delle Ricerche – Instituto di Studi sulle Societa del Mediterraneo, 2013.

United States Department of Agriculture. "USDA National Nutrient Database for Standard Reference." http://www.ars.usda.gov/Services/docs.htm?docid=8964.

United States Environmental Protection Agency. "Fuel Mix for U.S. Electricity Generation," 2015. http://www.epa.gov/energy/learn-about-energy-and-environment#Fuel.

– "Radiation Protection." http://www.epa.gov/radiation/radionuclides/thorium.html#affecthealth.

United States, Energy Information Administration. "Canada: International Energy Data and Analysis, 2013." http://www.eia.gov/beta/international/country.cfm?iso=CAN.

United States Secretary of the Interior. *Ninth Census: The Statistics of the Wealth and Industry of the United States, 1870.* Washington: Government Printing Office, 1872.

University of Calgary Environmental Science Program. "Impacts of Airborne Pollution from the Sour Gas Industry on Southern Alberta." ENSC 502 Project, 2003–04. http://www.ucalgary.ca/envirophys/sgimpacts2.

Urquhart, M.C. *Gross National Product, Canada, 1870–1926: The Derivation of the Estimates.* Montreal and Kingston: McGill-Queen's University Press, 1993.

Urquhart, M.C., and K.A.H. Buckley, eds. *Historical Statistics of Canada.* Toronto: Macmillan, 1965.

Usher, Peter J. "Northern Development, Impact Assessment and Social Change." In *Anthropology, Public Policy and Native Peoples in Canada*, ed. Noel Dyck and James B. Waldram, 98–130. Montreal and Kingston: McGill-Queen's University Press, 1993.

Vale, Thomas. "The Pre-European Landscape in the United States." In *Fire, Native Peoples and the Natural Landscape*, ed. Thomas R. Vale, 1–40. Washington DC: Island, 2002.

Van Huizen, Phil. "Flooding the Border: Development, Politics and the Environmental Controversy in the Canada-US Skagit Valley." PhD diss., University of British Columbia, 2013.

Van Wyck, Peter C. "The Highway of the Atom: Recollections along a Route." *Topia* 7 (Spring 2002): 99–115.

Vernon, James. *Hunger: A Modern History.* Cambridge, MA: Belknap of Harvard University Press, 2007.

Vickers, Daniel. *Farmers and Fishermen: Two Centuries of Work in Essex County, Massachusetts, 1630–1850.* Chapel Hill: University of North Carolina Press, 1994.

Walder, Ronald H. "The Utilization of Wood as an Energy Resource in Ontario." MA thesis, Wilfrid Laurier University, 1982.

Waldram, James B. *As Long as the Rivers Run: Hydroelectric Development and Native Communities in Western Canada.* Winnipeg: University of Manitoba Press, 1988.

– "Hydro-electric Development and the Process of Negotiation in Northern Manitoba, 1960–1977." *Canadian Journal of Native Studies* 4, no. 2 (1984): 205–39.

Walker, David F. "Transportation of Coal into Southern Ontario, 1871–1921." *Ontario History* 63 (1971): 15–30.

Walker, J.F., and D.R. Stevens. *Pulpwood Skidding with Horses: Efficiency of Technique,* 2nd ed. Montreal: Woodlands Section, Canadian Pulp and Paper Association, and Pulp and Paper Research Institute of Canada, 1947.

Wallace, Frederick William. *Wooden Ships and Iron Men: The Story of the Square-Rigged Merchant Marine of British North America.* London: Hodder and Stoughton, 1924.

Warde, Paul. *Energy Consumption in England and Wales, 1560–2000.* Naples: Consiglio Nazionale delle Ricerche: Istituto di Studi sulle Società del Mediterraneo, 2007.

– "Fear of Wood Shortage and the Reality of the Woodland in Europe, c. 1450–1850." *History Workshop Journal* 62, no. 1 (Autumn 2006): 28–57.

Watkins, Larry. *The Forest Resources of Ontario, 2011: Geographic Profiles of Natural Resources.* Sault Ste. Marie, ON: Forest Evaluation and Standards Section, Forests Branch Ontario Ministry of Natural Resources.

Watkins, M.H. "A Staple Theory of Economic Growth." In *Approaches to Canadian Economic History,* ed. W.T. Easterbrook and M.H. Watkins, 49–73. Ottawa: Carleton University Press, 1986.

Weaver, John. *The Great Land Rush and the Making of the Modern World, 1650–1900.* Montreal and Kingston: McGill-Queen's University Press, 2003.

Weisman, Alan. *The World without Us.* New York, Harper Perennial, 2007.

Wentworth, Edward N. "Dried Meat: Early Man's Travel Ration." *Agricultural History* 30, no. 1 (1956): 2–10.

Western Business and Industry. "Petroleum in Western Canada." May 1953.

Western Ontario Gazetteer and Directory, 1898–99. Ingersoll, ON: Ontario Publishing & Advertising, 1899.

Weyler, Rex. *Greenpeace*. Vancouver: Raincoast Books, 2004.

White, John H. Jr. "Steam in the Streets: The Grice and Long Dummy." *Technology and Culture* 27, no. 1 (January 1986): 106–9.

Williams, James. *Energy and the Making of Modern California*. Akron, OH: University of Akron Press, 1997.

Williams, Judith. *Clam Gardens: Aboriginal Mariculture on Canada's West Coast*. Vancouver, BC: New Star Books, 2006.

Williams, Michael. *Americans and Their Forests: A Historical Geography*. Cambridge: Cambridge University Press, 1989.

– *Deforesting the Earth: From Prehistory to Global Crisis*. Chicago: University of Chicago Press, 2003.

Williamson, Harold F., and Arnold P. Daum. *American Petroleum Industry*. Vol. 1, *The Age of Illumination*. Evanston, IL: Northwestern University Press, 1959.

Windfacts. "Why Wind Works." http://windfacts.ca/why-wind-works.

Windsor, J.E., and J.A. McVey. "Annihilation of Both Place and Sense of Place: The Experience of the Cheslatta T'En Canadian First Nation within the Context of Large-scale Environmental Projects." *Geographical Journal* 171, no. 2 (2005): 146–65.

Wlasiuk, Jonathan Joseph. "Refining Nature: Standard Oil and the Limits of Efficiency, 1863–1920." PhD diss., Case Western Reserve University, 2011.

Wood, David. *Making Ontario*. Montreal and Kingston: McGill-Queen's University Press, 2000.

Woodstock Review, 16 October 1874.

Woolliams, Nina G. *Cattle Ranch: The Story of the Douglas Lake Cattle Company*. Vancouver: Douglas and McIntyre, 1979.

World Energy Council. *Energy Resources, Hydropower*, 2011. http://www.iea.org/topics/renewables/subtopics/hydropower/.

World Nuclear Association. "Brief History of Uranium Mining in Canada." http://www.world-nuclear.org/info/Country-Profiles/Countries-A-F/Appendices/Uranium-in-Canada-Appendix-1-Brief-History-of-Uranium-Mining-in-Canada/.

– "Nuclear Power in Canada." http://www.world-nuclear.org/info/Country-Profiles/Countries-A-F/Canada-Nuclear-Power/.

– "Thorium." http://www.world-nuclear.org/info/Current-and-Future-Generation/Thorium/.

– "Uranium in Canada." http://www.world-nuclear.org/info/Country-Profiles/Countries-A-F/Canada--Uranium/.

World Wind Energy Association. *2015 Small Wind World Report.* www.uwea. com.ua/files/s w w R_2015.pdf.

Wrigley, E.A. *Continuity, Chance and Change: The Character of the Industrial Revolution in England.* Cambridge: Cambridge University Press, 1988.

– *Energy in the English Industrial Revolution.* Cambridge: Cambridge University Press, 2010.

– *Poverty, Progress, and Population.* Cambridge: Cambridge University Press, 2004.

– "Reflections on the History of Energy Supply, Living Standards and Economic Growth." *Australian Economic History Review* 33, no. 1 (1993): 3–21.

Wynn, Graeme. *Canada and Arctic North America: An Environmental History.* Santa Barbara, CA: ABC-CLIO, 2007.

– *Timber Colony: A Historical Geography of Early Nineteenth Century New Brunswick.* Toronto: University of Toronto Press, 1981.

Young, Edward. *Labor in Europe and America: A Special Report on the Rates of Wages, the Cost of Subsistence, and the Conditions of the Working Classes in Great Britain, France, Belgium, Germany, and Other Countries of Europe, also in the United States and British America.* Philadelphia: S.A. George, 1875.

Yudkin, John. "Some Basic Principles of Nutrition." In *The Making of the Modern British Diet,* ed. Derek Oddy and Derek Miller, 196–201. London: Croom Helm, 1976.

Zercher, Frederick K. "The Economic Development of the Port of Oswego." PhD diss., Syracuse University, 1935.

Contributors

GEORGE COLPITTS teaches environmental history at the University of Calgary. His publications include *Game in the Garden: A Human History of Wildlife to 1940* (UBC Press, 2002) and *Pemmican Empire: Food, Trade and the Last Bison Hunts in the North American Plains, 1780–1882* (Cambridge University Press, 2015).

JENNY CLAYTON is a British Columbia historian, independent scholar, and sessional lecturer at the University of Victoria. She completed her dissertation in 2009, "Making Recreational Space: Citizen Involvement in Outdoor Recreation and Park Establishment in British Columbia, 1900–2000," and has written and researched in the field of British Columbia social and environmental history. As a public historian, Jenny was website coordinator and a writer for the *Victoria's Chinatown* website and created interpretive materials for Government House in Victoria, BC.

JOANNA DEAN is an associate professor at Carleton University, where she teaches Canadian environmental history and animal history. She has written extensively on the history of street trees in Ottawa, and is co-editing *Animal Metropolis: Histories of Human-Animal Relations in Urban Canada*, with Christabelle Sethna and Darcy Ingram, forthcoming with University of Calgary Press.

COLIN DUNCAN taught environmental and modern British history at Queen's University and McGill University. He came to specialize on agricultural/legal aspects of environmental history after writing a multidisciplinary study: *The Centrality of Agriculture: Between Humankind and the Rest of Nature*. In retirement he has prepared a general global environmental history entitled "The Great Disconnect" (as yet unpublished).

MATTHEW EVENDEN is a professor of geography and serves as the associate dean, Research and Graduate Studies, in the Faculty of Arts, University of British Columbia. He has researched in the fields of energy and environmental history, with an emphasis on rivers and power. His latest book is *Allied Power: Mobilizing Hydro-Electricity during Canada's Second World War* (University of Toronto Press, 2015).

JACK LITTLE is a professor emeritus in the History Department at Simon Fraser University. He has written widely on Canadian rural, social, cultural, and political history, with an emphasis on Quebec. His most recent book is *Patrician Liberal: The Public and Private Life of Sir Henri-Gustave Joly de Lotbinière, 1829–1908* (University of Toronto Press, 2013). Forthcoming is *Fashioning the Canadian Landscape: Collected Essays* (University of Toronto Press).

JOSHUA MACFADYEN is an assistant professor of environmental humanities in the School of Historical, Philosophical, and Religious Studies and the School of Sustainability at Arizona State University. His work uses digital history methods such as historical geographic information systems to examine the social and ecological problems of energy in Canadian and US agriculture, particularly during the transition from traditional to modern agro-ecosystems. Before taking up his post at Arizona State, he was a postdoctoral fellow with the Sustainable Farm Systems project at the University of Saskatchewan and prior to that with the Network in Canadian History and Environment at Western University in London, Ontario.

JONATHAN PEYTON is an assistant professor in the Department of Environment and Geography, University of Manitoba, Winnipeg. His work lies at the intersection of environmental geography and political ecology and develops a cultural and historical approach. He has explored a series of resource development conflicts in northwest British Columbia, a region that is the site of intense mining exploration and controversy over energy projects. He situates current conflicts against the legacies of previous megaprojects – both failed and realized – to understand their social and environmental side effects and their legacies for future developments. He is the author of several articles and a monograph, *Unbuilt Environments: Legacies of Post-War Development and Extractive Economies in Northwest BC* (UBC Press). His research is on northern energy megaprojects and extractive economies in subarctic North America.

ERIC SAGER is a professor of history at the University of Victoria. He was a member of the Atlantic Canada Shipping Project at Memorial University of Newfoundland (1976–82). He has studied aspects of family history and labour history, and was director of the Canadian Families Project (1996–2002). He is working on the history of equality and inequality in Canada.

RUTH SANDWELL is a professor at the University of Toronto, where she teaches in the Department of History and the Department of Curriculum, Teaching and Learning at the Ontario Institute for Studies in Education. She publishes in the fields of rural and energy history. Her most recent book is *Canada's Rural Majority, 1870–1940: Households, Environments, Economies* (University of Toronto Press, 2016).

LAUREL SEFTON MACDOWELL is an emeritus professor of history in the Department of Historical Studies, University of Toronto Mississauga. She specialized in Canadian working-class history and North American environmental history. She is author of *"Remember Kirkland Lake": The Gold Miners' Strike 1941–42* (2nd ed., University of Toronto Press, 2001), *Renegade Lawyer: The Life of J.L. Cohen* (University of Toronto Press, 2001), and *An Environmental History of Canada* (UBC Press, 2012).

PHILIP VAN HUIZEN is an L.R. Wilson Assistant Professor in the Wilson Institute for Canadian History at McMaster University. His research on cross-border river and energy history has been published in journals like the *Pacific Historical Review* and *BC Studies.* His doctoral dissertation, "Flooding the Border: Development, Politics and Environmental Controversy in the Canadian-US Skagit Valley" (UBC, 2013), won best dissertation awards from the Canadian Studies Network and the American Historical Association.

ANDREW WATSON is a post-doctoral fellow working on the Sustainable Farm Systems project in the Department of History and the Historical GIS Laboratory at the University of Saskatchewan. He is also a collaborator on the London's Ghost Acres project, which uses digital methods to explore the history of nineteenth-century global commodity trading. He defended his dissertation, "Poor Soils and Rich Folks: Household Economies and Sustainability in Muskoka, 1850–1920," in May 2014.

His research is focused on the history of coal in Canada during the inter-war years.

LUCAS WILSON is a Toronto lawyer with an interest in the history of animal welfare. He studied history at Queen's University, Kingston.

Index

Page numbers in italics refer to the definition of a technical word or phrase.